Lecture Notes in Mathematics

Edited by A. Dold and B. Eckmann

Subseries: Scuola Normale Superiore, Pisa
Adviser: E. Vesentini

T0183944

992

Harmonic Analysis

Proceedings of a Conference
Held in Cortona, Italy, July 1 – 9, 1982

Edited by Giancarlo Mauceri, Fulvio Ricci and Guido Weiss

Springer-Verlag
Berlin Heidelberg New York Tokyo 1983

Editors

Giancarlo Mauceri
Università di Genova, Istituto di Matematica
16132 Genova, Italy

Fulvio Ricci
Politecnico di Torino, Dipartimento di Matematica
Corso Duca degli Abruzzi, 10129 Torino, Italy

Guido Weiss
Department of Mathematics, Washington University
St. Louis, Missouri 63130, USA

AMS Subject Classifications (1980): 42 A 20, 42 A 45, 42 A 50, 42 B 20, 42 B 25, 42 B 30, 42 A 50, 43 A 80, 43 A 85, 43 A 90, 46 A 15

ISBN 3-540-12299-0 Springer-Verlag Berlin Heidelberg New York Tokyo
ISBN 0-387-12299-0 Springer-Verlag New York Heidelberg Berlin Tokyo

Printed in Germany

Printing and binding: Beltz Offsetdruck, Hemsbach/Bergstr.
2146/3140-543210

-Introduction-

These are the Proceedings of the Conference in Harmonic Analysis that was held at the Palazzone of the Scuola Normale Superiore in Cortona in July 1982 and was supported by the National Science Foundation and the Consiglio Nazionale delle Ricerche. These Institutions have been supporting in the last three years the cooperation between a group of American and Italian harmonic analysts. As part of this activity, two conferences in Harmonic Analysis were organized: the first at the Scuola Normale Superiore di Pisa in April 1980 and the second at the University of Minnesota, Minneapolis, in April 1981.
The Proceedings of the Pisa Conference appeared as Supplemento ai Rendiconti del Circolo Matematico di Palermo n.1, 1981, and those of the Minneapolis Conference were published as Springer Lecture Notes 908.
As in the previous meetings, the Cortona Conference attracted the interest of a large number of harmonic analysts from all over the world, who contributed significantly to its success. Most of the articles in these Proceedings contain original results that were presented at the Conference, except for two survey articles by R.Blei and M.H.Reimann.

We wish to express our gratitude to the Scuola Normale Superiore di Pisa and to its Director, Professor Edoardo Vesentini, for hosting the Conference at the Palazzone and providing us with all its facilities.
We are also grateful to Maria Cristina Mauceri Cowling for her help during the Conference.

Giancarlo Mauceri
Fulvio Ricci
Guido Weiss

LIST OF PARTICIPANTS

A.Alesina	Università di Milano
A.Baernstein	Washington University in St.Louis
M.Baronti	Università di Genova
J.J.Benedetto	University of Maryland
R.Blei	University of Connecticut
S.Bloom	Siena College
A.Bonami	Université de Paris-Sud, Orsay
C.Bondioli	Università di Pavia
G.Brown	University of New South Wales
F.Cazzaniga	Università di Milano
C.Cecchini	Università di Genova
P.Cifuentes	Washington University in St.Louis
J.L.Clerc	Université de Nancy I
R.R.Coifman	Yale University
L.Colzani	Università di Milano
M.Cowling	Università di Genova
L.De Michele	Università di Milano
A.Dooley	University of New South Wales
E.Fabes	University of Minnesota
R.Fefferman	University of Chicago
A.Figà-Talamanca	Università di Roma
G.Gaudry	Flinders University
J.Gilbert	University of Texas
S.Giulini	Università di Milano
P.Greiner	University of Toronto
A.Hulanicki	Polish Academy of Sciences, Wrocław
I.Inglis	Università di Milano
A.Iozzi	Università di Roma
R.Johnson	University of Maryland
C.Karanikas	University of Crete
C.Kenig	University of Minnesota
A.Korányi	Washington University in St.louis
G.Kuhn	Università di Milano
R.Kunze	University of California at Irvine
P.Lemarié	Ecole Normale Superieure, Paris V
H.Leptin	Universität Bielefeld
J.Lewis	University of Illinois at Chicago Circle
S.Madan	Université de Paris-Sud, Orsay
A.M.Mantero	Università di Genova
G.Mauceri	Università di Genova

Y.Meyer	Ecole Polytechnique, Palaiseau
C.Nebbia	Università di Roma
V.Nestoridis	University of Athens
M.Pagliacci	Università di Perugia
D.Phong	Columbia University
M.A.Picardello	Università di Roma
D.Poguntke	Universität Bielefeld
D.Poornima	Université de Paris-Sud, Orsay
E.Prestini	Università di Milano
H.M.Reimann	Universität Bern
F.Ricci	Politecnico di Torino
C.Sadosky	Howard University
R.Scaramuzzi	Yale University
S.Semmes	Institute Mittag-Leffler and Washington University
P.Sjögren	Göteborg University
P.M.Soardi	Università di Milano
F.Soria	Washington University in St.Louis
E.M.Stein	Princeton University
M.Taibleson	Washington University in St.Louis
J.Torrea	Universidad Autónoma de Madrid
G.Travaglini	Università di Milano
N.Varopoulos	Université de Paris-Sud, Orsay
A.de la Villa	Universidad Complutense, Madrid
S.Wainger	University of Wisconsin
G.Weiss	Washington University in St.Louis
E.Wilson	Washington University in St.Louis
A.Zappa	Università di Genova

LIST OF LECTURES

E.M.Stein	Singular integrals related to the $\bar{\partial}$-Neumann problem
A.Korányi	Kelvin transform and harmonic polynomials on the Heisenberg group
A.Bonami	Multipliers of Sobolev spaces
S.Wainger	Applications of L^p estimates to the non-linear Klein-Gordon equation
G.Mauceri	Maximal operators and surfaces of vanishing curvature
R.Blei	Fractional dimensions in n-dimensional spaces
C.Sadosky	Vector valued inequalities of Marcinkiewicz-Zygmund and Grothendieck type for generalized Toeplitz kernels
P.Sjögren	On the maximal operator for the Mehler kernel
C.Kenig	Diffusion processes with singular transition probability densities
E.Fabes	Littlewood-Paley estimates in PDE
H.M.Reimann	Invariant differential operators on hyperbolic space
A.M.Mantero	Poisson integrals and uniformly bounded representations of the free group
M.A.Picardello	Harmonic analysis on graphs
N.Varopoulos	Transient groups and winding Brownian motions
A.Dooley	Contractions of Lie groups
J.E.Gilbert	Cauchy-Riemann systems on symmetric spaces
R.A.Kunze	A kernel for Cauchy-Riemann systems
R.R.Coifman	Non-linear harmonic analysis
Y.Meyer	The solution of A.P.Calderón's conjectures
C.Cecchini	Non-commutative L^1 spaces and conditional expectations
A.Baernstein	Aleksandrov's inner functions in the unit ball in C^n
E.Wilson	Structure of isometry groups of homogeneous Riemannian manifolds
H.Leptin	Radial functions on nilpotent groups

TABLE OF CONTENTS

UN NOUVEL ESPACE FONCTIONNEL ADAPTE A L'ETUDE
DES OPERATEURS DEFINIS PAR DES INTEGRALES SINGULIERES

R. R. COIFMAN, Y. MEYER et E. M. STEIN

L'objet de ce travail est de donner une nouvelle démonstration du théorème suivant :

Théorème 1 : Soit $A : \mathbb{R} \to \mathbb{C}$ une fonction lipschitzienne : $A'(x) = a(x) \in L^{\infty}(\mathbb{R})$. Alors pour tout entier $k \in \mathbb{N}$, l'opérateur T_k défini par le noyau v.p.$(A(x) - A(y))^k (x-y)^{-k-1}$ est borné sur $L^2(\mathbb{R}; dx)$ et la norme d'opérateur de T_k ne dépasse pas $C\|a\|_{\infty}^k (1+k)^4$; C est une constante numérique.

La preuve du théorème 1 que l'on peut trouver dans [2] comportait trois parties (notées (α), (β), (γ)) que nous allons rappeler

(α) la formule de représentation de Mc Intosh fournit

$$T_k = \text{v.p.} \int_{-\infty}^{+\infty} [(I + itD)^{-1} M]^k (I + itD)^{-1} \frac{dt}{t} \quad \text{où} \quad D = -i \frac{d}{dx}$$

et où $M : L^2(\mathbb{R}) \to L^2(\mathbb{R})$ est l'opérateur de multiplication ponctuelle par $a(x) = A'(X)$.

(β) Si $P_t = (I + t^2 D^2)^{-1}$ et $Q_t = t\,D\,P_t$, l'étude de T_k se ramène à celle d'une fonctionnelle quadratique adaptée à savoir

$$G_k[f](x) = \left(\int_0^{\infty} |Q_t (MP_t)^k f(x)|^2 \frac{dt}{t} \right)^{1/2}$$

et il s'agit de montrer que

$$\| G_k[f] \|_2 \leq C(1+k) \|f\|_2 \|a\|_{\infty}^k$$

(γ) la démonstration de cette inégalité fondamentale utilisait un calcul pseudo-différentiel adapté à la multiplication par les fonctions $a(x) \in L^{\infty}(\mathbb{R})$ ainsi que des majorations obtenues grâce à certaines "mesures de Carleson".

La nouvelle démonstration suit la même organisation générale. Nous donnons une nouvelle démonstration de la formule de McIntsoch. La partie (β) reste inchangée mais, grâce à un nouvel espace fonctionnel (l'espace des "tentes"), le calcul pseudo-différentiel de (γ) devient plus simple et plus élégant.

Enfin on peut se restreindre, tout au long de la démonstration, au cas où $a(x) \in C_o^\infty(\mathbb{R})$. Une fois vérifiée $\|T_k\| \leq 10^6 \|a\|_\infty^k (1+k)^4$ dans ce cas, le théorème de Calderón, Cotlar et Zygmund fournit une inégalité analogue pour les opérateurs définis par les noyaux tronqués $(A(x) - A(y))^k (x-y)^{-k-1} \chi\{\varepsilon \leq |x-y| \leq R\}$ où $R > \varepsilon > 0$. Il suffit pour conclure d'approcher $a \in L^\infty(\mathbb{R})$, dans la topologie $\sigma(L^\infty, L^1)$, par une suite $a_j \in C_o^\infty(\mathbb{R})$ et de passer à la limite dans les estimations relatives aux noyaux tronqués.

1. LA FORMULE DE REPRESENTATION DE McINTOSH

Une fois pour toutes $a \in C_o^\infty(\mathbb{R})$ est à valeurs complexes, M est l'opérateur de multiplication ponctuelle par la fonction $a(x)$. On pose $D = -i \frac{d}{dx}$ et l'opérateur $(I + itD)^{-1}$ a pour symbole $(1 + it\xi)^{-1}$ et pour noyau $\frac{1}{|t|} k_t(x-y)$ où $k_t(u) = k(\frac{u}{t})$ et $k(u) = 0$ si $u < 0$, $k(u) = e^{-u}$ si $u \geq 0$. Nous dirons qu'un opérateur T est défini par le noyau-distribution $K(x,y)$ si $K(x,y)$ est une distribution appartenant à $\mathcal{S}'(\mathbb{R}^n \times \mathbb{R}^n)$ définissant $T : \mathcal{S}(\mathbb{R}^n) \to \mathcal{S}'(\mathbb{R}^n)$ par $(Tf,g) = (K(x,y), f(x)g(y))$; $(.,.)$ exprime la dualité entre $\mathcal{S}'(\mathbb{R}^n)$ et $\mathcal{S}(\mathbb{R}^n)$ dans le premier cas et la dualité entre $\mathcal{S}'(\mathbb{R}^n \times \mathbb{R}^n)$ et $\mathcal{S}(\mathbb{R}^n \times \mathbb{R}^n)$ dans le second cas.

Proposition 1 : Avec les notations ci-dessus; pour tout entier $m \geq 0$ et tout nombre réel $t \neq 0$, le noyau-distribution de l'opérateur $(I + it\,D)^{-1} [M(I+it\,D)^{-1}]^k$ est

$$|t|^{-m-1} (m!)^{-1} (\text{sign}(x-y))^m (A(x) - A(y))^m k_t(x-y).$$

La preuve est simple. Posons $\mathcal{L}_t = (I + it\,D)^{-1} [M(I+it\,D)^{-1}]^k$. Le noyau de $\mathcal{L}_t$ est

$$L_t(x,y) = \int \cdots \int |t|^{-m-1} k_t(x-x_1) a(x_1) k_t(x_1-x_2) a(x_2) \cdots k_t(x_m-y) dx_1 \cdots dx_m .$$

Pour calculer cette intégrale, on distingue les deux cas $t > 0$ et $t < 0$. Si $t > 0$ le domaine d'intégration $S(x,y) \subset \mathbb{R}^m$ est défini par $x > x_1 > x_2 > \cdots > x_m > y$ et l'on a $k_t(x-x_1) k_t(x_1-x_2) \cdots k_t(x_m-y) = k_t(x-y)$. Si $t < 0$ le domaine d'intégration devient $x < x_1 < x_2 < \cdots < x_m < y$ et l'identité subsiste. On a donc, dans les deux cas

$$L_t(x,y) = |t|^{-m-1} k_t(x-y) J(x,y) \quad \text{où}$$

$$J(x,y) = \int \cdots \int_{S(x,y)} a(x_1)\ldots a(x_m)\,dx_1\ldots dx_m$$

Pour calculer $J(x,y)$, on observe que le groupe S_m des permutations de $\{1,\ldots,m\}$ agit sur $\mathbb{R}^m$ (par permutation des coordonnées). Les ensembles $\sigma\{S(x,y)\}$, $\sigma \in S_m$, forment une partition mesurable du cube $[x,y]^m$. Il vient donc $m!J(x,y) = (\int_{[x,y]} a(t)dt)^m$. La proposition 1 est démontrée.

Proposition 2 : Posons, pour $0 < \varepsilon < R$,

(1) $$T_k^{\varepsilon,R} = \int_{\varepsilon \leq |t| \leq R} (I+itD)^{-1}[M(I+itD)^{-1}]^k \frac{dt}{t} .$$

Alors pour toute fonction $f \in L^2(\mathbb{R})$, on a

(2) $$\| T_k f - T_k^{\varepsilon,R} f \|_2 \to 0 \quad \text{quand} \quad \varepsilon \to 0 \text{ et } R \to +\infty .$$

Pour le voir, on pose $\theta_k(u) = \frac{1}{k!}\int_0^{|u|} s^k \exp(-s)ds$ et l'on a

$$\int_{\varepsilon \leq |t| \leq R} |t|^{-k-1} k_t(x-y) \frac{dt}{t} = k!|x-y|^{-k-1}\operatorname{sign}(x-y)\{\theta_k(\frac{x-y}{\varepsilon}) - \theta_k(\frac{x-y}{R})\} .$$

Grâce à la proposition 1, le noyau de l'opérateur $T_k^{\varepsilon,R}$ est

$$\frac{(A(x) - A(y))^k}{(x-y)^{k+1}} \{\theta_k(\frac{x-y}{\varepsilon}) - \theta_k(\frac{x-y}{R})\} .$$

Il s'agit, puisque $\theta_k : [0,+\infty[\to [0,1]$ est une fonction croissante, d'une version tronquée (près et loin de la singularité) du noyau $(A(x) - A(y))^k(x-y)^{-k-1}$. La troncation usuelle correspondrait à remplacer θ_k par la fonction indicatrice de l'intervalle $[-1,1]$ et notre troncation est donc un "procédé de sommation" dominé par la troncation usuelle.

Puisque $a(x) \in C_0^\infty(\mathbb{R})$, le noyau $(A(x) - A(y))^k(x-y)^{-k-1}$ est un noyau de Calderón-Zygmund. Ceci sans cercle vicieux au niveau du théorème 1 ; (2) résulte du théorème de Calderón-Cotlar-Zygmund.

2. L'ESPACE DES TENTES

Nous présenterons la version de l'espace des tentes qui est adaptée à nos besoins ; il s'agit de l'espace $T_{2,1}$.

Définition 1 : Soit $\mathbb{R}^2_+$ le demi-plan supérieur : $(x,t) \in \mathbb{R}^2_+$ signifie donc $x \in \mathbb{R}$ et $t > 0$ et soit $f : \mathbb{R}^2_+ \to \mathbb{C}$ une (classe de) fonction mesurable pour la mesure dxdt. Posons

$$S(f)(x) = \left(\iint_{|x-y| \leq t} |f(y,t)|^2 \frac{dy\, dt}{t^2}\right)^{1/2}.$$

Nous définissons l'espace $T_{2,1}$ comme l'ensemble des (classe de) fonctions mesurables $f : \mathbb{R}^2_+ \to \mathbb{C}$ telles que $S(f) \in L^1(dx)$ et posons $\|f\|_{T_{2,1}} = \|S(f)\|_1$.

Le théorème fondamental concernant l'espace $T_{2,1}$ est l'existence d'une décomposition atomique. Mais il importe évidemment de définir les atomes.

Pour tout intervalle $I \subset \mathbb{R}$, nous désignerons par $\hat{I}$ la "tente au dessus de I" définie par $\hat{I} = \{(x,t)\ ; 0 < t < d(x,I^c)\}$ où $d(x,I^c)$ est la distance de x au complémentaire de I. En d'autres termes $(x,t) \in \hat{I}$ signifie $[x-t,x+t] \subset I$.
Si $\Omega \subset \mathbb{R}$ est une partie ouverte, $\hat{\Omega}$ est la réunion des $\hat{I}$ où $I \subset \Omega$.

Définition 2 : Une fonction mesurable $a : \mathbb{R}^2_+ \to \mathbb{C}$ est un atome (pour l'espace $T_{2,1}$) s'il existe un intervalle compact $I \subset \mathbb{R}$ tel que

(3) $\qquad a(x,t)$ soit portée par $\hat{I}$

(4) $$\left(\iint_{\hat{I}} |a(x,t)|^2 \frac{dx\, dt}{t}\right)^{1/2} \leq |I|^{-1/2}.$$

Nous allons vérifier que tout atome appartient à $T_{2,1}$ et que sa norme ne dépasse pas $\sqrt{2}$.

Nous utiliserons, à cet effet, la remarque évidente suivante.

Lemme 1 : Pour toute fonction mesurable $f : \mathbb{R}^2_+ \to \mathbb{C}$, on a

(5) $$\|S(f)\|_{L^2(dx)} = \sqrt{2}\,\|f\|_{L^2(\mathbb{R}^2_+\ ;\ dx\, dt/t)}.$$

C'est facile et laissé au lecteur.

Lemme 2 : Si a est un atome, $\|S(a)\|_1 \leq \sqrt{2}$.

En effet la fonction S(a) est portée par l'intervalle I et donc $\|S(a)\|_1 \leq |I|^{\frac{1}{2}} \|S(a)\|_2 \leq \sqrt{2}$.

Théorème 2 : Les deux propriétés suivantes d'une (classe de) fonction mesurable $f : \mathbb{R}_+^2 \longrightarrow \mathbb{C}$ sont équivalentes

(6) $$f \in T_{2,1}$$

(7) il existe une suite $a_k(x,t)$ d'atomes, $k \in \mathbb{N}$, et une suite λ_k de coefficients telles que $\sum_0^\infty |\lambda_k| < +\infty$ et que $f(x,t) = \sum_0^\infty \lambda_k a_k(x,t)$.

De plus la norme de f dans $T_{2,1}$ et la borne inférieure des sommes $\sum_0^\infty |\lambda_k|$ sont équivalentes.

Naturellement l'implication (7) ⇒ (6) résulte du lemme 2. Pour démontrer la réciproque, nous aurons besoin d'une version locale de l'identité (5).

Soient I un intervalle ouvert borné, $E \subset I$ une partie mesurable, χ_E la fonction indicatrice de E, χ_E^* la fonction maximale (non centrée) de Hardy et Littlewood de χ_E définie par $\chi_E^*(x) = \sup_{x \in J \subset I} \frac{|E \cap J|}{|J|}$ et $\Omega \subset I$ l'ensemble ouvert des x tels que $\chi_E^*(x) > \frac{1}{2}$. On a évidemment $E \subset \Omega$ (à un ensemble de mesure nulle près).

Proposition 2 : Désignons par K l'ensemble $\hat{I} \setminus \hat{\Omega}$. Avec les notations précédentes, pour toute fonction mesurable $f : \mathbb{R}_+^2 \to \mathbb{C}$, on a

(8) $$\left(\iint_K |f(y,t)|^2 \frac{dy\, dt}{t} \right)^{1/2} \leq \left(\int_{I \setminus E} (S(f))^2 dx \right)^{1/2}.$$

Pour démontrer la proposition 2, on désigne par $Z \subset I \times K$ l'ensemble des (x,y,t) vérifiant $x \in I \setminus E$ et $|x - y| \leq t$ cependant que $(y,t) \in K$. On a alors

Lemme 3 : Pour tout $(y,t) \in K$, la mesure de la tranche correspondante des x tels que $(x,y,t) \in Z$ dépasse t.

En effet si $(y,t) \in K$ c'est que $[y - t, y + t] \in I$ mais que cet intervalle (noté Y) n'est pas entièrement contenu dans Ω. Il existe donc $x_o \in Y$ tel que $\chi_E^*(x_o) \leq 1/2$ ce qui implique évidemment $|E \cap Y| \leq \frac{1}{2} |Y| = t$. Désignons par E^c l'ensemble $I \setminus E$. On a $|E^c \cap Y| \geq t$ ce qu'il fallait démontrer.

Considérons alors $\sigma = \iiint_Z |f(y,t)|^2 \frac{dy\,dx\,dt}{t^2}$ et appliquons le théorème de Fubini au calcul de cette intégrale triple. Fixons $(y,t) \in K$ et intégrons en x ; le lemme 3 fournit $\sigma \geq \iint_K |f(y,t)|^2 \frac{dy\,dt}{t}$. Fixons maintenant $x \in I \setminus E$ et intégrons en (y,t) en retenant seulement la contrainte $|x-y| \leq t$. Il vient $\sigma \leq \int_{I\setminus E} (S(f))^2 dx$ ce qui termine la démonstration de la proposition 2.

Naturellement si E étant fixé (ainsi que I), on remplace Ω par un ouvert $\Omega_1 \supset \Omega$, K est remplacé par $K_1 \subset K$ et la conclusion de la proposition 2 ne change pas. On pourra donc remplacer, dans la définition de Ω, χ_E^* par n'importe quelle fonction $\theta(x) \geq \chi_E^*$.

Revenons à la démonstration atomique.

Pour $k \in \mathbb{Z}$, on désigne par E_k l'ensemble des $x \in \mathbb{R}$ tels que $S(f)(x) > 2^k$ et l'on a évidemment $\sum_{-\infty}^{+\infty} 2^k |E_k| \leq 2\|S(f)\|_1$. On désigne par $\Omega_k \subset \mathbb{R}$ l'ensemble ouvert défini par $\chi_{E_k}^*(x) > 1/2$; Ω_k est la réunion des intervalles ouverts disjoints $I(k,j)$. On a $E_{k+1} \subset \Omega_{k+1} \subset \Omega_k$ et l'on pose $\Omega_{k+1}^{(j)} = \Omega_{k+1} \cap I(k,j)$. On pose

$$\Delta(k,j) = \hat{I}(k,j) \setminus \hat{\Omega}_{k+1}^{(j)} \tag{9}$$

et l'on a alors

<u>Proposition 3</u> : <u>Les ensembles</u> $\Delta(k,j)$ <u>forment une partition mesurable du support (mesurable) de</u> $f(x,t)$ <u>et si l'on pose</u> $a_{k,j}(x,t) = f(x,t)\,\chi_{\Delta(k,j)}$ <u>la série</u> $\sum_k \sum_j a_{k,j}(x,t)$ <u>est la décomposition atomique (non encore normalisée) de</u> $f(x,t)$.

Remarquons d'abord que, pour k fixé, $\bigcup_j \Delta(k,j) = \hat{\Omega}_k \setminus \hat{\Omega}_{k+1}$; cela implique, en appelant Ω la réunion des Ω_k, $\bigcup_k \bigcup_j \Delta(k,j) = \hat{\Omega}$. On conclut en remarquant que $S(f)(x) = 0$ sur Ω^c ce qui implique que $f(x,t) = 0$ hors de $\hat{\Omega}$.

Il reste à calculer $\sigma(k,j) = \left(\iint_{\hat{I}(k,j)} |a_{k,j}(x,t)|^2 \frac{dx\,dt}{t}\right)^{1/2}$.

On pose, à cet effet, $E(k,j) = I(k,j) \cap E_{k+1}$ et l'on applique la proposition 2 à $I = I(k,j)$, $E = E(k,j)$ en observant que $\Delta(k,j)$ est alors contenu dans l'ensemble K de la proposition 2. On a $\sigma(k,j) \leq \left(\int_{I(k,j)\setminus E(k,j)} (S(f))^2 dx\right)^{1/2} \leq 2^{k+1} |I(k,j)|^{1/2} = \lambda(k,j)\,|I(k,j)|^{-1/2}$.

On a posé $\lambda(k,j) = 2^{k+1}\,|I(k,j)|$ et il reste à montrer que $\sum_k \sum_j \lambda(k,j) < +\infty$. L'inégalité de type faible pour la fonction maximale donne $|\Omega_k| \leq C|E_k|$ et l'on

a

$$\sum_j \lambda(k,j) = 2^{k+1} \quad |\Omega_k| \quad \text{et donc} \quad \sum_k \sum_j \lambda(k,j) \leq 4C \, \|S(f)\|_1 \quad .$$

3. L'ESPACE DES TENTES ET L'ESPACE H^1-ATOMIQUE

Soit $\psi \in L^1(\mathbb{R})$ une fonction vérifiant $\int_{-\infty}^{+\infty} \psi(x)\,dx = 0$, $\int_{-\infty}^{+\infty} |\hat{\psi}(\xi)|^2 \frac{d\xi}{|\xi|} < +\infty$ et $|\psi(x)| \leq |x|^{-2}$ pour $|x| \geq 1$.

Pour tout $t > 0$, posons $\psi_t(x) = \frac{1}{t}\,\psi\,(\frac{x}{t})$ et désignons par Q_t l'opérateur de convolution avec ψ_t . Nous nous proposons de de démontrer le théorème suivant.

Théorème 3 : Soit $f(x,t) = f_t(x)$ une fonction appartenant à l'espace $T_{2,1}$ que l'on considérera comme une fonction de x indéxée par t.

Alors $h(x) = \int_0^\infty Q_t f_t \frac{dt}{t}$ appartient à l'espace $H^1(\mathbb{R})$ de Stein et Weiss.

La démonstration est, en principe, immédiate. La décomposition atomique de l'espace $T_{2,1}$ fournit la décomposition atomique (ou moléculaire) de h dans l'espace H^1.

Entrons dans les détails. Soit $c(\psi)$ le plus grand des deux nombres $\left(\int_0^\infty |\hat{\psi}(\xi)|^2 \frac{d\xi}{\xi}\right)^{1/2}$ et $\left(\int_0^\infty |\hat{\psi}(-\xi)|^2 \frac{d\xi}{\xi}\right)^{1/2}$. On a alors

Lemme 4 : Si $f \in L^2(\mathbb{R}; dx)$, on a

$$\left(\int_0^\infty \|Q_t f\|_2^2 \frac{dt}{t}\right)^{1/2} \leq c(\psi)\|f\|_2 \; .$$

C'est immédiat par la formule de Plancherel.

Lemme 5 : Si $f_t(x)$ est une fonction mesurable sur $\mathbb{R}_+^2$ et si $\int_0^\infty \|f_t\|_2^2 \frac{dt}{t} < +\infty$, alors l'intégrale singulière $\int_0^\infty Q_t(f_t) \frac{dt}{t}$ est convergente et l'on a

$$\left\|\int_0^\infty Q_t(f_t) \frac{dt}{t}\right\|_2 \leq c(\psi) \left(\int_0^\infty \|f_t\|_2^2 \frac{dt}{t}\right)^{1/2} \tag{10}$$

Oublions les problèmes de convergence et posons $h(x) = \int_0^\infty Q_t(f_t)\frac{dt}{t}$. Calculons le produit scalaire $I = \int h(x)\overline{u}(x)\,dx$ lorsque $\|u\|_2 \leq 1$. L'adjoint de l'opérateur Q_t est l'opérateur de convolution avec $\tilde{\psi}_t$ où $\tilde{\psi}(x) = \overline{\psi}(-x)$. Le lemme 4 s'applique donc à Q_t^* et l'on a

$$|I| \leq \left(\iint_{\mathbb{R}^2_+} |f_t(x)|^2 \frac{dxdt}{t}\right)^{1/2} \left(\iint_{\mathbb{R}^2_+} |Q_t^* u|^2 \frac{dxdt}{t}\right)^{1/2} \leq$$

$$c(\psi) \left(\iint_{\mathbb{R}^2_+} |f_t(x)|^2 \frac{dxdt}{t}\right)^{1/2}$$

On peut penser que l'inégalité (10) est mauvaise ; en effet si $f_t(x)$ ne dépend pas de t, elle ne redonne pas l'estimation attendue.

Dans le cas particulier où $\hat{\psi}(\xi) = \frac{\xi}{1+\xi^2}$, il existe une inégalité plus subtile (que nous utiliserons dans la preuve du théorème 1).

Lemme 6 : Dans le cas particulier où $\psi(x) = \frac{i}{2}$ sign x $e^{-|x|}$, on a si $f(x,t) \in \mathcal{S}(\mathbb{R}^2)$ et $f(x,0) = 0$,

$$(11) \quad \left\| \int_0^\infty Q_t f_t \frac{dt}{t} \right\|_2 \leq c\left(\int_0^\infty \|Q_t f_t\|_2^2 \frac{dt}{t}\right)^{1/2} + c\left(\int_0^\infty \left\|t \frac{\partial}{\partial t} f_t\right\|_2^2 \frac{dt}{t}\right)^{1/2} .$$

Ce résultat est un lemme de "presque-orthogonalité" des morceaux de l'intégrale de gauche. Le "défaut de presque -orthogonalité" est dû au comportement, peut-être aberrant, de $f_t(x)$ par rapport au paramètre t et le terme d'erreur du membre de droite est le prix payé lorsque la dépendance en t est irrégulière.

Pour démontrer cette inégalité, on appelle P_t l'opérateur de convolution avec $\varphi_t(x) = \frac{1}{t}\varphi(\frac{x}{t})$ où $\varphi(x) = \frac{1}{2}e^{-|x|}$ et l'on pose $A_t = -Q_t + 2P_tQ_t = Q_tB_t$ où $B_t = -I + 2P_t$. Alors on a $t\frac{\partial}{\partial t}A_t = Q_t - 8Q_t^3$ comme un calcul immédiat le montre.

Finalement $h(x) = \int_0^\infty Q_t f_t \frac{dt}{t} = h_1(x) - 8h_2(x)$ où

$$h_1(x) = \int_0^\infty \left(\frac{\partial}{\partial t}A_t\right)f_t\, dt = -\int_0^\infty A_t\left(t\frac{\partial}{\partial t}f_t\right)\frac{dt}{t} \quad \text{et} \quad h_2(x) = \int_0^\infty Q_t^3 f_t \frac{dt}{t}$$

(on a supposé que $f_0(x) = 0$, que $f_\infty(x) = 0$ et que la dépendance en t de f_t permet l'intégration par parties).

On applique alors le lemme 5. Pour calculer $\|h_1\|_2$, on écrit $A_t(t\frac{\partial}{\partial t}f_t) = Q_t B_t (t\frac{\partial}{\partial t}f_t)$ et l'on observe que la norme de l'opérateur B_t : $L^2(\mathbb{R}) \to L^2(\mathbb{R})$ est égale à 1. De même pour calculer $\|h_2\|_2$, on écrit $Q_t^3 = Q_t(Q_t^2 f_t)$ et l'on observe que $\|Q_t\| = 1/2$. Ceci termine la preuve du lemme 6.

Revenons au théorème 3. Il est agréable de commencer la preuve en supposant, en outre, que ψ a un support compact. Par exemple supposons que ψ soit port par [-1,1]. Utilisons la décomposition atomique de $T_{2,1}$. Si a(x,t) est un tel atome, vérifions que $\alpha(x) = \int_0^\infty Q_t\, a_t \frac{dt}{t}$ est un atome de $H^1(\mathbb{R})$.

On a, grâce au lemme 5, $\|\alpha\|_2 \leq c(\psi)\left(\int_0^\infty \|a_t\|_2^2 \frac{dt}{t}\right)^{1/2} \leq C|I|^{-1/2}$.

D'autre part l'intégrale de α est nulle. Enfin α est portée par l'intervalle double de I. Donc α est un 2-atome au sens de [1].

Il reste à compléter la démonstration dans le cas général où ψ n'a pas un support compact. On utilise alors la décomposition moléculaire de $H^1(\mathbb{R})$.

Une molécule centrée en x_0 est une fonction $m(x) \in L^1 \cap L^2$ vérifiant $\int_{-\infty}^{+\infty} m(x)\,dx = 0$ et

(12) $$\left(\int_{-\infty}^{+\infty} |m(x)|^2 dx\right)^{1/2} \left(\int_{-\infty}^{+\infty} |m(x)|^2 (x-x_0)^2 dx\right)^{1/2} \leq 1.$$

La norme dans $H^1(\mathbb{R})$ d'une molécule ne dépasse pas une constante absolue C.

<u>Lemme 7</u> : <u>Si</u> a(x,t) <u>est un atome de</u> $T_{2,1}$ <u>basé sur un intervalle</u> I <u>de centre</u> x_0 , <u>alors</u> $\alpha(x) = \int_0^\infty Q_t\, a_t \frac{dt}{t}$. <u>est une molécule centrée en</u> x_0 .

On a, comme ci-dessus, $\int \alpha(x)\,dx = 0$ et $\|\alpha\|_2 \leq C|I|^{-1/2}$. Il reste à montrer que $\left(\int (x-x_0)^2 |\alpha(x)|^2 dx\right)^{1/2} \leq C|I|^{1/2}$. Pour cela on observe que $\iint |a(x,t)|\,dxdt \leq |I|$ pour un atome a(x,t) basé sur I. On a ensuite

$$\int_{|x-x_0| \leq 10|I|} (x-x_0)^2 |\alpha(x)|^2 dx \leq 100|I|^2 \|\alpha\|_2^2 \leq C|I|.$$

Enfin si $|x-x_0| > 10|I|$, on a $|\psi_t(x-y)| \leq \frac{Ct}{(x-x_0)^2}$ pour tout $y \in I$ et donc $|\psi_t * a_t(x)| \leq \frac{Ct}{(x-x_0)^2} \int |a_t(y)|\,dy$. Cela implique

$$|\alpha(x)| \leq \frac{C}{(x-x_0)^2} \iint |a_t(y)|\,dydt \leq C|I|(x-x_0)^{-2} .$$

Finalement $$\int_{|x-x_0| \geq 10|I|} (x-x_0)^2 |\alpha(x)|^2 dx \leq C|I|.$$

Le théorème 3 est démontré.

4. RETOUR AU THEOREME DE CALDERÓN DE 1965

Nous allons interpréter un célèbre théorème de Calderón à la lumière du théorème 3. Soient f, g et h trois fonctions holomorphes dans $\mathfrak{I}m\, z > 0$, nulles à l'infini et reliées par l'équation fonctionnelle $h'(z) = f'(z)\, g(z)$. Alors si f et g appartiennent à l'espace de Hardy H^2, h appartient à H^1 et c'est là le théorème en question.

Posons, si a est un nombre réel et si f est holomorphe dans $\mathfrak{I}m\, z > 0$, $\sigma(f)(a) = \left(\iint_{|x-a|<y} |f'(x+iy)|^2 dx\, dy \right)^{1/2}$. Il est bien connu que $\|\sigma(f)\|_2 = \frac{1}{\sqrt{2}} \|f\|_2$ si f appartient à H^2.

<u>Lemme 8</u> : <u>Si</u> f <u>et</u> g <u>appartiennent à</u> H^2, $F(x,y) = y\, f'(x+iy) g(x+iy)$ <u>appartient à l'espace</u> $T_{2,1}$.

En effet on a, en désignant par $\Gamma(a)$ le secteur $y > |x-a|$,

$\sup_{z \in \Gamma(a)} |g(z)| \leq 3g^*(a)$ et donc

$$SF(a) = \left(\iint_{\Gamma(a)} |f'(x+iy) g(x+iy)|^2 dx\, dy \right)^{1/2} \leq 3g^*(a)\, \sigma f(a) \in L^1 .$$

Soit $\psi \in \mathcal{S}(\mathbb{R})$ une fonction dont la transformée de Fourier $\hat{\psi}(\xi)$ vaut $\xi e^{-\xi}$ si $\xi \geq 0$. Alors pour toute fonction holomorphe $h : \mathbb{R}^2_+ \to \mathbb{C}$ ayant un comportement raisonnable à l'infini et sur l'axe réel, on a

$$h(x) = -4i \int_0^\infty Q_y(y\, h'(x+iy)) \frac{dy}{y} \tag{13}$$

(Q_y est défini à l'aide de ψ comme ci-dessus).

Le théorème de Calderón affirme que la fonction holomorphe $h : \mathbb{R}^2_+ \to \mathbb{C}$, définie par $h(i\infty) = 0$ et $h'(z) = f'(z)g(z)$ appartient à H^1.

Notre point de vue donne la décomposition atomique de cette fonction : $yh'(x+iy) = yf'(x+iy)g(x+iy)$ appartient à $T_{2,1}$ et, en vertu du théorème 3, la fonction $h(x)$ de (13) appartient à H^1 .

5. GENERALISATION DU THEOREME DE CALDERÓN

Oublions les fonctions holomorphes pour construire un opérateur bilinéaire transformant un couple de fonctions à peu près arbitraires en une fonction de $H^1(\mathbb{R})$.

Pour cela, appelons φ une fonction de classe C^2 sur $\mathbb{R}$ vérifiant les inégalités $|\varphi(x)| \leqslant \frac{1}{1+x^2}$, $|\varphi'(x)| \leqslant C(1+x^2)^{-3/2}$ et $|\varphi''(x)| \leqslant C(1+x^2)^{-2}$. A l'aide de cette fonction, on définit l'opérateur P_t par $\mathcal{F}(P_t f)(\xi) = \varphi(t\xi)\hat{f}(\xi)$ ($\mathcal{F}f = \hat{f}$ est la transformée de Fourier de f et t est positif). Enfin $D = -i\frac{d}{dx}$.

Nous noterons $L^2_E(\mathbb{R})$ l'espace de Banach des fonctions $f : \mathbb{R} \to E$ mesurables et telles que $\int_{-\infty}^{+\infty} \|f(x)\|_E^2 dx < +\infty$. Les deux choix de l'espace de Banach E seront $E = L^\infty(0,+\infty)$ et $E = L^2[(0,+\infty) ; \frac{dt}{t}]$.

<u>Théorème 4</u> : <u>Soit</u> $f(x,t) = f_t(x)$ <u>une fonction appartenant à</u> $L^2_E(\mathbb{R})$ <u>pour</u> $E = L^2[(0,+\infty);\frac{dt}{t}]$ <u>et</u> $g(x,t) = g_t(x)$ <u>une fonction appartenant à</u> $L^2_F(\mathbb{R})$ <u>où</u> $F = L^\infty(0,+\infty)$. <u>Alors</u> $D\int_0^\infty (P_t f_t)(P_t g_t)dt \in H^1(\mathbb{R})$.

Supposons dans un premier temps que la fonction φ , de classe C^2, soit portée par l'intervalle [-1,1]. Rappelons que le spectre d'une distribution tempérée S est, par définition, le support de la distribution $\hat{S}$.

Alors le spectre de $P_t f_t$ est contenu dans $[-\frac{1}{t}, \frac{1}{t}]$ et il en est de même pour celui de $P_t g_t$. Le spectre du produit de ces fonctions est contenu dans [-2/t, 2/t] . Appelons ψ une fonction de $\mathcal{S}(\mathbb{R})$ dont la transformée de Fourier soit égale à ξ sur l'intervalle [-2,2] . Alors la transformée de Fourier de ψ_t vaut $t\xi$ sur [-2/t, 2/t] ce qui implique $tD\{(P_t f_t)(P_t g_t)\} = Q_t\{(P_t f_t)(P_t g_t)\}$

<u>Lemme 9</u> : <u>Sous les hypothèses du théorème</u> 4, <u>le produit</u> $(P_t f_t)(P_t g_t)$ <u>appartient à l'espace</u> $T_{2,1}$.

Appelons K(x) la fonction dont $\varphi(\xi)$ est la transformée de Fourier. Alors $|K(x)| \leqslant C(1+x^2)^{-1}$ et, en appelant p_t le semi-groupe de Poisson, on a $|(P_t f_t)(P_t g_t)| \leqslant C(p_t(|f_t|)(p_t(|g_t|))$. On peut donc oublier les valeurs absolues et supposer immédiatement que P_t est le noyau de Poisson.

On a, si $|x-y| \leqslant t$, $|P_t f_t(y)| \leqslant 3f_t^*(x)$ et de même pour $|P_t g_t(y)|$. Alors, si $\gamma(x) = \sup_{t>0} |g_t(x)|$, on a

$$\iint_{|x-y|\leq t} |P_t f_t(y)|^2 \, |P_t g_t(y)|^2 \frac{dydt}{t^2})^{1/2} \leq \sqrt{18}\, \gamma^*(x) (\int_0^\infty (f_t^*(x))^2 \frac{dt}{t})^{1/2} .$$

Pour démontrer que cette fonction appartient à $L^1(\mathbb{R})$, il suffit de vérifier que chaque facteur appartient à $L^2(\mathbb{R})$ ce qui est immédiat.

Lorsque φ est portée par $[-1,1]$, le théorème 4 découle donc du théorème 3.

Le second cas que nous allons envisager est celui où $\varphi \in C^2$ est portée par $[-T,T]$ avec $T \geq 1$ et où $|\varphi(x)| \leq 1$, $|\varphi'(x)| \leq T^{-1}$ et $|\varphi''(x)| \leq T^{-2}$. On s'efforcera alors de contrôler, en fonction de T, la norme de h dans $H^1(\mathbb{R})$.

En reprenant le raisonnement précédent il vient

$$D\{(P_t f_t)(P_f g_t)\} = \frac{T}{t} Q_{t/T} \{(P_t f_t)(P_t g_t)\}.$$

Considérons alors la fonction $\tilde{\varphi}(\xi) = \varphi(T\xi)$ qui est portée par $[-1,1]$ et l'opérateur $\tilde{P}_t$ dont le symbole est $\tilde{\varphi}(t\xi) = \varphi(tT\xi)$. On a donc $P_t = \tilde{P}_{t/T}$ et $|\tilde{\varphi}(\xi)| \leq 1$, $|D\tilde{\varphi}(\xi)| \leq 1$, $|D^2 \tilde{\varphi}(\xi)| \leq 1$.

Finalement

$$D\int_0^\infty (P_t f_t)(P_f g_t)dt = T\int_0^\infty Q_{t/T} \{(\tilde{P}_{t/T} f_t)(\tilde{P}_{t/T} g_t)\}\frac{dt}{t} =$$

$$T\int_0^\infty Q_t\{(\tilde{P}_t f_{tT})(\tilde{P}_t g_{tT})\}\frac{dt}{t} = h(x).$$

Remarquons que f_{tT} et f_t ont même norme dans $L^2_E(\mathbb{R})$ et que, de même g_{tT} et g_t ont même norme dans $L^2_F(\mathbb{R})$. On a donc $\|h\|_{H^1} \leq C\,T\,\|f_t\|_{L^2_E} \|g_t\|_{L^2_F}$ où C est une constante numérique.

Pour finir la preuve du théorème 4, on écrit $\varphi(\xi) = \sum_0^\infty 4^{-k} \varphi_k(\xi)$ où φ_k est portée par $[-2^k, 2^k]$ est de classe C^2 et vérifie $|\varphi_k(\xi)| \leq C$, $|\varphi_k'(\xi)| \leq C2^{-k}$, $|\varphi_k''(\xi)| \leq C4^{-k}$.

Alors $h(x) = D\int_0^\infty (P_t f_t)(P_t g_t)dt = \sum_j \sum_k 4^{-j-k} h_{j,k}(x)$ et l'on a

$\|h_{j,k}\|_{H^1} \leq C(2^j + 2^k)\, \|f_t\|_{L^2_E} \|g_t\|_{L^2_F}$ ce qui permet de sommer la série.

6. UN CALCUL PSEUDO-DIFFERENTIEL AVEC MULTIPLICATION PAR DES FONCTIONS DE $L^\infty(\mathbb{R})$.

Nous allons désormais supposer que $P_t = \frac{I}{I + t^2D^2}$ et que $Q_t = t\,D\,P_t$.

Théorème 5 : *Soit $g_t(x)$ une fonction de $L^2_F(\mathbb{R};dx)$ où $F = L\,(0,+\infty)$: $\sup_{t>0} |g_t(x)| \in L^2(\mathbb{R};dx)$ et soit $b(x)$ une fonction de BMO. Alors le commutateur*

$$\Delta(x,t) = P_t\,\{(b(x))(Q_t\,g_t(x)\} - Q_t\{(b(x))(P_t\,g_t(x))\}$$

vérifie

$$\left(\int_0^\infty \|\Delta(x,t)\|_2^2 \frac{dt}{t}\right)^{1/2} \leq C\|b\|_{BMO} \quad \|\sup_{t>0} |g_t(x)|\|_2 .$$

Pour le voir, on est amené à calculer le produit scalaire, dans $L^2(\mathbb{R}^2_+; \frac{dxdt}{t})$, entre $\Delta(x,t)$ et une fonction de test $f(x,t)$ de norme 1. On utilise le fait que si $u(x) \in L^2(\mathbb{R};dx)$ et $v(x) \in L^2(\mathbb{R};dx)$, on a $\int (P_t u)v\,dx = \int u(P_t v)dx$ tandis que $\int (Q_t u)v\,dx = -\int u\,(Q_t v)dx$. Il vient finalement

$$\iint_{\mathbb{R}^2_+} \Delta(x,t)f(x,t)\,\frac{dxdt}{t} = \int_{-\infty}^{+\infty} h(x)b(x)dx$$

où $$h(x) = \int_0^\infty (P_t f_t)(Q_t g_t)\frac{dt}{t} + \int_0^\infty (Q_t f_t)(P_t g_t)\frac{dt}{t} = D\int_0^\infty (P_t f_t)(P_t g_t)dt \in H^1(\mathbb{R})$$

grâce au théorème 4.

Corollaire 1 : *Désignons par $|||u|||_2$ la norme d'une fonction mesurable $u(x,t)$ dans $L^2(\mathbb{R}^2_+;\frac{dxdt}{dt})$. Alors si $\|b(x)\|_\infty \leq 1$ et si $g_*(x) = \sup_{t>0} |g_t(x)| \in L^2(\mathbb{R})$, on a, avec les notations du théorème 5,*

$$(14) \qquad |||Q_t\,\{(b(x))(P_t g_t(x))\}|||_2 \leq |||Q_t g_t|||_2 + C\|g_*\|_2$$

Pour le voir, on observe simplement que si $L_t : L^2(\mathbb{R}) \to L^2(\mathbb{R})$ est un opérateur linéaire continu de norme 1 et si $u(x,t) = u_t(x)$, on a $|||L_t\,u_t|||_2 \leq |||u_t|||_2$ grâce au théorème de Fubini.

Corollaire 2 : *Avec les hypothèses et notations précédentes, désignons par M l'opérateur de multiplication ponctuelle par $b(x)$. Alors pour toute fonction $f \in L^2(\mathbb{R})$ et tout entier $k \geq 0$, on a*

$$\text{(15)} \qquad |||Q_t(MP_t)^k f|||_2 \leq C(1+k)\|f\|_2 .$$

Si k = 0, cette inégalité n'est autre que le lemme 4. Pour traiter le cas général, on raisonne par récurrence en posant $g_t(x) = (MP_t)^{k-1}f$. Puisque le noyau de P_t est positif, $|g_t(x)| \leq (P_t)^{k-1}|f| \leq f^*(x)$. On a utilisé le fait que les fonctions de $L^1(\mathbb{R})$ qui sont ≥ 0, paires, décroissantes sur $[0, +\infty[$ et d'intégrale égale à 1 forment un semi-groupe pour la convolution.

En appliquant (14) il vient donc

$$|||Q_t(MP_t)^k f|||_2 \leq |||Q_t(MP_t)^{k-1}f|||_2 + C\|f\|_2$$

ce qu'il fallait démontrer.

<u>Corollaire 3</u> : <u>Supposons que l'opérateur linéaire</u> $L_t: L^2(\mathbb{R}) \to L^2(\mathbb{R})$ <u>soit continu, de norme</u> ≤ 1 <u>et dépende mesurablement de</u> t. <u>Soient</u> $b_1(x)$ <u>et</u> $b_2(x)$ <u>deux fonctions de</u> $L^\infty(\mathbb{R})$ <u>de norme</u> ≤ 1 <u>et soient</u> M_1 <u>et</u> M_2 <u>les opérateurs de multiplication ponctuelle par</u> $b_1(x)$ <u>et</u> $b_2(x)$. <u>Alors pour tout entier</u> $p \geq 0$ <u>et tout entier</u> $q \geq 0$, <u>l'opérateur</u>

$$\text{(16)} \qquad \mathcal{L}_{p,q} = \int_0^\infty (P_t M_1)^p Q_t L_t Q_t (M_2 P_t)^q \frac{dt}{t}$$

<u>est continu sur</u> $L^2(\mathbb{R})$ <u>et sa norme ne dépasse pas</u> $C(1+p)(1+q)$.

Pour le voir, on suppose $f \in L^2(\mathbb{R})$, $g \in L^2(\mathbb{R})$, $\|f\|_2 \leq 1$, $\|g\|_2 \leq 1$ et l'on calcule le produit scalaire $\langle \mathcal{L}_{p,q} f, g \rangle$ dans $L^2(\mathbb{R})$. Il vient, en appelant M_1^* l'adjoint de M_1,

$$\langle \mathcal{L}_{p,q} f, g\rangle = \int_0^\infty \langle L_t Q_t (M_2 P_t)^q f, Q_t (M_1 P_t)^p g\rangle \frac{dt}{t}$$

ce qui entraîne $|\langle \mathcal{L}_{p,q} f, g\rangle| \leq |||L_t Q_t (M_2P_t)^q f|||_2 \; |||Q_t(M_1^* P_t)^p g|||_2 \leq$ $|||Q_t(M_2P_t)^q f|||_2 \; |||Q_t(M_1^*P_t)^j g|||_2 \leq C(1+p)(1+q)$.

7. <u>RETOUR AU THEOREME 1</u>

Pour démontrer le théorème 1, il suffit maintenant de vérifier que l'opérateur T_k est la somme d'au plus $4(1+k)^2$ opérateurs $\mathcal{L}_{p,q}$ avec $p+q \leq k$.

Pour le voir, on utilise la formule de représentation de McIntosh et, grâce à $(I + itD)^{-1} = P_t - iQ_t$, on peut développer le produit

$(I+it\,D)^{-1}[M(I+it\,D)^{-1}]^k$ en 2^{k+1} "mots" $Y_{o,t}\,M\,Y_{1,t}\ldots M\,Y_{k,t}$ où $Y_{j,t}\in\{P_t,Q_t\}$. Cela conduit à écrire T_k comme une somme de 2^{k+1} opérateurs élémentaires.

Ce procédé a deux défauts que nous allons ensuite corriger. D'une part le nombre des opérateurs élémentaires est trop élévé : 2^{k+1} au lieu de $4(k+1)^2$. D'autre part les opérateurs élémentaires ne sont pas tous des opérateurs $\mathcal{L}_{p,q}$ car certains ne contiennent qu'une fois l'opérateur Q_t. Voici comment publier ces défauts.

Remarquons d'abord que le mot ne contenant jamais Q_t disparaît puisque P_t est une fonction paire de t alors que $\frac{1}{t}$ est impaire.

Etudions maintenant chacun des $k+1$ mots contenant une seule fois l'opérateur Q_t. On les étudie par la technique du lemme 6. Après intégration par parties, on a $t\frac{\partial}{\partial t}P_t = -2Q_t^2$ ce qui ramnène chacun des $k+1$ opérateurs élémentaires correspondants à s'écrire comme une somme de k opérateurs $\mathcal{L}_{p,q}$.

Etudions enfin les opérateurs élémentaires où Q_t apparaît au moins deux fois. Nous allons les regrouper en "paquets" notés $\pi(p,q)$ où $0\leqslant p<q\leqslant k$; le paquet $\pi(p,q)$ contient tous les mots où le premier j tel que $Y_{j,t}=Q_t$ est $j=p$ et le dernier j tel que $Y_{j,t}=Q_t$ est q. La somme de tous les opérateurs élémentaires du "paquet" $\pi(p,q)$ est donc

$$\text{v.p.}\int_{-\infty}^{-\infty}(P_tM)^p\,Q_t\,\{M(I+it\,D)^{-1}\ldots(I+it\,D)^{-1}M\,\}Q_t(M\,P_t)^{k-q}\,\frac{dt}{t}\;.$$

C'est un opérateur $\mathcal{L}_{p,q}$. Le nombre de paquets est $k(k+1)$. La preuve du théorème 1 est terminée.

REFERENCES

[1] R. R. COIFMAN and G. WEISS : Extensions of Hardy spaces and their use in analysis. Bull. Amer. Math. Soc. 83 (1977) 569-645.

[2] R. R. COIFMAN, A. McINTOSH et Y. MEYER : L'intégrale de Cauchy définit un opérateur borné sur L^2 pour les courbes lipschitziennes. Annals of Mathematics 116 (1982) (A paraître).

Centre de Mathématiques
Ecole Polytechnique
91128 Palaiseau Cedex
France

APPLICATION OF CARLESON MEASURES TO PARTIAL DIFFERENTIAL EQUATIONS AND FOURIER MULTIPLIER PROBLEMS

R. Johnson

Carleson measures were introduced as a means of characterizing measures for which solutions of the Dirichlet problem satisfied particular a priori estimates. In the solution of the corona problem [4], Carleson applied the condition to certain families of discrete measures. We shall study several larger families of Carleson measures and by using both the a priori estimates and duality, give other estimates for solutions of partial differential equations. We use the extended notion of a Carleson measure as considered by Amar-Bonami [1] (see also Barker [3], Duren [11]) and thus, begin by making precise the connection between the various definitions and evolving a workable criterion with which to decide if various measures are generalized Carleson measures. This leads to several equivalent ways to characterize known function spaces. Using the criterion, we compute explicit families of Carleson measures and deduce a priori estimates. For some of these families, the balayage of the Carleson measure has close connections with the Fourier multiplier problem for radial multipliers. We give several examples, including a new proof of the Sobolev mapping theorem.

Notation. We will be working on the Cartesian product of the half-line $R^+ = \{0 < t < \infty\}$ with a space of homogeneous type

X (for the definition see §1), which is $X = R^n$ from §2 forward. We denote

$$R^+ \times X = \{(t,x) | t \in R^+, x \in X\},$$

$$R_+^{n+1} = \{(t,x) | t \in R^+, x \in R^n\},$$

$$[0,\delta] \times Q = \{(t,x) | 0 \le t \le \delta, x \in Q\}.$$

$Q(x,\rho)$ denotes the ball with center x and radius ρ; in R^n, it has measure $|Q|$. A space of homogeneous type is equipped with a measure μ, which in the case of R^n is Lebesgue measure, and thus it is possible to consider $L^p(X)$ and $L^{p,q}(X)$, the Lorentz space. Here

$$L^p(X) = \{f | \quad f \text{ is } \mu\text{-measurable and } \int |f|^p d\mu < +\infty\},$$

for $1 \le p < \infty$, with the usual norm, while

$$L^\infty(X) = \{f | \quad f \quad \mu\text{-measurable and } \operatorname{ess\,sup} |f(x)| < +\infty\},$$

again with the usual norm. We can also define the distribution function of a measurable function

$$m_f(x) = \mu\{x | |f(x)| > \lambda\}$$

and the decreasing rearrangement

$$f^*(t) = \inf\{\lambda | m(\lambda) \le t\}.$$

The Lorentz space is $L^{p,q}(X) = \{f | (\int_0^\infty (f^*(t) t^{1/p})^q \frac{dt}{t})^{1/q} < +\infty\}$, with the same change as above for $q = \infty$.

There are two other function spaces we need. Recall that the $L_{p,\lambda}$ spaces are defined as the set of measurable functions on R^n (or measures if $p = 1$) for which there exists a constant A such that for every ball $Q(x,\rho)$, there is a constant $a_{x,\rho}$ with the property that

$$\int_{Q(x,\rho)} |f(x)-a|^p dx \leq A\rho^\lambda.$$

One must take $1 \leq p \leq \infty$, $0 \leq \lambda < n + p$. For $\lambda = 0$, this is just L^p if $1 < p < \infty$. For $p = 1$, we obtain the bounded measures $M(R^n)$. For $0 \leq \lambda < n$, the constant may be taken equal to zero and for $\lambda = n$, we obtain the John-Nirenberg space BMO, also characterized as

$$\{f| \ \frac{1}{|Q|} \int_Q |f(x)-f_Q|^p dx \leq A^p\},$$

for all cubes Q with sides parallel to the coordinate axes. We have denoted

$$f_Q = \frac{1}{|Q|} \int_Q f(x)dx.$$

For $\lambda > n$, $L_{p,\lambda} = \dot{B}_{\infty\infty}^{\frac{\lambda-n}{p}}$ is the homogeneous Besov space of function satisfying

$$|f(x)-f(y)| \leq A|x-y|^{\frac{\lambda-n}{p}}.$$

The homogeneous Besov spaces are naturally embedded in the Besov scale, $\dot{B}^s_{pq}$, defined for $0 < s < 1$ by the analogue of the above condition and for others by the shift operator. One has

$$\dot{B}^s_{pq} = \{f \mid \left(\int_{R^n} \left(\frac{\|f(\cdot+h)-f(\cdot)\|_p}{|h|^s}\right)^q \frac{dh}{|h|^n}\right)^{1/q} = \|f|\dot{B}^s_{pq}(R^n)\| < +\infty\},$$

with the usual modification for $q = \infty$. The shift operator R^α is densely defined on S', the space of tempered distributions, by the Fourier transform formula

$$\widehat{R^\alpha f}(\xi) = c_\alpha |\xi|^{-\alpha}\hat{f}(\xi).$$

Our convention for the Fourier transform is

$$\hat{f}(\xi) = \int e^{-i<x,\xi>}f(x)dx,$$

for $f \in L^1(R^n)$. The general definition of the Besov spaces follows from the observation that R^α is an isometric isomorphism which allows the definition of

$$\dot{B}^{\alpha+\beta}_{pq}(R^n) = R^\alpha \dot{B}^\beta_{pq}(R^n).$$

A kernel [1] on the space of homogeneous type X is a function $(t,x,y) \to P_t(x,y)$, measurable on $R^+ \times X^2$ such that for any $1 \le q \le \infty$, $P_t(x,\cdot) \in L^q(\mu)$. For any $f \in L^q(\mu)$,

$$P_tf(x) = \int_X P_t(x,y)f(y)d\mu(y).$$

Examples are the Poisson Kernel, $P(x,y) = c_n y(|x|^2+y^2)^{-\frac{n+1}{2}}$, on R_+^{n+1} and the Gauss-Weierstrass Kernel, $W(x,t) = (4\pi t)^{-n/2}\exp(-|x|^2/4t)$. Particularly important are the function χ_A, the characteristic function of a set and the measure δ_P, the Dirac measure at $P \in X$ defined by

$$<\delta_P, f> = f(P),$$

for f a continuous function with compact support on X.

We shall also need the family of maximal functions

$$M_\alpha f(x) = \sup_{Q \ni x} \frac{1}{|Q|^{1-\alpha/n}} \int_Q |f(t)|dt,$$

for $x \in R^n$ and the supremum is taken over all cubes (or balls) centered at x of Lebesgue measure $|Q|$. If $\alpha = 0$, $M_0 f = Mf$ is the Hardy-Littlewood maximal function.

The operation of convolution is denoted by $*$ and thus,

$$f*g(x) = \int_{R^n} f(x-y)g(y)dy.$$

§1. Generalized Carleson measures on a space of homogeneous type

Amar and Bonami have introduced the idea of an α-Carleson measure by means of a condition which differs from the usual condition, in that the appropriate condition must be verified on a tent over an arbitrary open set Ω. We will first explore the connections of this with the usual Carleson measure conditions. Let us recall some definitions.

Definition 1.1. A measure μ on R_+^{n+1} is an α-Carleson measure if there exists a constant C such that for every cube Q in R^n with side δ,

$$\mu([0,\delta] \times Q) \leq C|Q|^\alpha.$$

This definition is, for $\alpha > 1$, due to Duren [11].

The definition of Amar-Bonami is given in an arbitrary space of homogeneous type X (Coifman-Weiss [7]) provided with a metric ρ and a measure μ. We set

$$B(x,t) = \{y \in X | \rho(x,y) < t\},$$

and assume that there exists a constant A such that

$$\mu(B(x,2t)) \leq A\mu(B(x,t)).$$

For any open set $\Omega \subseteq X$, set

$$T(\Omega) = \{(t,x) \in R^+ \times X) | B(x,t) \subseteq \Omega\}.$$

Definition 1.2. A measure w on $R^+ \times X$ is a generalized α-Carleson measure if there is a constant C such that for every open set Ω,

$$|w(T(\Omega))| \le C|\Omega|^{\alpha}.$$

The space of Carleson measures is denoted V^{α}.

Stegenga [20] in studying the space of pointwise multipliers on the space $L^p_\alpha = \{f \mid \hat{f}(\xi) = (1+|\xi|^2)^{-\alpha/2}\hat{\varphi}(\xi),$ for some $\varphi \in L^p(R^n)\}$ introduced yet another generalization of Carleson measures. (Stegenga worked in the disc; for R^n it is more useful to modify his definition slightly.)

Definition 1.3. Let α be a real number. μ is an (α,p)-Carleson measure on R^{n+1}_+ if there exists a constant C_1 such that for every g with, $\hat{g}|\xi|^{n\alpha} = \hat{\varphi}$ for some φ in L^p, and u the Poisson integral of g,

$$u(x,y) = \int_{R^n} P(x-z,y)g(z)dz,$$

we have the inequality

$$\int |u(x,y)|^p d\mu(x,y) \le C^p \int |\varphi(x)|^p dx.$$

The normalizations are a little different in the three cases. In the case of Definition 1.1 and 1.2, $\alpha = 1$ corresponds to the usual Carleson measures, while for Definition 1.3, $\alpha = 0$ corresponds to usual Carleson measures. Our first observation is that for indices larger than one all definitions are equivalent.

Theorem 1.3. Let $\alpha \le 0$, $\mu \ge 0$. The following are equivalent:

(1) There is a constant C such that for every cube Q, $\mu(Q \times [0,\delta]) \le C|Q|^{1-p_1\alpha}$.

(2) There is a constant C such that for every open set Ω

$$\mu(T(\Omega)) \le C|\Omega|^{1-p_1\alpha}$$

(3) There is a $p_1 \ge 2$ and a C such that

$$\int |u(x,y)|^{p_1} d\mu(x,y) \le C^{p_1} \int |\varphi(x)|^{p_1} dx,$$

whenever $u(x,y) = P_y * g$, where $\hat{g}(\xi) = |\xi|^{-n\alpha}\hat{\varphi}(\xi)$, for some $\varphi \in L^{p_1}$.

(4) For every p, $1 < p < \infty$, there is a constant C (which might depend on n and p, but not on the function g) such that

$$\left(\int |u(x,y)|^{p(1-p_1\alpha)} d\mu(x,y)\right)^{\frac{1}{p(1-p_1\alpha)}} \le C\left(\int |g(x)|^p dx\right)^{1/p},$$

for every $g \in L^p(R^n)$.

(5) There exists a p, $1 < p < \infty$, and a constant C such that for every $g \in L^p(R^n)$

$$\left(\int |u(x,y)|^{p(1-p_1\alpha)} d\mu(x,y)\right)^{1/p(1-p_1\alpha)} \le C\left(\int |g(x)|^p dx\right)^{1/p}.$$

Various parts of the proof are implicit in Amar-Bonami [1] or in Stegenga [20]. We will prove the result by showing that $3 \to 5 \to 1 \to 2 \to 3$, and then handle (4) separately.

Proof. 3 → 5. If $\alpha = 0$, (3) is the same as (5), and we may assume $\alpha < 0$, and write

$$\varphi(x) = R^{-n\alpha}g(x)$$

$$= C\int \frac{g(z)}{|x-z|^{n+n\alpha}}\,dz.$$

Sobolev's embedding theorem says that if $g \in L^r(R^n)$, where $\frac{1}{r} + \alpha = \frac{1}{p_1}$, $\varphi \in L^{p_1}$ and (3) will hold. For this r, $r = \frac{p_1}{1-p_1\alpha}$, and (5) becomes

$$\left(\int |u(x,y)|^{r(1-p_1\alpha)}\,d\mu\right)^{1/r} \le C^{p_1}\left(\int |\varphi(x)|^{p_1}dx\right)^{1/p_1}$$

$$\le C^{p_1}\left(\int |g(x)|^r dx\right)^{1/r}.$$

5→1: Given a cube Q, take $g(x) = \chi_Q$ and observe that there is a constant depending only on n such that

$$u(x,y) \ge d_n \quad \text{on} \quad Q \times [0,\delta],$$

and (1) follows from (5).

1 → 2: (The proof below works for any space of homogeneous type allowing one to show the equivalence of (1) and (2) for any space of homogeneous type.)

First we observe that for any set Ω, if we put $d(x,{}^c\Omega) = d(x) = \inf\{\rho(x,y) \mid y \in {}^c\Omega\}$, then we can easily characterize $T(\Omega)$.

Proposition 1.4. $T(\Omega) = \{(x,t) | x \in \Omega, 0 < t < d(x)\}$.

Note that, in fact, $B(x,d(x)) \subseteq \Omega$ for any $x \in \Omega$. We apply Whitney's lemma ([7], p.70) in the form which works for spaces of homogeneous type. Given an open subset Ω of X, there is a sequence of balls $Q(x_i,\rho_i)$ such that $\Omega = \cup Q(x_i,\rho_i)$ and a fixed h with the property that $Q(x_i,h\rho_i)$ meets ${}^c\Omega$. We then have

$$T(\Omega) \subseteq \bigcup_j (Q(x_j(1+h),\rho_j) \times [0,(1+h)\rho_j]),$$

and

$$\mu(T(\Omega)) \leq \sum_j \mu(Q_j \times [0,(1+h)\rho_j]) \leq \sum_j [\mu(Q_j)]^{1-p_1\alpha} \leq [\sum_j \mu(Q_j)]^{1-p_1\alpha},$$

since $\alpha \leq 0$. Each point of Ω is in at most M of the balls and we conclude

$$\mu(T(\Omega)) \leq C|\Omega|^{1-p_1\alpha}.$$

2→3: The argument, due to Stein, which worked for $p_1 = 2$ [16] also works in this case and replaces an earlier longer proof of the author's. We note that

$$\left(\int_0^\infty \int_{R^n} |u(x,y)|^{p_1} y^{-p_1 n\alpha} dx \frac{dy}{y}\right)^{1/p_1}$$

$$= \|g|\dot{B}^{n\alpha}_{p_1p_1}\| = \|R^{n\alpha}\varphi|\dot{B}^{n\alpha}_{p_1p_1}\| = \|\varphi|\dot{B}^{0}_{p_1p_1}\|,$$

and since $p_1 \geq 2$. $\dot{B}^0_{p_1 p_1} \supseteq \dot{F}^0_{p_1 2} = L^{p_1}$ and if $\varphi \in L^{p_1}$,

$$\|\varphi|\dot{B}^0_{p_1 p_1}\| \leq C\|\varphi|L^{p_1}\|.$$

Thus, it suffices to prove that

$$\iint |u(x,y)|^{p_1} \, d\mu(x,y) \leq C^{p_1}\int_0^\infty\int_{R^n} |u(x,y)|^{p_1} y^{-p_1 n\alpha} dx\frac{dy}{y} .$$

Divide $R^{n+1}_+ = \cup Q^k_j$, where if $(x,y) \in Q^k_j$, $2^k \leq y \leq 2^{k+1}$ and $\{Q^k_j | -\infty < j < \infty\}$ represents a partitioning of the strip $2^k \leq y \leq 2^{k+1}$ into cubes. Then

$$\iint |u(x,y)|^{p_1} d\mu(x,y) = \sum_{j,k}\int_{Q^k_j} |u|^{p_1} d\mu \leq \sum_{j,k} (\sup|u|)^{p_1}\mu(Q^k_j).$$

By assumption, $\mu(Q^k_j) \leq C_1[2^{kn}]^{1-p_1\alpha}$, and we estimate $\sup_{Q^k_j}|u|$ using the fact that u is harmonic. There is a universal constant (10 will probably work) such that for every x in Q^k_j, $B_{2^{k-1}}(x) \subseteq hQ^k_j \times [0,h\delta^k_j]$, where hQ^k_j is a cube with the same center as Q^k_j and whose side length is expanded to $h\delta^k_j$. The mean value theorem for harmonic functions gives

$$|u(x)|^{p_1} \le c_1 \frac{1}{2^{k(n+1)}} \int_{B_{2^{k-1}}} |u(z,y)|^{p_1} dzdy$$

$$\le \frac{c_1}{2^{k(n+1)}} \int_{hQ_j^k} \int_0^{h\delta_j^k} |u(z,y)|^{p_1} dzdy,$$

and we obtain

$$\int |u|^{p_1} d\mu \le C \sum_j \sum_k 2^{kn(1-p_1\alpha)} 2^{-k(n+1)} \int_{hQ_j^k} \int_0^{h\delta_j^k} |u|^{p_1} dzdy$$

$$= C_1 \sum_j \sum_k \int_{hQ_j^k} \int_0^{h\delta_j^k} |u|^{p_1} y^{-np_1\alpha} dz \frac{dy}{y}$$

$$= C_2 \int_0^\infty \int_{R^n} |u(x,y)|^{p_1} y^{-np_1\alpha} dx \frac{dy}{y} \le C_2 \int |\varphi(x)|^{p_1} dx.$$

Finally, it is clear that (4) implies (5). Moreover (2) implies (4) by the argument in Barker [3] since the exponent is larger than 1.

The two possible definitions are also equivalent for $\alpha = 0$ since both require that w be a finite measure; however, for $0 < \alpha < 1$, they are not equivalent. To see this, consider R_+^{n+1} and special Carleson measures of the form $fdx \otimes \delta_{t_0}$ where δ_{t_0} is the Dirac measure at t_0 on the real line. The condition $|w([0,\delta] \times Q)| \le C|Q|^\alpha$ is equivalent to the condition

$$\int_Q |f(y)|dy \le C|Q|^\alpha,$$

which is the Morrey condition $f \in L_{1,n\alpha}$, and the constant C can be chosen independent of t_0. However, since

$$|w_{t_0}(T(\Omega))| = \int_{\{x\in\Omega|d(x)\ge t_0\}} |f(x)|dx,$$

the following theorem shows that we cannot have a uniform bound $|w_{t_0}(T(\Omega))| \le C|\Omega|^\alpha$.

<u>Theorem 1.5</u>. The following are equivalent on R^n.

(1) For every open set Ω, $\int_\Omega |f(x)|dx \le C|\Omega|^\alpha$.

(2) For every measurable set E, $\int_E |f(x)|dx \le C|E|^\alpha$.

(3) For every compact set K, $\int_K |f(x)dx \le C|K|^\alpha$.

(4) $f \in L^{p,\infty}(R_n)$, $\frac{1}{p} = 1 - \alpha$.

<u>Proof</u>. The equivalence of (1) through (3) follows because Lebesgue measure is a regular inner measure.

(4) $\Rightarrow$ (1): $\int_E |f(x)|dx \le \int_0^{|E|} f^*(t)dt \le \|f\|_{p,\infty} \int_0^{|E|} t^{-1/p}dt$

$$= \|f\|_{p,\infty} \frac{|E|^{1-1/p}}{1-1/p}.$$

(1) → (4): Let $E_\lambda = \{x \mid |f(x)| > \lambda\}$ and it is clear that $f \in L^{p,\infty}$.

Remark. Condition (2) was the original definition of $L^{p,\infty}$ [13].

In view of the above-cited connection with the $L_{p,\lambda}$ scale, it seems appropriate to remark that strengthening the usual $L_{p,\lambda}$ condition on balls to require that it holds for arbitrary open sets gives a scale that is better behaved with respect to interpolation.

Corollary 1.6. $$\{\mu \mid |\mu(\Omega)| \le C|\Omega|^\alpha\} = \begin{cases} M(R^n), & \alpha = 0 \\ L^{p,\infty}(R^n), & \frac{1}{p} = 1 - \alpha,\ 0 < \alpha < 1 \\ L^\infty(R^n), & \alpha = 1 \end{cases}$$

The other consequence is the non-equivalence of the two definitions for $0 < \alpha < 1$.

Theorem 1.7. $V^\alpha(R_+^{n+1}) \subsetneq \{w \mid |w([0,\delta] \times Q)| \le C|Q|^\alpha$, for every ball $Q\}$.

When the index is less than 1, the conditions bifurcate.

Theorem 1.8. Let $0 < \alpha < \dfrac{1}{p_1}$, $\mu \ge 0$. The following are equivalent:

(1) There is a constant C such that for every open set Ω,

$$\mu(T(\Omega)) \le C|\Omega|^{1-p_1\alpha}.$$

(2) For every p, $1 < p < \infty$, there is a constant C such

that

$$\left(\iint |u(x,y)|^{p(1-p_1\alpha)} d\mu(x,y)\right)^{\frac{1}{p(1-p_1\alpha)}} \leq C\|g\|_{(p,p(1-p_1\alpha))},$$

for every $g \in L^{p,p(1-p_1\alpha)}(R^n)$.

(3) There is a p such that $1 < p < \infty$ and a constant C such that for every $g \in L^{p,p(1-p_1\alpha)}(R^n)$

$$\left(\iint |u(x,y)|^{p(1-p_1\alpha)} d\mu(x,y)\right)^{\frac{1}{p(1-p_1\alpha)}}$$

$$\leq C\|g\|_{(p,p(1-p_1\alpha))}.$$

<u>Theorem 1.9</u>. Let $0 < \alpha < 1/p_1$, $p_1 > 1$, $\mu \geq 0$. The following are equivalent:

(1) There is a constant C such that for every open set Ω,

$$\mu(T(\Omega)) \leq C\dot{B}_{n\alpha,p_1}(\Omega),$$

where $\dot{B}_{n\alpha,p_1}(\Omega)$ is the capacity defined by

$$\dot{B}_{n\alpha,p_1}(\Omega) = \inf\{\|f\|_{p_1}^{p_1} \mid f \geq 0, R^{n\alpha}f \geq 1 \text{ on } \Omega,\ f \in L^{p_1}\}.$$

(2) There is a constant C such that

$$\iint |u(x,y)|^{p_1} d\mu(x,y) \leq C^{p_1}\|g|\dot{F}^{n\alpha}_{p_1 2}\|^{p_1},$$

for every $g \in \dot{F}^{n\alpha}_{p_1 2}$.

The proof of Theorem 1.8 is easy. We will prove $1 \to 2 \to 3 \to 1$. That (1) $\to$ (2) follows from the inclusion

$$\{(x,y) \mid |u(x,y)| > \lambda\} \leq T\{x \mid |Mg(x)| > \lambda\},$$

and hence, if (1) holds,

$$\mu\{(x,y) \mid |u(x,y)| > \lambda\} \subseteq \|\mu\| \, |\{x \mid |Mg(x)| > \lambda\}|^{1/q},$$

with $1/q = 1 - p_1\alpha$, and we have

$$u_\mu^*(t) \leq (Mg)^*((t/\|\mu\|)^q),$$

which gives the estimate

$$\begin{aligned} \int |u(x,y)|^{p(1-p_1\alpha)} d\mu(x,y) &= \int_0^\infty u_\mu^*(t)^{p(1-p_1\alpha)} dt \\ &\leq \int_0^\infty (1-p_1\alpha)\|\mu\| [(Mg)^*(s)s^{\frac{1}{p}}]^{p(1-p_1\alpha)} \frac{ds}{s} \\ &= (1-p_1\alpha)\|\mu\| \, \|Mg\|^{p(1-p_1\alpha)}_{(p,p(1-p_1\alpha))}, \end{aligned}$$

and (2) follows. Clearly, (2) $\to$ (3) and (3) $\to$ (1) by the argument of Theorem 1.1. If $g = \chi_\Omega$, for Ω an open set with $|\Omega| < +\infty$, we have $u(x,y) \geq d_n$ on $T(\Omega)$ and (1) follows.

The proof of Theorem 1.9 requires an auxiliary lemma.

Lemma 1.10. The measure μ satisfies (2) of Theorem 1.9 iff there exists C such that

$$\iint |v(x,y)|^{p_1} d\mu(x,y) \le C\int |f(x)|^{p_1} dx,$$

for $g = R^{n\alpha}f$, $f \ge 0$ in $L^{p_1}(R^n)$ and v the Poisson integral of g.

Proof. If $g = R^{n\alpha}f$, $\|g|\dot{F}^{n\alpha}_{p_1,2}\|^{p_1} = \|f|\dot{F}^{0}_{p_1,2}\|^{p_1} = \|f|L^{p_1}\|^{p_1}$.

If the above condition holds for every $f \ge 0$, since the Poisson kernel is nonnegative, it holds for every $f \in L^{p_1}$ and thus by the above identity, for every $g \in \dot{F}^{n\alpha}_{p_1,2}$,

$$\left(\iint |v(x,y)|^{p_1} d\mu(x,y)\right)^{1/p_1} \le C\,\|g(x)|\dot{F}^{n\alpha}_{p_1,2}\|.$$

Conversely, if this condition holds, the identity shows that

$$\left(\iint v^{p_1} d\mu\right)^{1/p_1} \le C\left(\int |f(x)|^{p_1} dx\right)^{1/p_1},$$

holds for every $f \ge 0$.

Proof of Theorem 1.9. First, (2) → (1) because if $f \ge 0$ and $g = R^{n\alpha}f \ge 1$ on Ω, $v(x,y) = \int P(x-z,y)g(z)dz$ satisfies by Lemma 1.10,

$$\iint |v(x,y)|^{p_1} d\mu(x,y) \le C\int |f(x)|^{p_1} dx,$$

and also since $g \ge \chi_\Omega$, $v(x,y) \ge d_n$ on $T(\Omega)$, which gives

$$d_n^{p_1}\mu(T(\Omega)) \le C\|f\|_{p_1}^{p_1}.$$

Taking the infimum over all such f gives (1).

We will show that (1) → (2) by showing that the condition of Lemma 1.10 is satisfied. Note that $M(R^{n\alpha}f) \le R^{n\alpha}(Mf)$ since f and the kernel of $R^{n\alpha}f$ are both nonnegative. This allows us to estimate, with

$$A_t = \{(x,y) \mid |u(x,y)| > t\},$$

from the inclusion

$$A_t \subseteq T(\{x \mid Mg(x) > t\}),$$

where g is the initial value of v,

$$\begin{aligned}\int |v(x,y)|^{p_1} d\mu(x,y) &= \int_0^\infty \mu(A_t)dt^{p_1} \\ &\le C\int_0^\infty \dot{B}_{n\alpha,p_1}(M(R^{n\alpha}f) > t)dt^{p_1} \\ &\le C\int_0^\infty \dot{B}_{n\alpha,p_1}(R^{n\alpha}(Mf) > t)dt^{p_1} \\ &\le C\int (Mf)^{p_1}dx,\end{aligned}$$

by the result of Dahlberg [9], Maz'ja [18]. The familiar estimate for the Hardy-Littlewood maximal function completes the proof.

Nevertheless, in the case $0 < \alpha < 1$, we can give a criterion which only requires that we look at cylinders, but over an arbitrary open set.

Theorem 1.11. If there exists a constant C such that for any open set Ω with $\delta_0 = \sup_{x \in \Omega} d(x) < +\infty$,

$$|w([0,\delta_0] \times \Omega)| \le C\mu(\Omega)^\alpha,$$

then $w \in V^\alpha$.

Proof. If $\delta_0 < +\infty$, we have $T(\Omega) \subseteq [0,\delta_0] \times \Omega$ and the Carleson condition is satisfied. If $\delta_0 = +\infty$, then there is a sequence x_n such that $d(x_n) \to \infty$. Since $B(x,d(x_n)) \subseteq \Omega$, this implies $\mu(\Omega) = \infty$ and hence, the condition is automatically satisfied.

It also follows that $\mu(\Omega) \ge b_n \delta_0^n$ for $X = R^n$.

We shall later discuss the balayages of Carleson measures in V^α and use the fact that these balayages are in $L^{p,\infty}$. For some purposes it is useful to have the balayage in L^p. The class of measures must be restricted in order to have this obtained.

Definition 1.12. Let $P_t^0(x,y) = \frac{1}{\mu(B(x,t))} \chi_{B(x,t)}(y)$, and set

$$S_w(y) = \int P_t^0(x,y)dw(t,x).$$

One calls W^α the space of measures on $R^+ \times X$ such that the area function $S_{|w|}$ associated to $|w|$ belongs to $L^p(d\mu)$,

$\frac{1}{p} = 1 - \alpha, \quad 0 \le \alpha \le 1.$

As remarked by Amar-Bonami, $S_w(y)$ is equal to, up to a constant factor,

$$\int_{\Gamma(y)} \frac{1}{\mu(B(y,t))} \, dw(t,x),$$

where $\Gamma(y)$ is the cone $\{(t,x) | \rho(x,y) \le t\}$. When $w = t|\nabla u|^2 dt dx$ in $R^+ \times R^n$, where u is the Poisson integral of a function f on R^n, $S_w(y) = Sf(y)^2$, where S is the Lusin area function of f [21].

§2. Families of Carleson measures in R_+^{n+1}

With the criterion of Theorem 1.11 in hand, it is easy to construct families of Carleson measures. Since the a priori estimates will depend on the Carleson measure norm, where possible we provide a norm bound.

Proposition 2.1. $\delta_{(\xi_0,t_0)} \in V^\alpha$ for every $\alpha \geq 0$ and

$$\|\delta_{(\xi_0,t_0)}\|_{V^\alpha} = \frac{1}{(b_n t_0^n)^\alpha}.$$

Proof. If $(\xi_0,t_0) \notin T(\Omega)$, $\delta_{(\xi_0,t_0)}(T(\Omega)) = 0$ and we are done. If $(\xi_0,t_0) \in T(\Omega)$, $B(\xi_0,t_0) \subseteq \Omega$ and $|\Omega| \geq b_n t_0^n$. This gives the estimate

$$|\delta_{(\xi_0,t_0)}(T(\Omega))| \leq \left(\frac{|\Omega|}{b_n t_0^n}\right)^\alpha,$$

with equality when $\Omega = B(\xi_0,t_0)$.

Corollary 2.2. If $\sum |a_j| t_j^{-n\alpha} < +\infty$, $\sum a_j \delta_{(\xi_j,t_j)} \in V^\alpha$.

Remark. An easy computation shows that $S_\delta(y) = t_0^{-n} \chi_{B(\xi_0,t_0)}(y)$, which belongs to every L^p space and thus for $0 \leq \alpha \leq 1$, $\delta_{(\xi_0,t_0)} \in W^\alpha$. The same applies for the Corollary.

Proposition 2.3. If $0 < \alpha \le 1$, $\{fdx \otimes \delta_{t_0} | t_0 > 0\}$ is a bounded family in V^α if and only if $f \in L^{p,\infty}(R^n)$, $\frac{1}{p} = 1 - \alpha$, and

$$\|fdx \otimes \delta_{t_0}\|_{V^\alpha} \le \|f\|_{p,\infty}.$$

Proof. Consider $[0,\delta_0] \times \Omega$. If $t_0 > \delta_0$, there is nothing to prove. If $t_0 < \delta_0$, $|(fdx \otimes \delta_{t_0})([0,\delta_0] \times \Omega) = \int_\Omega |f(x)|dx$, and the result follows from Theorem 1.5.

Proposition 2.4 . If $0 < \alpha \le 1$, $\{fdx \otimes \delta_{t_0}\}$ is a bounded family in W^α if and only if $f \in L^p(R^n)$, $\frac{1}{p} = 1 - \alpha$.

In this case,

$$S_{|w_{t_0}|}(y) = \frac{1}{b_n t_0^n} \int_{B(y,t_0)} |f(z)|dz,$$

is a competitor in the definition of the Hardy-Littlewood maximal function and the required estimates follow. The appropriate substitute for $\alpha = 0$ is that for any $\mu \in M(R^n)$, $\mu \otimes \delta_{t_0}$ forms a bounded family in $V^0(R_+^{n+1})$. Although the Hardy-Littlewood maximal function is not in L^1, the functions $S_{|w_{t_0}|}$ are and $fdx \otimes \delta_{t_0}$ forms a bounded family in W^0 if and only if $f \in L^1$.

Proposition 2.5. Given a measurable function $t(x)$, as f ranges over a bounded subset $L^{p,\infty}$, $\frac{1}{p} = 1 - \alpha$, $1 < p < \infty$,

$f(x)\delta_{t(x)}dx$ forms a bounded set of measures in V.

Proof. Denoting $\mu_f(x,y) = f(x)\delta_{t(x)}dx$, it follows that

$$|\mu|(T(\Omega)) = \int_\Omega\int_0^{\delta(x)} |f(x)|\delta_{t(x)}dx$$

$$= \int_{\{x\in\Omega|t(x)\le\delta(x)\}} |f(x)|dx,$$

and since t is measurable and δ is continuous, we obtain

$$|\mu|(T(\Omega)) \le \|f\|_{p,\infty}|\Omega \cap \{x|t(x) \le \delta(x)\}|^\alpha \le \|f\|_{p\infty}|\Omega|^\alpha.$$

Proposition 2.6. For any family of measures such that the mapping $t \to \mu_t(\Omega)$ is measurable for every open set Ω, and $\mu_t \in M(R^n)$ with $\|\mu_t(\cdot)\|_M \le C$, then if $\alpha > 0$, $\{\mu_t(\cdot)t^{n\alpha}dt/t\}$ forms a bounded family in V^α.

Proof. For an open set Ω such that $\delta_0 < \infty$, we have

$$|w([0,\delta_0] \times \Omega)| \le \int_0^{\delta_0} t^{n\alpha}\mu_t(\Omega)\frac{dt}{t} \le C\int_0^{\delta_0} t^{n\alpha}\frac{dt}{t} \le C'|\Omega|^\alpha.$$

Two particular cases of this are $\mu(\cdot,t) = \delta_{\xi_0}$ in which case $C = 1$ is independent of ξ_0 and $\mu(\cdot,t) = \delta_{\xi(t)}$, where $\xi: (0,\infty) \to R^n$ is measurable.

The function $S_{\delta_{\xi_0}t^{n\alpha}dt/t}(y) \sim |\xi_0-y|^{\alpha-1} = |\xi_0-y|^{-1/p}$ does not belong to L^p for any p which shows that $\delta_{\xi_0}t^{n\alpha}\frac{dt}{t} \notin W^\alpha$.

Coifman-Meyer [6] have shown that for suitable restrictions on ψ, $|a*\psi_t|^2 dx\,\frac{dt}{t}$ defines a Carleson measure ($\alpha=1$). In fact, one has a more general result. Let $\psi \in L^1(R^n)$ be a function such that $|\psi(x)| \le A(1+|x|)^{-(n+1)}$, $\int \psi(x)dx = 0$. We set $\psi_t(x) = t^{-n}\psi(x/t)$.

Proposition 2.7. If $a \in L^{r,\infty}$, $|a*\psi_t(x)|^2 dx\,\frac{dt}{t} \in V^{1-2/r}$; if $a \in L^r(R^n)$, $|a*\psi_t|^2 dx\,\frac{dt}{t} \in W^{1-2/r}$, $2 < r < \infty$.

Proof. We define an operator by

$$T(a)(x,t) = |a*\psi_t(x)|^2 dx\,\frac{dt}{t}\ .$$

The Corollary of ([6], Lemma 1, p. 148) shows that

$$\int_0^\infty\int_{R^n} T(a)(x,t) = \int_0^\infty \|a*\psi_t\|_2^2\,\frac{dt}{t} \le C\|a\|_2^2 .$$

This shows that $T\colon L^2 \to V^0$. It is shown in Lemma 2 on the same page that $T\colon BMO \to V^1$. Since T is quasi-linear, it follows from interpolation, after linearizing, that

$$T\colon (L^2, BMO)_{\theta q} \to (V^0, V^1)_{\theta q}.$$

Amar and Bonami have shown that $(V^0,V^1)_{\theta\infty} = V^\theta$, while $(V^0,V^1)_{\theta r} = W^\theta$, $\frac{1}{r} = \frac{1-\theta}{2}$; the corresponding interpolation theorem for (L^2, BMO) is well-known and yields the proposition along with a norm bound.

Examples. (1) If we take $\psi_j(x) = \frac{-(n+1)C_n x_j}{(1+|x|^2)^{\frac{n+3}{2}}}$,

$(\psi_j)_t = t \frac{\partial P}{\partial x_j}(x,t)$, and if $\psi_0(x) = \frac{|x|^2+1-(n+1)C_n}{(|x|^2+1)^{\frac{n+3}{2}}}$,

$(\psi_0)_t = t \frac{\partial P}{\partial t}$, giving the familiar facts that if $a \in L^{r,\infty}$,

$t|\frac{\partial u}{\partial x_j}(x,t)|^2 dxdt \in V^{1-2/r}$, $t|\frac{\partial u}{\partial t}(x,t)|^2 dxdt \in V^{1-2/r}$ and, summing, $t|\nabla u|^2 dxdt \in V^{1-2/r}$.

More complicated choices of ψ would give $t^K \nabla^K u$.

(2) If we take $\psi_j(x) = -(x_j/2)(4\pi)^{-n/2} e^{-|x|^2/4}$, $(\psi_j)_t = t \frac{\partial W}{\partial x_j}(x,t^2)$. To obtain temperatures, we must change variables, but we leave this to the interested reader.

Proposition 2.8. (1) If $0 < \alpha < \frac{1}{p}$, and $\beta = 1 - \frac{1}{p} + \alpha$, then if $f \in L^{p,\infty}(R^n)$, $f(x)t^{n\alpha}dx \frac{dt}{t} \in V^\beta$.

(2) If $\alpha \geq \frac{1}{p}$ and β is as above, then if $f \in L_{1,n-n/p}(R^n)$, $f(x)t^{n\alpha}dx \frac{dt}{t} \in V^\beta$.

Proof. (1) follows from the estimate

$$\int_\Omega \int_0^{\delta_0} |f(x)| t^{n\alpha} dx \frac{dt}{t} \leq \|f\|_{p,\infty} |\Omega|^{1-1/p} \delta_0^{n\alpha}/n\alpha \leq C\|f\|_{p,\infty} |\Omega|^\beta .$$

In case (2), $\beta \geq 1$ and we need only check the condition on balls.

The balayage of a measure of this type will give new proof of Sobolev's theorem. However, if one wished to decide whether $f(x)t^{n\alpha}dx\,\frac{dt}{t} \in W^{\beta}$, one finds that $S_w(y) = R^{n\alpha}f(y)$. A non-circular proof can be constructed to show that if $f \in L^p(R^n)$, $0 < \alpha < \frac{1}{p}$, then $f(x)t^{n\alpha}dx\,\frac{dt}{t} \in W^{\beta}$.

Proposition 2.9. (1) If $f \in L^{p_0,\infty}(R^n)$, $\varphi(t) \in L^{p_1,\infty}(R^+)$ and $1 - 1/p_1 < \frac{n}{p_0}$, $1 < p_0, p_1 < +\infty$, $\beta = 1 - 1/p_0 + (1-1/p_1)1/n$, $f(x)\varphi(t)dxdt \in V^{\beta}$.

If $p_1 = 1$, $\mu \in M(R^+)$, $fdxd\mu \in V^{1-1/p_0}$. If $p = 1$, $\mu \in M(R^n)$, $d\mu\,\varphi(t)dt \in V^{(1-1/p_1)1/n}$.

(2) If $1 - 1/p_1 \geq \frac{n}{p_0}$, β as above and $f \in L_{1,n-n/p_0}$, $f(x)\varphi(t)dxdt \in V^{\beta}$.

Finally Proposition 1 of Amar-Bonami [1] can be used to provide another family of generalized Carleson measures. We generalize their result slightly.

Theorem 2.10. If $w_0 \in V^{\beta_0}$, $h \in L^{p,\infty}(w_0)$, then $hw_0 \in V^{\beta_0(1-1/p)}$. If $h \in L^1(w_0)$, $hw_0 \in V^0$ and if $h \in L^{\infty}(w_0)$, $hw_0 \in V^{\beta_0}$.

Proof. The same argument as before gives

$$\int_{T(\Omega)} |w|dxdt = \int_{T(\Omega)} hw_0 \le \int_0^{|w_0|(T(\Omega))} h^*(t)dt$$

$$\le \|h\|_{p,\infty} \int_0^{|w_0|(T(\Omega))} t^{-1/p} dt$$

$$\le \|h\|_{p,\infty} (C|\Omega|^{\beta_0})^{1-1/p}.$$

The modifications for $h \in L^1$ and L^∞ are clear.

Amar-Bonami considered the case $\beta_0 = 1$ and showed that the conditions are also sufficient for that case.

Since $dxdt \in V^{1+1/n}$, we deduce the following corollary.

Corollary 2.11. (1) If $h \in L^{n+1,\infty}(R_+^{n+1})$, $hdxdt \in V^1$.

(2) If $h \in L^{p,\infty}(R_+^{n+1})$, $hdxdt \in V^{(1+1/n)(1-1/p)}$, $1 < p < n + 1$.

(3) If $p \ge n + 1$, it suffices that for each cube with side on the $t = 0$ axis in R_+^{n+1},

$$\int_Q |h|dxdt \le C|Q|^{(1-1/p)},$$

in order that $hdxdt \in V^{(1+1/n)(1-1/p)}$.

Since $f(x)t^{-\lambda} \in L^{p,\infty}(R_+^{n+1})$ if and only if $\lambda = \frac{1}{p}$ and $f \in L^p$, we see that the Corollary does not include Proposition 2.8.

§3. Application to a priori estimates in R_+^{n+1}

As we have remarked in §1, the a priori estimate which follows for a Carleson measure takes the following form. For an L^p function g, if $u(x,y)$ is the Poisson integral of g,

$$u(x,y) = \int P(x-z,y)g(z)dz,$$

then if $w \in V^\alpha$,

$$\left(\int |u(x,y)|^{p\alpha} dw(x,y)\right) \le C\|w\|_{V^\alpha} \|g\|^{p\alpha}_{(p,p\alpha)}, \qquad 1 < p < \infty.$$

For $\alpha = 1$ this is the familiar estimate of Carleson [4] and for $\alpha > 1$ the extension of Duren [11] and Barker [3], if one notes that since $\alpha > 1$, $L^p \subseteq L^{(p,p\alpha)}$. For $p\alpha = 1$, it was proved by Amar-Bonami [1], and their proof was used in §1 to get the general inequality. The key to this proof was the set-theoretic inclusion, valid for harmonic functions,

$$\{(x,y) \mid |u(x,y)| > \lambda\} \subseteq T(\{x \mid Mg(x) > \lambda\}).$$

If ψ is an arbitrary function on R_+^{n+1}, and we define

$$\psi^*(x) = \sup_{|y-x|<t} |\psi(y,t)|,$$

it follows easily that

$$\{(x,y) \mid |\psi(x,y)| > \lambda\} \subseteq T\{x \mid \psi^*(x) > \lambda\},$$

which gives the estimate

$$\int_0^\infty\int_{R^n} |\psi(x,y)|^{p\alpha} dw(x,y) \le C\|w\|_{V^\alpha} \|\psi^*\|^{p\alpha}_{(p,p\alpha)},$$

due to Stein ($\alpha = 1$, [21]) and Barker ($\alpha > 1$, [3]). For this to be useful, an estimate of ψ^* in terms of $\psi(x,0) = g(x)$, is needed. This follows when $\psi^*(x) \le AMg(x)$, where M is the Hardy-Littlewood maximal function, or any other maximal function for which L^p estimates are known. The fact that Carleson measures can be used to control harmonic functions is now a consequence of the fact (Stein [21], 92-93) that for a harmonic function u with trace g,

$$|u(y,t)| \le A\left(1 + \frac{|y-x|}{t}\right)^n Mg(x),$$

and thus

$$u^*(x) \le 2^n AMg(x).$$

The method used by Stein can be used with several other kernels.

<u>Lemma 3.1</u>. For $g \in L^p$, $p \ge 1$, if

$$u(x,t) = \int W(x-\xi,t)g(\xi)d\xi,$$

where

$$W(x,t) = (4\pi t)^{-n/2}\exp(-|x|^2/4t),$$

$$|u(x-z,t)| \le A(1 + |z|^2/t)^n Mg(x).$$

Pf. The inequality is invariant under the dilatation $(x,z,t) \to (x\delta, z\delta, t\delta^2)$, so we may assume $t = 1$, and the kernel is $W(x,1) = (4\pi)^{-n/2}\exp(-|x|^2/4)$. It follows from Theorem 2, Stein [21] that

$$|u(x-z,1)| \leq A_z Mg(x),$$

where

$$A_z = \int Q_z(x)dx,$$

with Q_z the smallest decreasing radial majorant of $W(x,1)$. An easy computation gives

$$Q_z(x) = \begin{cases} \exp\left(-\frac{(|x|-|z|)^2}{4}\right), & |x| \geq |z| \\ 1, & |x| < |z|, \end{cases}$$

and hence,

$$A_z \leq d_n(1+|z|)^n.$$

<u>Corollary 3.2</u>. If u is a temperature with trace g, $w \in V^\alpha$

$$\int |u(x,t^2)|^{p\alpha} dw(x,t) \leq C\|w\|_{V^\alpha} \|g\|^{p\alpha}_{(p,p\alpha)}.$$

It has been shown by Aronson [2] that the fundamental solution of a uniformly parabolic equation of second order in divergence form satisfies an estimate

$$\Gamma(x,y;t,\tau) \le C_1(t-\tau)^{-n/2}\exp(-\mu_1|x-y|^2/t-\tau)$$

$$= C_1 W(\sqrt{2\mu_1}(x-y),t-\tau),$$

which gives a corresponding estimate for solutions of such equations with traces in $L^{(p,p\alpha)}$.

Our final example is the kernel, for λ complex,

$$K_1^\lambda(x,y,t) = C_n\left(\frac{t^2}{|x-y|^2+t^2}\right)^{\frac{\lambda n}{2}} t^{-n}$$

or, since they are of the same orders of magnitude

$$K_2^\lambda(x,y,t) = C_n\left(\frac{t}{|x-y|+t}\right)^{\lambda n} t^{-n}.$$

Lemma 3.3. For $g \in L^p$, $p \ge 1$, if $\operatorname{Re}\lambda > 1$,

$$u_j(x,t) = \int_{R^n} K_j^\lambda(x,y,t)g(y)dy,$$

then

$$|u_j(x-z,t)| \le A\left(1+\frac{|z|^2}{t^2}\right)^{n/2} Mg(x).$$

Pf. We give the proof in the case $j = 1$. The inequality is invariant under the map $(x,y,t) \to (x\delta,y\delta,t\delta)$, and it suffices to prove

$$|u(x-z,1)| \le A(1+|z|^2)^{n/2}Mg(x).$$

(For $\operatorname{Re}\lambda \ge 1 + 1/n$,

$$K^\lambda(x,1) \le K^{(1+1/n)}(x,1) = P(x,1)$$

and the result follows from the above-cited result for the Poisson integral. However, the argument below works for any $\operatorname{Re}\lambda > 1$.)

Once more the problem is to estimate

$$Q_z(x) = \sup_{|w|\geq|x|} |K_1^\lambda(w,z,1)|.$$

For small x, use the obvious estimate $|Q_z(x)| \leq 1$. For large x, $(|x| \geq 2|z|)$, the fact that $|w-z| \geq \frac{1}{2}|x|$ implies that

$$Q_z(x) \leq C|x|^{-(\operatorname{Re}\lambda)n},$$

and since $\operatorname{Re}\lambda > 1$, $Q_z \in L^1(R^n)$. To estimate

$$A_z = \int Q_z(x)dx,$$

for $|z| \leq 1$,

$$A_z \leq \int_{|x|\leq 2} Q_z(x)dx + \int_{|x|>2} Q_z(x)dx$$

$$\leq C2^n + C2^{-(\operatorname{Re}\lambda)n+n},$$

since $|x| \geq 2|z|$ in the last integral. For $|z| > 1$, the same splitting gives

$$A_z \leq C|z|^n + C|z|^{-(\operatorname{Re}\lambda)n+n} \leq C|z|^n.$$

This gives the estimate

$$A_z \leq C(1+|z|^n).$$

Corollary 3.4. For any $g \in L^{(p,p\alpha)}$, $p > 1$, if $\mathrm{Re}\,\lambda > 1$, $\mu \in V^\alpha$

$$\int |u_j(x,y)|^{p\alpha} d\mu(x,y) \leq C\|\mu\|_{V^\alpha} \|g\|^{p\alpha}_{(p,p\alpha)},$$

for $j = 1$ or 2.

A direct calculation shows that if $\gamma = \frac{n-1}{2}(\alpha - 1/2)$, $h = y^\gamma P^{1/2+\alpha}(x,y)$ is an eigenvector of the operator $L = y^2\left(\left(\frac{\partial}{\partial y}\right)^2 + \Delta\right)$ with eigenvalue $(\gamma+\alpha-1/2)(\gamma+\alpha+1/2)$, i.e.,

$$Lh = (\gamma+\alpha+1/2)(\gamma+\alpha-1/2)h.$$

Now

$$h(x,y) = y^{-\frac{n+1}{2}(\alpha-1/2)} K_1^\lambda(x,y),$$

with $\lambda = (1+1/n)(1/2+\alpha)$. Hence, if we form the eigenfunction of L with trace g and eigenvalue $(\gamma+\alpha+1/2)(\gamma+\alpha-1/2)$,

$$\begin{aligned} u(x,y) &= \int h(x-z,y)g(z)dz \\ &= y^\gamma \int P^{\frac{1}{2}+\alpha}(x-z,y)g(z)dz \\ &= y^{-\frac{n+1}{2}(\alpha-1/2)} \int K_1^\lambda(x-z,y)g(z)dz \\ &= y^{-\frac{n+1}{2}(\alpha-1/2)} u_1(x,y), \end{aligned}$$

allowing us to write the eigenfunction as a power of y times a function which can be controlled by a Carleson measure.

Next, we give examples of inequalities which can be deduced from applying the above corollaries.

Example 1. Using the fact that $\delta_{(x_0,y_0)} \in V^{\alpha}$ for all $\alpha > 0$, we obtain

$$|u(x_0,y_0)|^{p\alpha} \leq C y_0^{-n\alpha} \|g\|^{p\alpha}_{(p,p\alpha)},$$

or,

$$|u(x_0,y_0)| \leq C y_0^{-n/p} \|g\|_{(p,p\alpha)},$$

for any harmonic function of the form $u = P_y * g$. Taking $\alpha = 1$ gives the familiar estimate for harmonic functions with L^p trace. Since α is arbitrary, f may be in a larger Lorentz class; if we use interpolation, we may take $f \in L^{p,\infty}(R^n)$ [11].

The same estimate applied to the heat equation gives

$$|u(x_0,y_0^2)|^{p\alpha} \leq C\, y_0^{-n\alpha} \|g\|_{(p,p\alpha)},$$

or

$$|u(x_0,y_0)| \leq C\, y_0^{-n/2p} \|g\|_{(p,p\alpha)},$$

which is again of the correct order of magnitude, and we can see that this estimate, with a different C will be valid for solutions of a uniformly parabolic equation of second order.

When applied to the Laplace-Beltrami operator L, we see that if u is an eigenfunction such that

$$y^{\frac{n+1}{2}(\alpha-1/2)} u$$

has $L^{(p,p\alpha)}$ trace g,

$$|u(x_0,y_0)|\,y_0^{\frac{n+1}{2}(\alpha-1/2)} \le C\, y_0^{-n/p} \|g\|_{(p,p\alpha)},$$

which we may also expect to be of the correct order of magnitude.

Example 2. Since $dx \otimes \delta_{y_0} \in V^1$, we obtain

$\int |u(x,y_0)|^p dx \le \int |g(x)|^p dx,$ for harmonic functions,

$\int |u(x,t_0^2)|^p dx \le \int |g(x)|^p dx,$ for temperatures,

and

$$y_0^{\frac{n+1}{2}(\mathrm{Re}\,\alpha-1/2)} \left(\int |u(x,y_0)|^p dx\right)^{1/p} \le \left(\int |g(x)|^p dx\right)^{1/p},$$

for an eigenfunction u with $y^{\frac{n+1}{2}(\alpha-1/2)} u$ having L^p trace g.

This inequality can be used to give estimates of Hardy-Littlewood and Flett, since

$$\int |u(x_0,t_0^q)|^r dx = \int |u(x,t_0^q)|^p |u(x,t_0)|^{r-p} dx$$

$$\leq C_1(t_0)^{r-p}\left(\int |u(x,t_0^q)|^p dx\right)\left(\int |g(x)|^p dx\right)^{r-p}$$

$$\leq C_1(t_0)^{r-p} C_2(t_0)\left(\int |g(x)|^p dx\right)^{r},$$

and using the appropriate values of C_1 and C_2, we obtain

$\|u(\cdot,t_0)\|_r \leq C\, t_0^{-n(1/p-1/r)}\|g\|_p$, for harmonic functions,

$\|u(\cdot,t_0)\|_r \leq C\, t_0^{-\frac{n}{2}(1/p-1/r)}\|g\|_p$, for temperatures,

and

$$\|u(\cdot,t_0)\|_r \leq C\, t_0^{-n(1/p-1/r)-\frac{n+1}{2}(\mathrm{Re}\,\alpha-1/2)}\|g\|_p,$$

for eigenfunctions such that $y^{\frac{n+1}{2}(\alpha-1/2)}u$ has L^p trace g.

Example 3. Since $t^{\lambda n}dx\,\frac{dt}{t} \in V^{1+\lambda}$, we obtain for $r/p_0 = 1+\lambda$,

$$\left(\int_0^\infty\int_{R^n} |u(x,t)|^r t^{\lambda n} dx\,\frac{dt}{t}\right)^{1/r} \leq C\|g\|_{(p_0,p_0(1+\lambda))},$$

for harmonic functions u,

$$\left(\int_0^\infty\int_{R^n} |u(x,t)|^r t^{\frac{\lambda n}{2}} dx\,\frac{dt}{t}\right)^{1/r} \leq C\|g\|_{(p_0,p_0(1+\lambda))},$$

for a temperature u and

$$\left(\int_0^\infty \int_{R^n} |u(x,t)|^r t^{\lambda n + \frac{n+1}{2}(\mathrm{Re}\,\alpha - 1/2)} dx \frac{dt}{t}\right)^{1/r} \le C\|g\|_{(p_0, p_0(1+\lambda))}$$

for eigenfunctions of the Laplace-Beltrami operator.

This result (for harmonic functions) is due to Flett [14] (see also Barker [3] and Lemma 5 of [12].

Our final example gives another proof of the equivalence of H^p and L^p for $p > 1$.

Example 4. Since $\delta_{t(x)}dx \in V^1$, we obtain, for $g \in L^p(R^n)$,

$$\left(\int |u(x,t(x)|^p dx\right)^{1/p} \le A\|g\|_p, \qquad p > 1.$$

However, we saw from Example 1 that for each $x \in R^n$, $|u(x,t)| \to 0$ as $t \to +\infty$. Hence, $\sup_{t>0}|u(x,t)| = u(x,t(x))$ for some function $t(x)$, and we obtain the estimate,

$$\|u^+\|_p = \left(\int |u^+(x)|^p dx\right)^{1/p} \le A\|g\|_p, \qquad p > 1$$

which implies $g \in H^p$.

In each of these estimates we applied the basic estimate to a particular Carleson measure. Now we will discuss the result of applying the estimate to families of Carleson measures. By varying the functions in the family and using duality, we will obtain sharper estimates. The sharpest possible result says that $|u|^{p\alpha} \in (V^\alpha)'$; however, at the time of the first draft of this paper only an atomic description of this space was known. Now using ideas of [5], we can give an explicit description of this product. However, interesting inequalities also follow by considering special families.

<u>Theorem 3.5.</u> For any $g \in L^{(p,q)}(R^n)$, $0 < q < \infty$.

$$\left(\int_0^\infty [(\sup_x |u(x,y)|)y^{n/p}]^q \frac{dy}{y}\right)^{1/q} \le C\|g\|_{(p,q)}, \quad u \text{ harmonic},$$

and

$$\left(\int_0^\infty [\sup_x |u(x,y)|y^{n/2p}]^q \frac{dy}{y}\right)^{1/q} \le C\|g\|_{(p,q)}, \quad u \text{ temperature},$$

where C depends only on n, p and q.

Proof. Since (Proposition 2.6) $\delta_{\xi(t)} t^{n\alpha} \frac{dt}{t} \in V^\alpha$, if we choose $\alpha = q/p$, we obtain

$$\left(\int_0^\infty\int_{R^n} |u(\xi(t),t)|^q t^{nq/p} \frac{dt}{t}\right)^{1/q} \le C\|g\|_{(p,q)},$$

and since ξ is arbitrary, the results follow.

This gives a significant improvement of Example 1 for which the integrand was shown to be bounded. Theorem 3.5 is an embedding theorem. Whereas Example 1 showed $L^p \subseteq \dot{B}^{-n/p}_{\infty\infty}$, we now obtain

<u>Corollary 3.6.</u> $L^{p,q} \subseteq \dot{B}^{-n/p}_{\infty q}$.

We also obtain an estimate for the particular eigenfunctions of $y^2\left((\frac{\partial}{\partial y})^2 + \Delta\right)$ identified in this section.

Theorem 3.7. For any $g \in L^{(p,q)}(R^n)$, $0 < q < \infty$, $\operatorname{Re}\gamma > +1/2$, if u is the eigenfunction of $y^2\left((\frac{\partial}{\partial y})^2 + \Delta\right)$ with eigenvalue $\gamma(\gamma-1)$ and trace g discussed after Corollary 3.4, then

$$\left(\int_0^\infty [\sup_x |u(x,y)|\, y^{\gamma+n/p-1}]^q \frac{dy}{y}\right)^{1/q} \le C\|g\|_{(p,q)}.$$

Pf. The condition on γ quarantees that if we write $\gamma - 1 = \frac{n+1}{2}(\alpha-1/2)$, $\operatorname{Re}((1+1/n)(1/2+\alpha)) > 1$ and Corollary 3.4 can be applied to the measure $\delta_{\xi(y)}y^{nq/p}\frac{dy}{y}$.

In our next sequence of theorems, the Carleson measures will be parametrized by $L^{p,\infty}$ and we will repeatedly use the following lemma, which can be found in [8], page 91.

Lemma 3.8. If g is a measurable function such that for any simple function f,

$$\left|\int_X fg\,d\mu\right| \le K\|f\|_{(p,\infty)},$$

with K independent of f, then $g \in L^{p',1}$ and

$$\|g\|_{(p',1)} \le cK.$$

The first result is well-known but we give it to illustrate how Lemma 3.8 will be used.

Theorem 3.9. For $g \in L^{(p,q)}$, $\{u(x,y_0) | y_0 > 0\}$ is bounded in $L^{(p,q)}$, where u is the harmonic function with trace g.

Proof. For any $f \in L^{p_0,\infty}$, $f dx \delta_{t_0} \in V^\alpha$, $1/p_0 = 1 - \alpha$, and its Carleson measure norm is bounded by $\|f\|_{(p_0,\infty)}$. We thus have

$$\int |u(x,y_0)|^{p\alpha} |f(x)| dx \le C \|f\|_{(p_0,\infty)} \|g\|^{p\alpha}_{(p,p\alpha)}.$$

By Lemma 3.8,

$$|u(x,y_0)|^{p\alpha} \in L^{p_0',1},$$

which implies that $u(x,y_0) \in L^{(p,p\alpha)}$ and

$$\|u(\cdot,y_0)\|_{(p,p\alpha)} \le C \|g\|_{(p,p\alpha)}.$$

Theorem 3.10. If $g \in L^{(p,q)}$, $1 < p < \infty$, $0 < q < \infty$, and $p < r$ and $r > q$, and if u is the harmonic function with trace g,

$$\left\| \left(\int_0^\infty (|u(x,y)| y^{n(1/p-1/r)})^q \frac{dy}{y} \right)^{1/q} \right\|_{(r,q)} \le C \|g\|_{(p,q)}.$$

Proof. We want to show that

$$G(x) = \int_0^\infty (|u(x,y)| y^{n(1/p-1/r)})^q \frac{dy}{y}$$

is in $L^{(r/q,1)}$. We apply Proposition 2.7 which says $f(x) y^{n\alpha} dx \frac{dy}{y} \in V^\beta$, with $\beta = 1 - 1/p_0 + \alpha$, for $f \in L^{p_0,\infty}$.

Hence,

$$\int_0^\infty \int_{R^n} |u(x,y)|^{p\beta} |f(x)| y^{n\alpha} dx \frac{dy}{y} \le C\|f\|_{(p_0,\infty)} \|g\|^{p\beta}_{(p,p\beta)}.$$

It follows from Lemma 3.8 that

$$\left(\int_0^\infty |u(x,y)|^{p\beta} y^{n\alpha} \frac{dy}{y} \right) \in L^{p_0',1}.$$

This expression is G if $p\beta = q$ and $n\alpha = n(\frac{1}{p} - \frac{1}{r})q$, and hence we will obtain the desired result if $p_0' = r/q$, which can be done for $r > q$. We must have $\alpha > 0$ which requires $r > p$.

Both of the above theorems hold for temperatures with the replacement of y by $y^{1/2}$. Theorem 3.10 can also be viewed as the case $\gamma = 1$ of a result on eigenfunctions of the operator $y^2\left((\frac{\partial}{\partial y})^2 + \Delta\right)$.

Theorem 3.11. If $g \in L^{(p,q)}$, $1 < p < \infty$, $0 < q < \infty$, $r > p$ and $r > q$, then if u is the eigenfunction of $y^2\left((\frac{\partial}{\partial y})^2 + \Delta\right)$ with eigenvalue $\gamma(\gamma-1)$ where $\mathrm{Re}\,\gamma > 1/2$,

$$\left\| \left(\int_0^\infty (|u(x,y)| y^{\gamma-1+n(1/p-1/r)})^q \frac{dy}{y} \right)^{1/q} \right\|_{(r,q)} \le C\|g\|_{(p,q)}.$$

There is also an estimate with an arbitrary weight in the y-variable.

Theorem 3.12. If $g \in L^{(p,q)}(R^n)$, $0 < q < \infty$, and $\varphi \in L^{(p_1,\infty)}(R^+)$ with $p_1 > q$, then for any r with $n(\frac{1}{p} - \frac{1}{r}) = \frac{1}{q} - \frac{1}{p_1}$,

$$\left\| \left(\int_0^\infty [|u(x,y)|\varphi(y)]^q dy \right)^{1/q} \right\|_{(r,q)} \leq C\|g\|_{(p,q)}.$$

Proof. We want to estimate the $L^{(r/q,1)}$ norm of

$$G(x) = \int_0^\infty |u(x,y)|^q \varphi(y)^q dy,$$

and thus we begin with the observation that by Proposition 2.8, if $\psi \in L^{p_2,\infty}(R^+)$, $f \in L^{p_0,\infty}(R^n)$,

$$\int_0^\infty \int_{R^n} |u(x,y)|^{p\beta} |f(x)| |\psi(y)| dx dy \leq C\|f\|_{(p_0,\infty)} \|\psi\|_{(p_2,\infty)} \|g\|^{p\beta}_{(p,p\beta)},$$

where $\beta = 1 - \frac{1}{p_0} + \frac{1}{n}(1 - \frac{1}{p_2})$. Since f is arbitrary, we have

$$\int_0^\infty \int_{R^n} |u(x,y)|^{p\beta} |\psi(y)| dy \in L^{p_0',1}.$$

This expression equals $G(x)$ if $p\beta = q$ and $\psi = \varphi^q$, and we get the theorem if $p_0' = r/q$. Since $\varphi^q \in L^{p_1/q,\infty}$, we must have $p_1 > q$ and $r > q$ follows since we need $1 < p_0 < \infty$. Substituting the values into the expression for β and we obtain the condition $n(\frac{1}{p} - \frac{1}{r}) = \frac{1}{q} - \frac{1}{p_1}$.

A similar argument dualizing over ψ shows that for any $f \in L^{p_0,\infty}(R^n)$,

$$\left(\int_{R^n}^{\infty} |u(x,y)|^q |f(x)|^q dx \right)^{1/q} \in L^{(r,q)}(0,\infty),$$

with $\frac{1}{p} - \frac{1}{q} + \frac{1}{p_0} = \frac{1}{nr}$, and suitable conditions on r, p_0. There is also an inequality for the other eigenfunctions, which we leave to the reader.

Proposition 2.10 implies a generalization of the result of Duren [11].

Theorem 3.13. For $g \in L^{(p,q)}(R^n)$, $0 < q \leq p(1+1/n)$, $u \in L^{(p(1+1/n),q)}(R_+^{n+1})$.

Proof. For $f \in L^{p_0,\infty}(R_+^{n+1})$, $f dx dt \in V^{(1+1/n)(1-1/p_0)}$, and then, we apply Carleson's inequality and Lemma 3.8.

Our final result is a variant of Theorem 3.10.

Theorem 3.14. If $g \in L^{(p,q)}$, $1 < p < \infty$, $r = (1+1/n)p$, $0 < q \leq p(1+1/n)$.

$$\left\| \left(\int_0^{\infty} (|u(x,y)| y^{1/r})^q \frac{dy}{y} \right)^{1/q} \right\|_r \leq C \|g\|_{(p,q)}.$$

Proof. With $\beta = (1+1/n)(1-1/p_0)$, it follows that

$$\int_0^{\infty} \int_{R^n} |u(x,y)|^{p\beta} |f(x)| y^{-1/p_0} dx dy \leq C \|f\|_{p_0} \|g\|_{(p,p\beta)}^{p\beta},$$

and since this holds for every $f \in L^{p_0}$,

$$\left(\int_0^\infty |u(x,y)|^{p\beta} y^{-1/p_0} dy\right) \in L^{p_0'}(R^n).$$

If we choose $p\beta = q$, then $p_0' = r/q$, we have

$$\left(\int_0^\infty |u(x,y)|^{q} y^{-1/p_0} dy\right)^{1/q} \in L^{r}(R^n),$$

and solving for p_0 gives the result.

§4. Fourier multipliers on R^n

Amar and Bonami have defined the balayage of a Carleson measure, by duality, as a function P^*w which satisfies

$$\int_0^\infty\int_{R^n} P_t g(x)dw(x,t) = \int_{R^n} g(y)(P^*w)(y)dy,$$

for $w \in V^\alpha$, $g \in L^{q,1}(R^n)$, $q = 1/\alpha$, $0 < \alpha < 1$. If p is the conjugate exponent of q, the function belongs to $L^{p,\infty}(R^n)$. For nonnegative g and w, we can apply Fubini and conclude that

$$P^*w(y) = \int_0^\infty\int_{R^n} P(y-z,t)dw(z,t) \tag{4.1}$$

Since $P^*w \in L^{p,\infty}(R^n)$, it is finite almost everywhere. For an arbitrary $w \in V^\alpha$, we write $w = w_1 - w_2$ with $w_j \geq 0$. It is easy to see that $w_j \in V^\alpha$ and hence we find that (4.1) holds for any $w \in V^\alpha$. If $\alpha \geq 1$, the balayage belongs to a space defined modulo polynomials. The balayage as defined above may be identically infinite but can be regularized with preservation of its good properties. We will first discuss some applications of the case $\alpha < 1$.

Theorem 4.1. The operator

$$g_\lambda^*f(x) = \left(\int_0^\infty\int_{R^n} \left(\frac{y}{|t|+y}\right)^{\lambda n} |\nabla u(x-t,y)|^2 y^{1-n}dtdy\right)^{1/2}$$

maps $f \in L^r(R^n) \to L^r(R^n)$ (and $f \in L^{r,\infty}(R^n) \to L^{r,\infty}(R^n)$) boundedly for any $r > 2$, $\operatorname{Re}\lambda > 1$.

Proof. $g_\lambda^* f(x)^2 = (K_1^\lambda)*(y|\nabla u|^2 dxdy)$, where u is the Poisson integral of f. For $f \in L^{r,\infty}(R^n)$, $r > 2$, Proposition 2.7 shows that $y|\nabla u|^2 dxdy \in V^{1-2/r}$, and then by the result discussed above, $g_\lambda^* f(x)^2 \in L^{r/2,\infty}(R^n)$. For $f \in L^r(R^n)$, $y|\nabla u|^2 dxdy \in W^{1-2/r}$ and the same argument yields $g_\lambda^* f \in L^r(R^n)$.

Remark. For $r = 2$, $f \in L^2$ implies $y|\nabla u|^2 dxdy \in V^0$ and thus $g_\lambda^* f \in L^2$ follows.

The other interesting family of measures is $\{f(x)\varphi(t)dxdt \mid f \in L^{p,\infty}(R^n), \varphi \in L^{p_1,\infty}(R_+)\}$, with p, p_1 chosen so that the family consists of Carleson measures of order less than one. It is easy to see that the balayage is

$$P*w(y) = \int_0^\infty \left(\int_{R^n} P(y-z,t) f(z) dz \right) \varphi(t) dt$$

$$= \int_0^\infty u(y,t)\phi(t)dt, \quad u \text{ harmonic extension of } f.$$

If we balayage by the Gauss-Weierstrass kernel, we obtain

$$W*\mu(y) = \int_0^\infty u(y,t^2)\phi(t)dt,$$

u temperature with initial value f, or balayaging by fractional powers of the Poisson kernel,

$$G^*_\rho\mu(y) = \int_0^\infty \left(\int_{R^n} P^{1/2+\alpha}(y-z,t)f(z)dz \right) \varphi(t)dt.$$

$$= \int_0^\infty t^{(\rho-1)} u_f(y,t)\varphi(t)dt,$$

where u is the eigenfunction of the operator L with eigenvalue $\rho(\rho-1)$ and trace f, for $\operatorname{Re}\rho > 1/2$.

If we now take the Fourier transform, a radial Fourier multiplier results.

<u>Theorem 4.2.</u> The operator T defined by

$$\widehat{Tf}(\xi) = m(|\xi|)\hat{f}(\xi),$$

where

$$m(|\xi|) = \int_0^\infty e^{-y|\xi|}\varphi(y)dy, \quad \text{with} \quad \varphi \in L^{p_1,\infty}(R^+)$$

sends $L^p \to L^q$, $1 < p < \min\left(2, \frac{n}{1-1/p_1}\right)$, $p_1 > 1$,

$$\frac{1}{q} = \frac{1}{p} - \frac{1}{n}\left(1 - \frac{1}{p_1}\right).$$

Proof. We will approximate T by $T_\varepsilon f$ where

$$\widehat{T_\varepsilon f}(x) = m_\varepsilon(|\xi|)\hat{f}(\xi),$$

with

$$m_\varepsilon(|\xi|) = \int_0^\infty e^{-(y+\varepsilon)|\xi|}\varphi(y)dy;$$

and show that

$$\|T_\varepsilon f\|_q \le C\|f\|_p,$$

with a bound independent of ε. In fact, we take

$$T_\varepsilon f(x) = P*(u(z,\varepsilon)\varphi(y)dzdy),$$

where $u(z,\varepsilon) = P_\varepsilon * f(z)$. For $f \in L^{p,\infty}$, $u(z,\varepsilon) \in L^{p,\infty}$ and

$$\|u(z,\varepsilon)\|_{p\infty} \le \|f\|_{p,\infty}.$$

Thus, $u\varphi dzdy \in V^\beta$, with $\beta = 1 - \frac{1}{p} + \frac{1}{n}(1-1/p_1)$, (and because of the assumptions on p, $\beta < 1$) and its norm there is bounded by $\|f\|_{p\infty}\|\varphi\|_{p_1,\infty}$. The remark above gives

$$T_\varepsilon f(x) = \int_0^\infty u(x,y+\varepsilon)\varphi(y)dy,$$

and

$$\|T_\varepsilon f\|_{q\infty} \le C\|f\|_{p\infty},$$

with C independent of ε, and an appeal to the Marcinkiewicz interpolation theorem completes the proof of boundedness. Next, we must show that

$$\widehat{T_\varepsilon f}(\xi) = m_\varepsilon(|\xi|)\hat{f}(\xi).$$

Since $f \in L^{p,\infty}$, $p < 2$, $e^{-(y+\varepsilon)|\xi|}\hat{f}$ is integrable. If we write,

$$T_\varepsilon f(x) = \int_0^\infty \left(\int e^{-(y+\varepsilon)|\xi|} \hat{f}(\xi) d\xi \right) \varphi(y) dy,$$

we need only justify an interchange in the order of integration. But, one estimates

$$\int_0^\infty \int_{R^n} e^{-(y+\varepsilon)|\xi|} |\hat{f}(\xi)| \varphi(y) dy d\xi \le \int_0^\infty \| e^{-(y+\varepsilon)|\xi|} \|_{p,1} \| \hat{f} \|_{p',\infty} \varphi(y) dy$$

$$= C \int_0^\infty (y+\varepsilon)^{-n/p} \varphi(y) dy$$

$$\le C \int_0^\infty (y+\varepsilon)^{-n/p} y^{-1/p_1} dy,$$

which is finite.

Example (1). Let $\varphi(t) = t^{\alpha-1}$, $0 < \alpha < 1$. It is easy to see that $\varphi \in L^{\frac{1}{1-\alpha},\infty}(R^+)$ and the theorem gives

$$T: L^p \to L^q, \quad \frac{1}{q} = \frac{1}{p} - \frac{\alpha}{n}, \quad 1 < p < \min(2, n/\alpha).$$

An easy computation gives $m(\xi) = C|\xi|^{-\alpha}$, and we have obtained the Riesz potential as the balayage of a Carleson measure and the estimate of Sobolev's theorem. The representation of the Riesz potential as a balayage has the advantage of providing some insight into the range of the Riesz potential. By Proposition 2[1], any L^q function is of the form $P*w$ for a measure $w \in W^{1-1/q}$ of the form

$$\sum \delta_{t_k} \otimes f_k(x)dx,$$

with $f_k \in L^q$. Those whose balayages give the same function as the balayage of $\{f(x)t^{\alpha}dx\frac{dt}{t}\}$ for $f \in L^p(R^n)$ will be in the range of the Riesz potential, and we obtain our next corollary.

<u>Corollary 4.3</u>. Range $R^{\alpha} = \{f \in L^q | f(x) = \sum_{k=1}^{\infty} u_k(x,t_k)$, with u_k harmonic with trace in L^q and $t_k > 0\}$.

The convergence is in L^q rather than pointwise, but the result shows the smoothing effect of the Riesz potential.

Example 2. Since m is the Laplace transform of φ, we note that [10]

$$\mathcal{L}\{s^{-1/2}(I_{1/2}(s) - L_{1/2}(s))\} = \begin{cases} \sqrt{\pi/2}, & t < 1 \\ 0, & t > 1 \end{cases}$$

and the Laplace transform is convergent for all t. Here I is a modified Bessel function and L_ν is the Struve function of order ν. For $\nu = 1/2$ they can be calculated explicitly and

$$s^{-1/2}(I_{1/2}(s) - L_{1/2}(s)) = \frac{\sqrt{2/\pi}}{s}(1-e^{-s}).$$

With this choice of φ, we see that $\varphi \in L^{1+\varepsilon}$ for every $\varepsilon > 0$. Thus, $S_0 f = c\chi_B(\xi)\hat{f}(\xi)$ has the property that $S_0 f$ is a balayage and we obtain the mapping theorem,

$$S_0\colon\ L^p \to L^q,\quad 1 < p < \min(2, n(1+1/\varepsilon)),$$

$\frac{1}{q} = \frac{1}{p} - \frac{\varepsilon}{n(1+\varepsilon)}$. Since $\varepsilon > 0$ is arbitrary, we obtain any $q > p$ for $1 < p < 2$ and hence,

$$S_0 \colon L^p \to L^q, \quad 1 < p < 2, \quad q > p,$$

but with a bound depending on q. By Fefferman's theorem that $S_0 \colon L^p \to L^p$ if and only if $p = 2$, the constants must blow up as $q \to p$.

Results about the Bochner-Riesz means follow since

$$L\left\{\frac{I_\nu(s)-L_\nu(s)}{s^\nu}\right\} = \begin{cases} (1-t)^{\nu-1/2}/2^{\nu-1}\sqrt{\pi}\,\Gamma(\nu+1/2), & t < 1 \\ 0, & t > 1 \end{cases}$$

for $\operatorname{Re}\nu > -1/2$ and the Laplace transform converges for all t (10, p.140). If we balayage with the Gauss-Weierstrass kernel we get $m(\xi) = c(1-|\xi|^2)_+^{\nu-1/2} = \begin{cases} c(1-|\xi|^2)^{\nu-1/2}, & |\xi| < 1, \\ 0, & |\xi| > 1. \end{cases}$

The same procedure works for a large class of multipliers.

Theorem 4.4. If $\varphi(s)$ is bounded at 0, and $\varphi(s)s$ is bounded at infinity, and φ has no other singularities, then

$$m(\xi) = \int_0^\infty e^{-y|\xi|}\varphi(y)dy$$

defines an operator $T_m \colon L^p \to L^q$ for any $1 < p < 2$, $q > p$.

Proof. The hypotheses guarantee that $\varphi \in L^{1+\varepsilon}(R^+)$ for every $\varepsilon > 0$. We then apply Theorem 4.2.

Examples of multipliers m obtained in this way are $\exp(-\alpha|\xi|)L_n(2\alpha|\xi|)$, $\alpha > 0$, L_n the nth Laguerre polynomial, $\chi_{|\xi|<a} - \chi_{a<|\xi|<2a}$, where χ denotes the characteristic function of a set, $\exp(ib|\xi|)\chi_{|\xi|<\frac{2n\pi}{b}}$, where $b > 0$ and n is an arbitrary positive integer and $\exp(i|\xi|)\chi_{|\xi|<\pi}$.

The Gauss-Weierstrass kernel gives similar results with $\xi \to |\xi|^2$ and $s^{-1}\varphi$ bounded at the origin and $s\varphi$ bounded at infinity.

As noted above, for $\alpha \geq 1$ the definition of $P*w$ by

$$P*w(x) = \int_0^\infty\int_{R^n} P(x-z,y)dw(z,y)$$

need not give a meaningful result. Indeed , if $w = |z|^{-n/p}t^{n/p}$ $dzdt \in V^1$, one sees that

$$P*w(x) = R^{n/p}(|x|^{-n/p}) \equiv +\infty.$$

The balayage must be regularized and thus, we consider

$$P*w(x) = \int_0^\infty\int_{R^n} [P(x-z,y) - P(z,y)\chi_{c_{T(\widetilde{B})}}]dw(z,y),$$

where $\widetilde{B} = B(0,K)$, for a suitably large K. We have

$$P*w(x) = \iint_{T(\tilde{B})} P(x,z,y)dw(z,y) + \iint_{{}^cT(\tilde{B})} [P(x-z,y)-P(z,y)]dw(z,y).$$

$$= F_1(x) + F_2(x).$$

The function F_1 is finite almost everywhere because

$$\int_{R^n} |F_1(x)|dx \le \iint_{T(\tilde{B})} |dw| \int |P(x-z,y)|dx$$

$$\le |w(T(\tilde{B})| \le C|\tilde{B}|^{\alpha} < +\infty.$$

For $|x| \le 1$, F_2 is finite everywhere by the estimate and arguments on page 31-33 of [1]. If $|x| \le 1$, $(z,y) \notin T(\tilde{B})$, then $|z| + y \ge K \ge K|x|$, and estimate (H3) for the Poisson kernel gives

$$|P(x-z,y) - P(z,y)| \le C|x|(y+|z|)^{-(n+1)}.$$

If $n+1 > n\alpha$, we can apply Lemma 2 of [1] and conclude

$$\iint_{{}^cT(\tilde{B})} |P(x-z,y) - P(z,y)||dw| \le C|x| \iint_{{}^cT(\tilde{B})} (|z|+y)^{-n+1}|dw|$$

$$\le C|x|K^{n\alpha-(n+1)}.$$

If $x \notin B$, $|x| \ge 1$, put x into the appropriate dyadic ring, say $2^{K_0} \le |x| \le 2^{K_0+1}$. Write

$${}^cT(\tilde{B}) = \bigcup_1^{\infty} (T(\tilde{B}_k) - T(\tilde{B}_{k-1})),$$

where $\tilde{B}_k = B(0,2^k K)$. For $k \geq K_0 + 1$, $y + |z| \geq K|x|$ on ${}^cT(\tilde{B}_{K-1})$ and we can argue as above to estimate

$$\iint_{{}^cT(B_k)} [P(x-z,y) - P(z,y)]dw(z,y).$$

What remains is $2^{K_0} \leq |x| \leq 2^{K_0+1}$, $\displaystyle\iint_{\bigcup_1^{K_0+1} (T(\tilde{B}_k)\setminus T(\tilde{B}_{k-1})} [P(x-z,y)-P(z,y)]dw(z,y).$

But this is a set of finite w-measure, away from the singularity of P and it is easy to show that it is finite almost everywhere.

We have justified the regularization for $1 \leq \alpha < 1 + 1/n$. In this range, the balayage (by the same argument as in [1]) belongs to $\dot{B}^{\beta}_{\infty\infty}$, $\beta < 1$. These Besov spaces are defined by first differences. For $\beta \geq 1$, second differences are needed and the regularization must be modified by subtracting the first order Taylor series terms. We will only give results in the range $1 \leq \alpha < 1 + 1/n$, where the balayage sends $V^1 \to BMO$ and $V^{\alpha} \to \dot{B}^{n(\alpha-1)}_{\infty\infty}$.

Theorem 4.5. (a) If $0 < \alpha < n$, $R^{\alpha}: L_{1,n-\alpha} \to BMO$.

(b) If $0 < \beta < n$, $\beta < \alpha < \beta + 1$, then $R^{\alpha}: L_{1,n-\beta} \to \dot{B}^{\alpha-\beta}_{\infty\infty}$.

Proof. For $f \in L_{1,n-\beta}$, $f(x)t^{\alpha}dx\,\frac{dt}{t} \in V^{1+\frac{\alpha-\beta}{n}}$, and the result follows, if we note that $\chi_{{}^cT(\tilde{B})} = \chi_{U_1} + \chi_{U_2} + \chi_{U_3}$, where

$$U_1 = \{(x,y) \mid x \in {}^c\tilde{B},\ y > 0\},$$

$$U_2 = \{(x,y) \mid x \in \tilde{B},\ y \geq K\},$$

$$U_3 = \{(x,y) \mid K-y \leq |x| \leq K,\ 0 < y \leq K\},$$

and hence,

$$P*(ft^{\alpha}dx\ \frac{dt}{t}) = \iint [P(x-z,y) - P(z,y)\chi_{U_1}]f(z)y^{\alpha}dz\ \frac{dy}{y}$$

$$+ \iint P(z,y)\chi_{U_2} f(z)dzy^{\alpha}\frac{dy}{y}$$

$$+ \iint P(z,y)\chi_{U_3} f(z)y^{\alpha}dz\ \frac{dy}{y}.$$

It is easy to see that the first term is the regularized Riesz potential of order α, which is known to be finite almost everywhere [15]. The next two terms are finite constants. The first is finite because it can be estimated by

$$\int_K^{\infty}\int_{\tilde{B}} |P(z,y)|\,|f(z)|y^{\alpha}dz\ \frac{dy}{y}$$

$$\leq \int_K^{\infty}\int_{\tilde{B}} y^{\alpha-n}|f(z)|dz\ \frac{dy}{y}$$

$$= \left(K^{\alpha-n}/(n-\alpha)\right)\int_{\tilde{B}} |f(z)|dz \leq \left(K^{\alpha-n}/(n-\alpha)\right)CK^{n-\alpha} \leq \frac{C}{n-\alpha}.$$

The second term is estimated by

$$\int_0^K \int_{K-y\le|z|\le K} |P(z,y)|\,|f(z)|\,y^{\alpha} dz\, \frac{dy}{y} \le (2/K^2)^{\frac{n+1}{2}} K^{\alpha} C K^{n-\alpha} \le C'.$$

A more classical proof can be found in [15].

REFERENCES

1. E. Amar and A. Bonami, Measures de Carleson d'ordre α et solutions au bord de l'equation $\bar{\partial}$, Bull. Soc. Math. France, 107 (1979), 23-48.
2. D. Aronson, Nonnegative solutions of linear parabolic equations, Ann. Suola Norm. Sup. Pisa. 22 (1968), 607-694.
3. S. R. Barker, An inequality for measures on a half-space, Math. Scand. 44 (1979), 92-102.
4. L. Carleson, Interpolation of bounded analytic functions and the corona problem, Ann. of Math. 76 (1962), 547-559.
5. R. Coifman, Lecture at Cortona, Summer, 1982.
6. R. Coifman and Y. Meyer, Au-delá des operateurs pseudo-differentieles, Asterisque, 57 (1978).
7. R. Coifman and G. Weiss, Analyse harmonique non-commutative sur certains espaces homogenes, Lecture Notes in Math, vol. 242, Springer-Verlag, Berlin and New York, 1971.
8. M. Cwikel, The dual of weak L^p, Ann. Inst. Fourier Grenoble, 25 (1975), 81-126.
9. B. Dahlberg, Regularity properties of Riesz potentials, Ind. Univ. Math. J. 28 (#2), 1979, 257-268.
10. G. Doetsch, Tabellen zur Laplace-transformation und anleitung zum gebrech, Springer-Verlag, Berlin, 1947.
11. P. L. Duren, Extension of a theorem of Carleson, Bull. Amer. Math. Soc., 75 (1969), 143-146.

12. C. Fefferman and E. Stein, Some maximal inequalities, Amer. J. Math., 93 (1971), 107-115.

13. C. Fefferman and E. Stein, H^p spaces of several variables, Acta Math., 129 (1972), 137-193.

14. T. Flett, On the rate of growth of mean values of holomorphic and harmonic functions, Proc. Lond. Math. Soc. (3) 20 (1970), 749-68.

15. R. Johnson and U. Neri, Remarks on Riesz potential, BMO and Lip (α,p) spaces, University of Maryland Tech. Report, TR76-25, 1976.

16. R. Johnson, Definition of generalized Carleson measures and applications, Proc. Conf. Harmonic Analysis, Torino-Melano, May-June 1982.

17. G. G. Lorentz, Some new functional spaces, Ann. of Math. 51 (1950), 37-55.

18. V. M. Maz'ja, Multipliers in spaces of differentiable functions Trudy Seminar S. Sobolev, 1979, #1, 37-90.

19. P. Sjogren, Weak L^1 characterizations of Poisson integrals, Green potentials and H^p spaces, Trans. Amer. Math. Soc. 233 (1977), 179-196.

20. D. Stegenga, Multipliers of the Dirichlet space, Ill. J. Math. 24 (1980), 113-139.

21. E. Stein, Singular integrals and differentiability properties of functions, Princeton University Press, Princeton, N.J., 1970.

Department of Mathematics
University of Maryland
College Park, MD 20740

ON THE MAXIMAL FUNCTION FOR THE MEHLER KERNEL.

Peter Sjögren

1. INTRODUCTION.

Let $Nu = -\Delta u + x\cdot \operatorname{grad} u$ be the well-known number operator for the quantum-mechanical harmonic oscillator in $\mathbb{R}^n$. In $\mathbb{R}^{n+1} = \{(x,t): x\in\mathbb{R}^n,\ t>0\}$, the initial-value problem

$$-\frac{\partial u}{\partial t} = Nu$$

$$u(x,0) = f(x)$$

is solved by

$$u(x,t) = e^{-tN}f(x) = M_\lambda f(x) = \int M_\lambda(x,y)f(y)dy$$

with $\lambda = e^{-t}$. Here

$$M_\lambda(x,y) = (2\pi(1-\lambda^2))^{-n/2}\exp\left(-\frac{(y-\lambda x)^2}{2(1-\lambda^2)}\right)$$

is the Lebesgue measure form of the Mehler kernel, and $(e^{-tN})_{t>0}$ is the Hermite semigroup, whose infinitesimal generator is $-N$. The n-dimensional Hermite polynomials

$$H_m(x) = \prod_1^n H_{m_i}(x_i),\quad m = (m_1,\dots,m_n)\in\mathbb{N}^n,$$

are defined so as to be orthogonal with respect to the canonical Gaussian measure γ, whose density is $\gamma(x) = (2\pi)^{-n/2}\exp(-|x|^2/2)$. In terms of these polynomials, M_λ is conveniently expressed:

$$M_\lambda \Sigma a_m H_m = \Sigma\lambda^{|m|} a_m H_m,\quad |m| = \Sigma m_i.$$

The operators M_λ are bounded and of norm 1 on L^p_γ, $1 \le p \le \infty$, and they are self-adjoint on L^2_γ. Further, they are given by a positive kernel and leave constant functions invariant. This makes the maximal theorem from semigroup theory (see Stein [3, III.3]) applicable. Hence, the operator

$$M^* f(x) = \sup_{0<\lambda<1} |M_\lambda f(x)|$$

is bounded on L^p_γ, $1 < p < \infty$. This works even in infinite dimension.

The one-dimensional case is studied by Muckenhoupt [1], who also shows that M^* maps L^1_γ into $L^{1,\infty}_\gamma$ (i.e., weak L^1_γ). We shall prove the same thing in arbitrary finite dimension. Of course, estimates for M^* imply convergence results for $M_\lambda f$ as $\lambda \to 1$ $(t \to 0)$.

The author is grateful to C. Borell for suggesting this maximal function.

Theorem. For each finite dimension n, the operator M^* is bounded from L^1_γ into $L^{1,\infty}_\gamma$.

We need some notation for the proof. If $D \subset \mathbb{R}^n \times \mathbb{R}^n$, we let $D^x = \{y: (x,y) \in D\}$ for $x \in \mathbb{R}^n$, and slightly abusively, $D^y = \{x: (x,y) \in D\}$ for $y \in \mathbb{R}^n$. By $c > 0$ and $C < \infty$, we denote various constants, and $f \sim g$ means $c \le f/g \le C$.

2. First part of the proof.

Notice that

$$\gamma(x) \sim \gamma(y) \quad \text{for} \quad |x-y| < C/|y| \tag{2.1}$$

when y stays away from 0. We first study M^* when x is near y in this sense, setting for $R > 0$

$N_R = \{(x,y) \in \mathbb{R}^n \times \mathbb{R}^n : |x| \le R \text{ and } |y| \le R, \text{ or } |y| \ge R/2 \text{ and } |x-y| \le R/|y|\}$

Lemma 1. The operator

$$f \to \sup_{0<\lambda<1} \left| \int_{N_R^x} M_\lambda(x,y) f(y)\,dy \right|$$

maps L^1_γ boundedly into $L^{1,\infty}_\gamma$, for any $R < \infty$.

Proof. We cover $\mathbb{R}^n$ with $B(0,R)$ together with a sequence of balls of type $B(z, C\,R/|z|)$, $|z| \ge R/2$, with bounded overlap, so that $(x,y) \in N_R$ implies that one of these balls contains x and y. Hence, it is enough to verify that the restriction of M^* to a ball of this type is uniformly of weak type (1,1) for γ or, equivalently in view of (2.1), for Lebesgue measure. Because of the bounded overlap, we may then add these estimates and obtain the lemma.

So take $g \ge 0$ in $L^1(B)$ (Lebesgue measure), with $B = B(z, C\,R/|z|)$. Now if $\sqrt{1-\lambda^2} \ge 1/|z|$ and $x, y \in B$, we can estimate $M_\lambda(x,y)$ by $(1-\lambda^2)^{-n/2} \le C|B|^{-1}$, $C = C(R,n)$, where $|\cdot|$ means Lebesgue measure. Hence,

$$\int_B M_\lambda(x,y) g(y)\,dy \le C|B|^{-1} \int_B g\,dy \le C g^*(x),$$

g^* denoting the Hardy-Littlewood maximal function. And if $\sqrt{1-\lambda^2} < 1/|z|$ and $x, y \in B$, then $|y-\lambda x|$ differs from $|x-y|$ by at most $(1-\lambda)|x| < C\sqrt{1-\lambda^2}$. Considering separately the cases when $|y-\lambda x|$ is or is not much larger than $\sqrt{1-\lambda^2}$, we see that

$$\exp\left(-\frac{|y-\lambda x|^2}{2(1-\lambda^2)}\right) \le C \exp\left(-\frac{c|y-x|^2}{1-\lambda^2}\right)$$

for some c. But then $\int M_\lambda(x,y) g(y)\,dy$ is bounded in B by a convolution of g with a normalized contraction of the integrable radial decreasing kernel $C\exp(-c|.|^2)$ and thus by $C\,g^*(x)$, see [4, III.2.2]. Lemma 1 follows since the case of $B(0,R)$ is similar.

Outside N_R, we shall estimate M^* by the operator defined by the pointwise sup kernel

$$M(x,y) = \sup_{0<\lambda<1} M_\lambda(x,y).$$

Lemma 2. For some R, the operator

$$f \mapsto \int_{\mathbb{R}^n \setminus N_R^x} M(x,y)f(y)dy$$

maps L^1_γ into $L^{1,\infty}_\gamma$.

This would clearly imply the theorem.

We must estimate M and need some notation. If $y \neq 0$, let $\eta = |y|$ and $e = y/\eta$, and set $x = \xi e + v$ where v is orthogonal to e. By a and A we mean, respectively, $\min(\xi,\eta)$ and $\max(\xi,\eta)$. Of course, $\xi_+ = \max(\xi,0)$.

Lemma 3. Given a small $\beta > 0$, we may choose R so that the following estimates hold when $(x,y) \notin N_R$ for some c and C depending only on β, R, and n.

(a) If $|x-y| \geq \beta \max(|x|,|y|)$, then

$$M(x,y) \leq C \min\left(1, e^{\xi_+^2/2 - \eta^2/2}\right).$$

(b) If $|x-y| < \beta \max(|x|,|y|)$ and $|v| < A-a$, then

$$M(x,y) \leq C\left(\frac{A}{A-a}\right)^{n/2} \exp\left(-\frac{cA|v|^2}{A-a}\right) \min(1, e^{\xi^2/2 - \eta^2/2}).$$

(c) If $|x-y| < \beta \max(|x|,|y|)$ and $|v| \geq A-a$, then

$$M(x,y) \leq C(A/|v|)^{n/2} \exp(-cA|v|) \min(1, e^{\xi^2/2 - \eta^2/2}).$$

Proof. For x and y fixed, $x \neq y$, it is easily seen that $M_\lambda(x,y)$ takes its sup in $0 < \lambda < 1$ for some λ_{max} in $[0,1[$. The derivative $\partial M_\lambda(x,y)/\partial\lambda$ equals a positive factor times

$$\begin{aligned} U &= n\lambda(1-\lambda^2) + (x - \lambda y)\cdot(y-\lambda x) \\ &= n\lambda(1-\lambda^2) - \lambda|v|^2 + (\xi-\lambda\eta)(\eta-\lambda\xi) = I - II + P. \end{aligned} \tag{2.2}$$

Here the last product is

$$P = (A-\lambda a)(a-\lambda A) = ((1-\lambda)a + A-a)((1-\lambda)A - (A-a)). \tag{2.3}$$

If we replace η and v by 0 here, we see that then $\lambda_{max} = (a/A)_+$ and so

$$\sup_{0<\lambda<1} \exp\left(-\frac{(\eta-\lambda\xi)^2}{2(1-\lambda^2)}\right) = \min\left(1,\ e^{\xi_+^2/2 - \eta^2/2}\right). \tag{2.4}$$

In case (a), we conclude from (2.2) that

$$\begin{aligned} U &\le n - (x-y)\cdot(y-x) + (1-\lambda)(|x| + |y|)^2 \\ &\le n - |x-y|^2 + 4(1-\lambda)\max(|x|,|y|)^2 < 0 \end{aligned}$$

if R is large and λ close to 1. Hence, λ_{max} is bounded away from 1, and (2.4) gives the estimate in (a).

In case (b), notice that $a > A/2$ and $A-a > RA^{-1}/2$ because $|x-y| > R/|y|$. We see from (2.3) that

$$P \sim (1-\lambda)^2A^2 \quad \text{for } 1-\lambda > 4(A-a)A^{-1}. \tag{2.5}$$

So if $1-\lambda$ is much larger than $(A-a)A^{-1}$, then $II < P$ and $U > 0$. And if $1-\lambda < (A-a)A^{-1}/2$, we see by estimating I and P that

$$U < n(A-a)A^{-1} - (A-a)^2/2 < 0$$

for suitable R. Hence, $1-\lambda_{max} \sim (A-a)A^{-1}$, and (b) follows from (2.4).

To prove (c), we may assume $|v| > \bar{B}(A-a)$ for any fixed B, since the contrary case is covered by the method of (b). Notice that $|v| > RA^{-1}/2$. It is enough to show that

$$1 - \lambda_{max} \sim |v|A^{-1}. \tag{2.6}$$

For $1 - \lambda < c|v|A^{-1}$, we have

$$I/II < 2n(1-\lambda)/|v|^2 < cA^{-1}/|v| < 1/2.$$

Since (2.5) remains valid, (2.6) follows if we can exclude $1-\lambda_{max} \leq 4(A-a)A^{-1}$. But $1-\lambda \leq 4(A-a)A^{-1}$ implies $P < C(A-a)^2 < II/2$, and thus $U < 0$, if B is large enough. This completes the proof of (c) and Lemma 3.

3. Proof of Lemma 2.

We introduce sets forming a disjoint partition of $\mathbb{R}^n \times \mathbb{R}^n \setminus N_R$ if $\beta > 0$ is small. Let $\alpha(x,y)$ denote the angle between non-zero x and y, satisfying $0 \leq \alpha(x,y) \leq \pi$, and define

$$D_1 = \{(x,y) \notin N_R\colon \xi \leq \eta, \text{ and } \alpha(x,y) \geq \pi/4\}$$

$$D_2 = \{(x,y) \notin N_R\colon \xi > \eta, \text{ and } |x-y| \geq \beta \max(|x|,|y|)\}$$

$$D_3 = \{(x,y) \notin N_R\colon |x-y| < \beta \max(|x|,|y|) \text{ or both } \xi \leq \eta \text{ and } \alpha(x,y) < \pi/4\}.$$

Take an $f \geq 0$ in L^1_γ. We write

$$\int_{\mathbb{R}^n \setminus N_R^x} M(x,y)f(y)dy = \int_{D_1^x} + \int_{D_2^x} + \int_{D_3^x}$$

and estimate these three terms.

The first two terms turn out to be in L^1_γ. For

$$\int \gamma(x)dx \int_{D_1^x} M(x,y)f(y)dy = \int f(y)dy \int_{D_1^y} M(x,y)\gamma(x)dx,$$

and the integral over D_1^y here can be estimated by $C\gamma(y)$ if we use Lemma 3(a) and the fact that $\xi_+^2 \leq |x|^2/2$ in D_1^y. As to D_2, we arrive similarly at the integral

$$\int_{D_2^y} e^{-|x|^2/2}\, dx \leq C \int_y^\infty e^{-\xi^2/2}\, d\xi \leq C\gamma(y).$$

Before dealing with D_3, we divide $\mathbb{R}^n$ into disjoint cubes Q_i centered at x_i, $i = 1,2,\ldots$, such that

$$c \min(1, 1/|x_i|) \leq \operatorname{diam} Q_i \leq \min(1, 1/|x_i|).$$

Choose the enumeration so that $|x_i|$ is nondecreasing in i.

If χ denotes the characteristic function of D_3, we set $M_3 = \chi M$. Since we do not want our kernel to vary too much within a Q_i, we let

$$\overline{M}_3(x,y) = \sup\{M_3(x',y)\colon x \text{ and } x' \text{ in the same } Q_i\}.$$

Notice that the estimates of Lemma 3 hold also for $\overline{M}_3$, with new constants. Clearly $\overline{M}_3 f(x) = \int \overline{M}_3(x,y) f(y)dy$ dominates $M_3 f(x)$ and is constant in each Q_i.

Given $\alpha > 0$, we shall construct a subset E of $\{x\colon \overline{M}_3 f(x) > \alpha\}$ such that

$$\gamma\{\overline{M}_3 f > \alpha\} \leq C\gamma(E) \tag{3.1}$$

and

$$U(y) \leq C\gamma(y) \text{ in } \mathbb{R}^n. \tag{3.2}$$

Here

$$U(y) = \int_E \overline{M}_3(x,y)\gamma(x)dx.$$

This would yield

$$\gamma\{M_3 f > \alpha\} \le C\gamma(E) \le C\alpha^{-1} \int_E \overline{M}_3 f(x)\gamma(x)dx$$

$$= C\alpha^{-1} \int f(y)U(y)dy \le C\alpha^{-1} \|f\|_{L^1_\gamma} ,$$

and thus complete the proof of Lemma 2. This method is similar to that used for Theorem 1 in [2].

The set E will be constructed as the union of certain Q_j, which will be selected inductively. To obtain (3.2), we must not select too many Q_j close to each other. Therefore, we associate with each Q_j a forbidden region F_j, defined as the union of those Q_i, $i > j$, which intersect the set $Q_j + K_j$, where K_j is the cone $\{x: \alpha(x,y) \le \pi/4$ for some $y \in Q_j\}$.

The first step of the construction consists of selecting Q_1 if and only if it intersects, and thus is contained in, $\{\overline{M}_3 f > \alpha\}$. At the ith step, Q_i is selected if and only if it intersects $\{\overline{M}_3 f > \alpha\}$ and is not forbidden, i.e., it is not contained in F_j for any Q_j already selected. Then E is defined as the union of those Q_i selected.

To verify (3.1), we observe that $\{\overline{M}_3 f > \alpha\}$ is contained in the union of those Q_j selected and the corresponding F_j. The Q_j selected of course have total γ-measure $\gamma(E)$. So (3.1) follows if we verify that $\gamma(F_j) \le C\gamma(Q_j)$. When $|x_j| \le C$, we have $\gamma(Q_j) \sim \gamma(\mathbb{R}^n)$, so assume the contrary. Let H_s be the hyperplane $\{x: x \cdot x_j/|x_j| = x_j + s\}$. Then $F_j \cap H_s$ is empty for $s \le -C/|x_j|$, and has $(n-1)$-dimensional Lebesgue measure at most $C \max(s, 1/|x_j|)^{n-1}$ for $s > -C/|x_j|$. On $F_j \cap H_s$, we see that

$$\gamma(x) \le e^{-(s+|x_j|)^2/2} \le e^{-|x_j|^2/2 - |x_j|s} .$$

Hence,

$$\gamma(F_j) \le C \int_{-C/|x_j|}^{\infty} \max(s,1/|x_j|)^{n-1} e^{-|x_j|^2/2 - |x_j|s} ds$$

$$\leq C|x_j|^{-n} e^{-|x_j|^2/2} \sim \gamma(Q_j),$$

and (3.1) follows.

To show (3.2), we fix y and may assume $|y| \geq R/2$ since $D_3^y = \emptyset$ and $U(y) = 0$ otherwise. Let S_y denote the support of $\overline{M}_3(\cdot,y)$, which is the union of those Q_j intersecting D_3^y. For $v \perp e = y/\eta$, $\eta = |y|$, we let $\ell = \ell_v$ denote the line $\{ s\,e+v\colon s \in \mathbb{R}\}$, and set

$$I(v) = \int_{\ell_v \cap E \cap S_y} \overline{M}_3(s\,e+v,y)\gamma(s\,e+v)ds$$

so that

$$U(y) = \int I(v)dv, \tag{3.3}$$

the integral taken over $e^{\perp} = \mathbb{R}^{n-1}$.

Assume z belongs to some $Q_j \subset E \cap S_y$. Then Q_j intersects D_3^y and so e is in K_j. Therefore, F_j includes any Q_i, $i > j$, intersecting the ray $\{z + te\colon t > 0\}$. It follows that $\ell_v \cap E \cap S_y$ is contained in an interval $J = \{ s\,e+v\colon \xi \leq s \leq \xi + C\min(1,1/|\xi|)$. The point $x = \xi e+v$ is in or near D_3^y. We shall estimate $I(v)$ by means of Lemma 3, and consider the same cases (a), (b), (c) as in this lemma. Let a and A be as there. Notice that the estimates for $M(x,y)$ of Lemma 3 still hold if we replace x by any point in J.

(a) Lemma 3(a) gives

$$I(v) \leq C\min(1,1/|\xi|)e^{\xi_+^2/2 - \eta^2/2 - |x|^2/2} \leq Ce^{-\eta^2/2 - |v|^2/2}. \tag{3.4}$$

(b) Here $A \sim a \sim \eta$ and $A-a > 1/\eta$. Lemma 3(b) gives

$$I(v) \leq C\eta^{-1}\left(\frac{A}{A-a}\right)^{n/2} \exp\left(-\frac{c\,A}{A-a}|v|^2\right)e^{-|x|^2/2} \min(1,e^{\xi^2/2 - \eta^2/2})$$

$$\leq C\eta^{-1}\left(\frac{\eta}{A-a}\right)^{n/2} \exp\left(-\frac{c\,\eta}{A-a}|v|^2\right)e^{-\eta^2/2}.$$

Varying $A-a$, we see that this expression takes its maximum when $A-a \sim \eta|v|^2$. Such a value of $A-a$ is compatible with $A-a > 1/\eta$ only when $\eta|v| > 1$, and otherwise the largest admissible value of the expression occurs when $A-a \sim 1/\eta$. In both cases, we get

$$I(v) \leq C\eta^{-1} \min(|v|^{-n}, \eta^n) \, e^{-\eta^2/2}. \tag{3.5}$$

(c) Here $|v| > 1/\eta$, and Lemma 3(c) gives

$$I(v) \leq C\eta^{-1}(\eta/|v|)^{n/2} \exp(-\eta|v|) \, e^{-\eta^2/2}.$$

Estimating $\exp(-\eta|v|)$ by $C(\eta|v|)^{-n/2}$, we see that (3.5) holds also in this case.

Applying now (3.4-5) to (3.3), we obtain (3.2), and the proof is complete.

References

1. Muckenhoupt, B., Poisson integrals for Hermite and Laguerre expansions. Trans. Amer. Math. Soc. 139(1969), 231-242.

2. Sjögren, P., Weak L^1 characterizations of Poisson integrals, Green potentials, and H^p spaces. Trans. Amer. Math. Soc. 233(1977), 179-196.

3. Stein, E.M., Topics in harmonic analysis related to the Littlewood-Paley theory. Princeton University Press, Princeton 1970.

4. Stein, E.M., Singular integrals and differentiability properties of functions. Princeton University Press, Princeton 1970.

University of Göteborg
Chalmers University of Technology
S-41296 Göteborg
Sweden

POINTWISE BEHAVIOUR OF SOLUTIONS TO SCHRÖDINGER EQUATIONS

by Michael G. Cowling

ABSTRACT. Suppose that H is a self-adjoint (possibly unbounded, but with dense domain) operator on the Hilbert space $\mathcal{H}$. Take ϕ in $\mathcal{H}$ and let $\psi(t)$ be given by the formula

$$\psi(t) = \exp(itH)\phi \qquad t \in \mathbb{R}.$$

Then $\lim_{t \to 0} \psi(t) = \phi$.
Suppose that $\mathcal{H}$ is the space $L^2(X)$, for some measure space X. It is reasonable to ask when $\psi(t)$ converges to ϕ pointwise almost everywhere. We show that if $|H|^\alpha \phi$ is in $L^2(X)$ for some α in $(1/2, +\infty)$, then pointwise convergence is verified.

To motivate our work, consider the following examples. If $\mathcal{H} = L^2(\mathbb{R})$, and

$$H\phi(x) = x\phi(x) \qquad x \in \mathbb{R}, \quad \phi \in \mathrm{Dom}(H),$$

then $$\psi(t,x) = \exp(itx)\phi(x) \qquad x \in \mathbb{R},$$

and $$\psi(t,x) \to \phi(x) \qquad \text{a.e.}\, x$$

as $t \to o$ for any ϕ in H. On the other hand, if

$$H\phi(x) = id\phi(x)/dx \qquad x \in \mathbb{R}, \quad \phi \in \mathrm{Dom}(H),$$

then $$\psi(t,x) = \phi(x-t) \qquad x \in \mathbb{R},$$

and for general ϕ in $L^2(\mathbb{R})$,

$$\psi(t,x) \not\longrightarrow \phi(x) \qquad \text{a.e.}\,x$$

as $t \to o$. If however we assume that $|H|^{\alpha}\phi \in L^2(\mathbb{R})$ for some α in $(1/2, +\infty)$, then this forces ϕ to be continuous, and so pointwise convergence is obvious.

More recent examples arise in work of L. Carleson [1] and B.E.J. Dahlberg and C.E. Kenig [3], in which the case where

$$H\phi(x) = d^2\phi(x)/dx^2 \qquad x \in \mathbb{R}$$

is treated. These authors show that $|H|^{\alpha}\phi$ in $L^2(\mathbb{R})$ is sufficient to guarantee pointwise convergence if and only if $\alpha \geqslant 1/4$.

Our approach to this problem is abstract. It is based on the ideas we present in fuller detail in [2]. In particular, we assume only that H is self-adjoint, and further, our results hold for any realisation of the Hilbert space H as $L^2(X)$.

THEOREM. Let H be a self-adjoint operator on $L^2(X)$, and let α be in $(1/2,1)$. Suppose that $\phi \in L^2(X)$, and $|H|^{\alpha}\phi \in L^2(X)$. Let $M\phi$ be defined by the formula

$$M\phi = \sup \{ |t|^{-\alpha} |(\exp(itH)-I)f| \ : t \in \mathbb{R} \}.$$

Then $M\phi$ is in $L^2(X)$, and

$$\| M\phi \|_2 \leq C(\alpha) \| \ |H|^{\alpha}\phi \|_2 .$$

Consequently $\exp(itH)\phi$ converges to ϕ pointwise almost everywhere as t tends to 0.

Proof. By spectral theory, there are projections P and Q which commute with H and with each other such that HP and HQ are positive definite, and

$$H = HP + HQ.$$

Then

$$\exp(itH)\phi - \phi = \exp(itHP)P\phi - P\phi + \\ + \exp(itHQ)Q\phi - Q\phi,$$

and

$$|t|^{-\alpha}|\exp(itH)\phi-\phi| \leqslant |t|^{-\alpha}|\exp(itHP)P\phi-P\phi| + |t|^{-\alpha}|\exp(itHQ)Q\phi-Q\phi|.$$

We shall treat the maximal function

$$\sup\{|t|^{-\alpha}|\exp(itHQ)Q\phi-Q\phi| : t \in \mathbb{R}^+\}.$$

The way is which the other terms are treated is analogous.

We recall two Mellin transform inversions:

$$\lambda^{-\alpha}(\exp(-i\lambda)-1) = \int_{\mathbb{R}} du\ c^-(u)\ \lambda^{iu} \qquad \lambda \in \mathbb{R}^+$$

where $\quad c^-(u) = (1/2\pi)\exp([i\alpha-u]\pi/2)\Gamma(-\alpha-iu) \qquad u \in \mathbb{R}$

and $\quad \lambda^{-\alpha}(\exp(i\lambda)-1) = \int_{\mathbb{R}} du\ c^+(u)\ \lambda^{iu} \qquad \lambda \in \mathbb{R}^+$

where $\quad c^+(u) = (1/2\pi)\exp([u-i\alpha]\pi/2)\Gamma(-\alpha-iu) \qquad u \in \mathbb{R}.$

Let c be given by the rule

$$c(u)=(1/2\pi)\exp(\pi|u|/2)|\Gamma(-\alpha-iu)| \qquad u \in \mathbb{R}.$$

Then $|c^-| \leqslant c$ and $|c^+| \leqslant c$, and by standard estimates for the Γ-function (see, e.g., E.C. Titchmarsh [4], p. 151)

$$\int_{\mathbb{R}} du\ c(u) < \infty.$$

By spectral theory (substituting λ for $-tHQ$),

$$t^{-\alpha}(\exp(itHQ)Q\phi - Q\phi)$$

$$= (-tHQ)^{-\alpha}(\exp(-itHQ)-I)\ (-HQ)^{\alpha}\ Q\phi$$

$$= \int_{\mathbb{R}} du\ c^-(u)\ (-tHQ)^{iu}\ (-HQ)^{\alpha}\ Q\phi$$

$$= \int_{\mathbb{R}} du\ c^-(u)\ t^{iu}\ (-HQ)^{iu+\alpha}Q\phi.$$

In the Banach lattice $L^2(X)$, it follows that

$$|t^{-\alpha}(\exp(itHQ)Q\phi - Q\phi)|$$

$$\leqslant \int_{\mathbb{R}} du\ c(u)\ |(-HQ)^{iu+\alpha}Q\phi|$$

This last integral is t-independent, and

$$\left\| \int_{\mathbb{R}} du\ c(u)\ |(HQ)^{iu+\alpha}\ Q\phi| \right\|$$

$$\leq \int_{\mathbb{R}} du\ c(u)\ \|(-HQ)^{iu+\alpha}\ Q\phi\|$$

$$= \left(\int_{\mathbb{R}} du\ c(u)\right) \|(-HQ)^{\alpha}\ Q\phi\|,$$

so it represents an element of $L^2(X)$. Then

$$\sup\{|t^{-\alpha}(\exp(itHQ)Q\phi - Q\phi)| : t \in \mathbb{R}^+\}$$

$$\leq \int_{\mathbb{R}} du\ c(u)\ |(-HQ)^{iu+\alpha}Q\phi|.$$

Similarly,

$$\sup\{|t|^{-\alpha}\ |\exp(itHQ)Q\phi - Q\phi| : t \in \mathbb{R}^-\}$$

$$\leq \int_{\mathbb{R}} du\ c(u)\ |(HQ)^{iu+\alpha}\ Q\phi|.$$

and

$$\sup\{|t|^{-\alpha}\ |\exp(itHP)P\phi - P\phi| : t \in \mathbb{R}\}$$

$$\leq \int_{\mathbb{R}} du\ c(u)\ |(HP)^{iu+\alpha}\ P\phi|.$$

We conclude that

$$M\phi = \sup \{ |t|^{-\alpha} \, |\exp(itH)\phi - \phi| \; : \; t \in \mathbb{R} \}$$

$$\leq \int_{\mathbb{R}} du \; c(u) \; [\, |(HP)^{iu+\alpha} \, P\phi| + |(-HQ)^{iu+\alpha} Q\phi| \,]$$

and

$$\|M\phi\| \leq \int_{\mathbb{R}} du \; c(u) \; [\, \|(HP)^{\alpha} \;\; P\phi\| + \|(-HQ)^{\alpha} \; Q\phi\| \,]$$

$$\leq [\int_{\mathbb{R}} du \; c(u)] \; \sqrt{2} \; \| \, |H|^{\alpha} \; \phi\| ,$$

as required. It is clear that $\exp(itH)\phi$ converges to ϕ at points where $M\phi$ is finite. □

COROLLARY. If H is as above, and $\alpha \in (1/2, +\infty)$, $\phi \in L^2(X)$ and $|H|^{\alpha}\phi \in L^2(X)$, then $\exp(itH)\phi$ converges to ϕ almost eveŕywhere.

Proof. If $\alpha < 1$, we are done. Otherwise, pick β in $(1/2, 1)$, and use the moment inequality.

$$\| \, |H|^{\beta} \; \phi\| \leq \|\phi\|^{1-\beta/\alpha} \; \| \, |H|^{\alpha} \; \phi\|^{\beta/\alpha} .$$

to show that $|H|^{\beta} \; \phi \in L^2(X)$, and then apply the theorem. □

REFERENCES

[1] L. Carlèson, Some analytic problems related to statistical mechanics, in Euclidean Harmonic Analysis, Lecture Notes in Math. 779 (1979), 5-45.

[2] M. Cowling, Harmonic analysis on semigroups, to appear, Annals of Math..

[3] B.E.J. Dahlberg and C.E. Kenig, A note on the almost everywhere behaviour of solutions to the Schrödinger equation, in Harmonic Analysis, Lecture Notes in Math. 908 (1982), 205-209.

[4] E.C. Titchmarsh, The Theory of Functions. Oxford Univ. Press, Oxford, etc., 1978.

MICHAEL COWLING
Istituto di Matematica,
Università di Genova,
Via L.B. Alberti 4,
16132 GENOVA, ITALY.

AN APPLICATION OF L^p ESTIMATES TO SCATTERING THEORY

Walter Strauss* and Stephen Wainger**

The purpose of this lecture is to give an idea of how the tools of "L^p-Harmonic Analysis" can be used in a "physical" problem.

We begin with some physical background. For our purposes any dynamic physical system is described by a state vector $\psi(t)$ which evolves in time in a definite way. For a system of n particles moving in 3-space $\psi(t)$ could be, of course, a $6 \cdot n$ dimensional vector giving the position and momentum of each particle. In ordinary quantum mechanics of one particle in a potential field $\psi(t) = \psi(t,x)$ would be a solution of a Schroedinger equation

$$i \frac{\partial \psi(t,x)}{\partial t} = \Delta_x \psi(t,x) + q(x)\psi(t,x) .$$

Of course in this case $\psi(t,x)$ would be for each t a vector in the Hilbert Space $L^2(R^3)$. Note that if $\psi(t,x)$ is known for any fixed time, t, it is known for all future times. This is a property of Schroedinger's equation. So $\psi(t,x)$ evolves in a definite manner (even though one associates a probabalistic interpretation to $\int_E |\psi(t,x)|^2 dx$ for any measurable set E.) We see further that $\psi(t,x)$ does not have a physical interpretation but only functionals of ψ have physical interpretation This is true in general: the state vector $\psi(t)$ need not have a physical interpretation, but every physical property of the physical system is supposed to be deducible from $\psi(t)$.

*Mathematics Department, Brown University, Providence, RI 02912.
**Department of Mathematics, University of Wisconsin, Madison, WI 53706.

The typing and preparation of this report is sponsored by the United States Army under Contract No. DAAG29-80-C-0041.

If a particle is free, that is, moving without interaction, its state vector $\psi(t)$ satisfies a linear equation $L\psi = 0$. Now suppose we have two particles, say type 1 and type 2, with corresponding linear operators L_1 and L_2. Then if the two particles did not interact (that is, were free), the physical system would be described by

$$L\psi = 0$$

where

$$L = \begin{pmatrix} L_1 & 0 \\ 0 & L_2 \end{pmatrix}$$

and

$$\psi = \begin{pmatrix} \psi_1 \\ \psi_2 \end{pmatrix} .$$

However, if interaction is also considered, and it must be for any interesting physical system, ψ would satisfy

$$(L + N)\psi = 0$$

where N expresses the interaction. In many problems N is assumed to be non-linear though its exact nature is not known. Thus we have a state vector ψ_P which describes a real "physical" system undergoing interaction that satisfies

$$(L + N)\psi_P = 0 .$$

We shall also consider the equation

$$L\psi_F = 0 .$$

ψ_F represents an idealized free system, that is, one not underoing interaction.

Specifically, we will take L to be the Klein Gordon operator

$$L\psi(t,x) = \frac{\partial^2\psi(t,x)}{\partial t^2} - \Delta_x\psi(t,x) + \psi(t,x)$$

and take N to be the very simple operator

$$N\psi(t,x) = |\psi(t,x)|^\gamma \operatorname{sgn} \psi(t,x)$$

where γ is a suitably restricted positive number. Here x is in R^3 and the state vector at time t is not just $\psi(t,\cdot)$ but should be conisdered as the pair $(\psi(t,\cdot), \frac{\partial\psi}{\partial t}(t,\cdot))$ of Cauchy data. This choice of L is a model for a free meson of unit mass. While the right N is not really known our particular choice is a simple choice which satisfies two criteria: that the system be invariant under Lorentz transformations and that the physical energy (which we shall get to later) be positive. The most commonly made choice in physics texts is $\gamma = 3$. In that case we can write $N\psi = \psi^3$. The physical interpretation of our equation is not very simple. This Klein-Gordon equation is the analogue for a meson of the system of 6 ordinary differential euqations which would classically describe an ordinary particle. Schroedinger's equation is the "quantized" version of these ordinary differential equations. For the case of the meson the Klein-Gordon equation is the "classical" description. Thus the true equation of the meson must be obtained by "quantizing" our equation. This "quantized" version involves an infinite number of degrees of freedom and is too complicated to describe here.

Thus we will consider the interacting system

$$\text{(1)}\qquad \frac{\partial^2\psi_p}{\partial t^2} - \Delta_x\psi_p + \psi_p + |\psi_p|^\gamma \operatorname{sgn} \psi_p = 0$$

and the free equation

$$\text{(2)}\qquad \frac{\partial^2\psi_F}{\partial t^2} - \Delta_x\psi_F + \psi_F = 0 .$$

The problem we consider is that of scattering. The idea of scattering is that at time $-\infty$ a physical system is in an initial free state. As time evolves the system undergoes interaction. Finally at time $+\infty$ we expect that the particle is in a final free state. The scattering operator is the mapping taking the initial free state to the final free state. We are interested in the existence of the scattering operator. So our problem is roughly the following: Given $\psi_F^{(1)}$ satisfying (2) find ψ_p satisfying (1) so that ψ_p in some sense agrees with $\psi_F^{(1)}$ at time $-\infty$. Then we wish to find $\psi_F^{(2)}$ satisfying (2) so that $\psi_F^{(2)}$ agrees with ψ_p at time $+\infty$.

In order to make a precise mathematical problem we need to specify the meaning of these asymptotic statements. We introduce the energy norm of a vector $\psi(t,x)$,

$$E(\psi(t,x)) = \{ \int_{R^3} \{\psi^2(t,x) + |\nabla_x\psi(t,x)|^2 + |\frac{\partial\psi(t,x)}{\partial t}|^2\}dx\}^{1/2} .$$

(This norm has the physical interpretation of energy.) The importance of the energy norm resides in the fact that, while $E\psi$ should vary with time, in fact $E\psi$ is constant if ψ satisfies (2).

We are now ready for the precise mathematical question. Let $E(\psi_F^1(t,x)) < \infty$ and satisfy

2) $$L\psi_F^1(t,x) = 0 \ .$$

Can we find

i) $\psi_p(t,x)$ such that

$$E(\psi_F^{(1)}(t,x) - \psi_p(t,x)) \to 0$$

as $t \to -\infty$ and such that

1) $$(L + N)\psi(t,x) = 0 \ ?$$

ii) Can we then find $\psi_F^{(2)}(t,x)$ satisfying

$$L\psi_F^{(2)}(t,x) = 0$$

such that

$$E(\psi_F^{(2)}(t,x) - \psi_p(t,x)) \to 0$$

as $t \to \infty$?

We then have the following

Theorem: (Strauss [S]). If $E(\psi_F^1(t,x))$ is sufficiently small and if $7/3 \leq \gamma \leq 3$, the answer to i) and ii) above is yes.

We shall only consider question i). (The treatment of ii) is similar.) To give an affirmative answer to question i) requires the construction of a solution to the equation 1). To do this we introduce a vector

$$\Phi(t) = \begin{pmatrix} \Phi_1(t) \\ \Phi_2(t) \end{pmatrix}$$

where we identify $\Phi_1(t)$ with $\psi(t)$ and Φ_2 with $\frac{\partial\psi}{\partial t}$. Then (1) looks like a first order differential euqation for Φ

3) $$\dot{\Phi} = L\Phi + P\Phi$$

where

$$L(\Phi) = \begin{pmatrix} 0 & 1 \\ \Delta_x - I & 0 \end{pmatrix}\Phi$$

and

$$P\Phi = \begin{pmatrix} 0 \\ -|\phi_1|^\gamma \operatorname{sgn} \phi_1 \end{pmatrix} .$$

Now by using the method of variation of parameters one obtains

4) $$\Phi(t) = T(t)f + \int_s^t T(t-\tau)P(\Phi)(\tau)dy$$

where

$$T(t) = \exp(tL)$$

or

$$\widehat{T(t)f}(\xi) = \begin{pmatrix} \cos t\beta & -\frac{\sin t\beta}{\beta} \\ \beta \sin t\beta & \cos t\beta \end{pmatrix}\hat{f}(\xi)$$

where $\beta = \sqrt{1+|\xi|^2}$, $f = \binom{f_1}{f_2}$. $T(t)f$ is a solution of the free equation

2') $$LT(t)f = 0$$

(2' is the translation of 2)). The condition of finite energy is that f_1, ∇f_1 and f_2 have finite L^2 norm. We say f is in X if this is the case. We are looking for solutions $\Phi(t)$ which are continuous functions of t with values in X; that is Φ is in $C(X)$.

So the solution $\Phi(t)$ of 4) represents the solution of our non-linear equation which at time s agrees with a given free solution, $T(t)f$. If s is finite this is really the ordinary Cauchy problem, and one can construct a solution of (4) by the Picard iteration procedure. But we want to solve (4) with $s = -\infty$, so that our true solution should agree with the given free solution at time $-\infty$. Now it seems difficult to use Picard iteration in $C(X)$ when $s = -\infty$ because $\| T(t)f \|_X$ will not decay, so iteration seems dubious.

The idea in principle is to divide the problem into two parts:

a) To find a solution of (4) with $s = -\infty$ in some appropriate auxillary space.

b) To use a "soft" argument to show that this solution is in $C(X)$.

Actually, this idea basically words. The auxillary space is something like L^p in space and time, with $p = \gamma + 1$. The soft part of the argument requires $p = \gamma + 1$ (at least in the spatial norm) because one presumably wishes to use the fact that

$$\int (\phi_1^2 + |\nabla\phi_1|^2 + \phi_2^2 + \tfrac{1}{\gamma+1}|\phi_1|^{\gamma+1})dx$$

doesn't change with time if $\Phi = \begin{pmatrix}\phi_1\\ \phi_2\end{pmatrix}$ is a solution of 4. This actually implies that the norm of $\Phi(t)$ of the solution to 4) is uniformly bounded in s. If we denote this solution by $\Phi_s(t)$ it follows that given any sequence $s_k \to -\infty$, there is a subsequence s_{k_j} such that $\Phi_{s_{k_j}}(t)$ converges weakly in X. If we have a good candidate from our construction in step a) for this weak limit, hopefully this solution will be in X and have all desired properties. This all works out.

Let us consider now the problem of trying to find a solution to (4) in some L^p space with $s = -\infty$, but, to simplify matters a little, consider instead the first component of (4). This is the integral equation

$$5) \qquad \phi(t) = S(t)f(t) + \int_{-\infty}^{t} S(t-\tau)|\phi(\tau)|^{\gamma}\mathrm{sgn}\ \phi(\tau)dy ,$$

where $f(t)$ is an L^2 function and

$$S(t)f(\xi) = \frac{\sin t\sqrt{1+|\xi|^2}}{\sqrt{1+|\xi|^2}}\hat{f}(\xi).$$

Our hope is to produce a solution of (5), essentially in $L^p(dx,dt)$ by Picard iteration. If there is any hope to do this, then we must at least be able to show the following:

I) If f is in $L^2(R^3)$, then

$$\| S(t)f \|_{L^p(dx,dt)} < \infty .$$

II. If $\phi(\tau) = \phi(\tau,x)$ is in $L^p(dx,d\tau)$ then so is

$$R\phi(t,x) = \int_{-\infty}^{t} S(t-\tau)|\phi(\tau,x)|^{\gamma}\mathrm{sgn}\ \phi(\tau,x)d\tau.$$

We shall first show that I) is essentially a restriction theorem of the Stein-Tomas type [T]. For

$$\widehat{S(t)f(\xi)} = \frac{\sin t\sqrt{1+|\xi|^2}}{\sqrt{1+|\xi|^2}}\hat{f}(\xi)$$

with f and hence $\hat{f}(\xi)$ in $L^2(R^3)$. So

$$S(t)f(x) = \int e^{i\xi\cdot x}\, e^{-it\sqrt{1+|\xi|^2}}\, \frac{\hat{f}(\xi)}{\sqrt{1+|\xi|^2}}\, d\xi$$

$$+ \text{ similar term}$$

$$= \int e^{i\xi\cdot x}\, e^{it\tau}\, \delta(\tau-\sqrt{1+|\xi|^2})\, \frac{\hat{f}(\xi)}{\sqrt{1+|\xi|^2}}\, d\xi\ .$$

Thus $S(t)f$ is the Fourier transform of a singular measure in R^4 supported on the surface $|\tau|^2 = |\xi|^2 + 1$, and so since $\hat{f}$ is in L^2, the dual statement of an appropriate restriction asserts $S(t)f$ is in $L^p(dx,dt)$ for appropriate p. This work was done by Strichartz [St].

Let us consider problem II). The condition that ϕ be in L^p is equivalent to saying that $u = |\phi|^\gamma \operatorname{sgn}\phi$ is in $L^{p/\gamma}$. Since $\gamma = p - 1$, $p/\gamma = p'$. So let

$$Wu = \int_{-\infty}^{t} S(t-\tau)u(\tau)d\tau\ .$$

We are then interested in showing

$$\text{6)}\qquad \|Wu\|_{L^p(dx,dt)} \le c\|u\|_{L^{p'}(dx,dt)}\ .$$

Now

$$\|Wu\|_{L^p(dx)} \le \int_{-\infty}^{t} \|S(t-\tau)u(\tau)\|_{L^p(dx)}\, d\tau$$

and $\|u(\tau)\|_{L^{p'}(dx)}$ is in $L^{p'}(dt)$. So if one could show

$$\text{7)}\qquad \|S(t-\tau)u(\tau)\|_{L^p(dx)} \le \frac{c}{|t-\tau|^\alpha}\, \|u(\tau)\|_{L^{p'}(dx)}$$

for appropriate α one could deduce 6) by the fractional integration Theorem. Estimate 7) was proved by Marshall, Strauss, and Wainger [MSW], the proof uses Stein's Complex interpolation between the (L^2,L^2) and

(H^1, BMO). Thus the machinery of Fefferman and Stein [FS] is involved in the argument. The analytic family of operators is defined by

$$S_z f(\xi) = (1+|\xi|^2)^{z/2} \frac{\sin t\sqrt{1+|\xi|^2}}{\sqrt{1+|\xi|^2}} \hat{f}(\xi)$$

We refer further details to the cited literature.

The same $L^p - L^q$ estimates have been used to advantage for several other problems in non-linear wave equations, notably by Glassey and Tsutsumi [GT], Kleinerman and Ponce [KP], and Shatah [Sh].

References

[FS] C. Fefferman and E. M. Stein, "H^p Spaces of several variables," Acta Math., Vol. 129 (1972) pp. 137-193.

[GT] R. Glassey and M. Tsutsumi "On uniqueness of weak solutions to some non-linear wave equations," J. Diff. Eqns. 46 (1982) pp. 409-425.

[KP] S. Kleinerman and G. Ponce, "Global, small amplitude solutions to nonlinear evolution equations," Comm. on Pure and Appl. Math., Vol. 36 (1983) pp.133-141.

[MSW] B. Marshall, W. Strauss, and S. Wainger, "L^p-L^q estimates for the Klein-Gordon equation", J. Math. Pures et Appl., Vol. 59 (1980) pp. 417-440.

[Sh] J. Shatah, "Global existence of small solutions to nonlinear evolution equations," J. Diff. Eqns., Vol. 46 (1982), pp. 409-425.

[S] W. Strauss, "Non linear scattering theory at low energy," Journal of functional analysis, Vol. 41 (1981) pp. 110-133.

[St] R. Strichartz, "Restrictions of Fourier transforms to quadratic surfaces," Duke Math. J., Vol. 44 (1977) pp. 705-714.

[T] P. Tomas, "Restriction theorems for the Fourier transform," Proc. of Symp. in Pure Math., Vol. 35, Part 1 (1979) pp. 111-119.

WS/SW/jvs

Elementary Characterizations of the Morrey-Campanato Spaces

by Svante Janson, Mitchell Taibleson*, and Guido Weiss*

The spaces $L(\alpha,q,s)$ are spaces of "smooth" functions on R^n, where α is non-negative, $1 \leq q \leq \infty$, and s is an integer that is at least as large as $[n\alpha]$, the greatest integer in $n\alpha$. For $\alpha = 0$ these spaces are variants of BMO (the space of functions bounded mean oscillation) and for $\alpha > 0$ they are variants of the homogeneous Besov-Lipschitz spaces with order of smoothness $n\alpha$. We denote by $\mathcal{P}^s$ the space of polynomials of degree at most s and by $B_q(h)$,

$$B_q(h) = \begin{cases} \left[\frac{1}{|B|}\int_B |h(x)|^q dx\right]^{\frac{1}{q}}, & 1 \leq q < \infty \\ \operatorname{ess\,sup}_{x\in B}|h(x)|, & q = \infty \end{cases},$$

where h is locally in L^q and B is a ball in R^n. We say that $f \in L(\alpha,q,s)$ if

$$\sup_B |B|^{-\alpha} \inf_{P\in\mathcal{P}^s} B_q(f-P) = \|f\|_{L(\alpha,q,s)} < \infty .$$

We may view $L(\alpha,q,s)$ as a function space or as a space whose elements are equivalence classes of functions modulo polynomials in $\mathcal{P}^s$. Adopting

*Research of these authors was supported in part by the National Science Foundation under Grant #MCS-7903122

this second point of view $L(\alpha,q,s)$, equipped with $\|\cdot\|_{L(\alpha,q,s)}$, is a Banach space. In this paper we give elementary proofs that if $\alpha > 0$ then $L(\alpha,q,s)$ and $L(\alpha,\infty,[n\alpha])$ are equivalent Banach spaces and if $\alpha = 0$ then $L(0,q_0,0)$ is equivalent to $L(0,q_1,s)$ provided $1 \leqq q_0, q_1 < \infty$.

§1. Introduction. For the balance of this paper we will view the $L(\alpha,q,s)$ spaces as spaces of functions. Our two principal results are:

Theorem 1. *If* $f \in L(\alpha,q,s)$, $\alpha \geqq 0$, $1 \leqq q \leqq \infty$, $s > n\alpha$, *then there is a polynomial* $P \in \mathcal{P}^s$ *such that* $f - P \in L(\alpha,q,s-1)$ *and* $\|f - P\|_{L(\alpha,q,s-1)} \leqq A\|f\|_{L(\alpha,q,s)}$ *where* $A > 0$ *is independent of* f .

Theorem 2. *If* $f \in L(\alpha,q,s)$, $\alpha > 0$, $1 \leqq q \leqq \infty$, $s \geqq [n\alpha]$ *then* f *is equal a.e. to a continuous function* $g \in L(\alpha,\infty,s)$ *and* $\|g\|_{L(\alpha,\infty,s)} \leqq A\|f\|_{L(\alpha,q,s)}$ *where* $A > 0$ *is independent of* f .

In the process of proving Theorem 2 we will obtain the more general result:

Theorem 3. *Let* $1 \leqq q \leqq \infty$ *and* k *be an integer such that* $0 \leqq k < n\alpha$, $s \geqq [n\alpha]$. *Then* $f \in L(\alpha,q,s)$ *if and only if* f *is equal a.e. to a function in* C^k *and for all* $\nu = (\nu_1,\ldots,\nu_n)$, $|\nu| = |\nu_1|+\ldots+|\nu_n| = k$, $D^\nu f \in L(\alpha - k/n, \infty, s-k)$. *Moreover,* $\|f\|_{L(\alpha,q,s)} \sim \sum_{|\nu|=k} \|D^\nu f\|_{L(\alpha - k/n,\infty,s-k)}$, (*where* $D^{\nu}f$ *represents the distribution derivative of* f *as well as the pointwise derivative*.)

Recall the result of John and Nirenberg [6] which implies the continuous inclusion $L(0,1,0) \subset L(0,q,0)$ if $1 \leqq q < \infty$. This is the analogue of Theorem 2 for $\alpha = 0$. The inclusions $\mathcal{P}^{s-1} \subset \mathcal{P}^{s}$, $s \geqq 1$ give the continuous embeddings $L(\alpha,q,s) \supset L(\alpha,q,s-1)$. The inequality $B_1(h) \leqq B_q(h)$ (which is a consequence of Hölder's inequality) gives the continuous embeddings $L(\alpha,q,s) \subset L(\alpha,1,s)$. Thus, from Theorem 1, Theorem 2 and these observations we obtain the equivalences among the spaces $L(\alpha,q,s)$ that were described in the opening paragraph.

A motivation for this work was to provide, by elementary means, a simplification in the definitions of the atomic Hardy spaces as well as in the identification of their duals (see [9]). Once the equivalences described above are established one may define the atomic Hardy spaces, $H^{p,q,s}$ directly, as continuous linear functionals on $L(1/p - 1,q',s)$ $(1/q + 1/q' = 1)$. It then follows that the dual of $H^{p,q,s}$ is $L(1/p - 1,q',s)$ by the arguments outlined in [9]. One can, of course, define the Hardy spaces as spaces of tempered distributions, but this approach creates some difficulties. Consider the problem of analyzing the "numerical" properties of $K*f$ where K is some kernel (for example, weak-L^p properties or estimates of associated maximal operators). With the results stated above one only needs to check that K has the requisite smoothness and then $K*f$ is defined as a function. When the

Hardy spaces are defined as spaces of tempered distributions it is usually necessary to work with a priori estimates and then use a limiting argument (note that the Poisson kernel is in every $L(\alpha,q,s)$ space but it is not in O_C, the class of convoluters on $\mathcal{S}'$).

We will complete this section with some historical remarks and notes to related results.

Observe that $L(0,1,0)$ is the space of functions of bounded mean oscillation (BMO) of John and Nirenberg [6]. If $\alpha > 0$ and $1 \leqq q \leqq \infty$, $L(\alpha,q,s)$ is closely related to one of the Morrey spaces studied by Campanato [2]. The chief differences are that Campanato did not consider the case $q = \infty$, his functions were defined on bounded open sets, and his space was normed differently (L^q-norm of the function was added to the norm defined above); such spaces are referred to as "non-homogeneous" in the literature. The Campanato notation for $L(\alpha,q,s)$ is $\mathcal{L}_s^{q,\lambda}$ with $\lambda = n+\alpha q$.

Certain aspects of Theorem 2 are not new. Grevholm [4] has used real interpolation to show that if $\alpha > 0$ and $q < \infty$ then $L(\alpha,q,s)$ coincides with a homogeneous Besov-Lipschitz space $B^{\alpha n}$ of functions defined as an intermediate space. The Grevholm notation for $L(\alpha,q,s)$ is $\mathcal{L}_{s+1}^{q,\lambda}$.

The $\alpha = 0$ case of Theorem 1 was announced earlier by Janson [5] where it is described as a result due to Campanato. This is so in the sense that it follows rapidly from Lemma (2.1) and versions of (2.1) occur often in [1].

Both Theorem 1 and Theorem 2 have been proven by Greenwald [3] in a companion paper to this one. He obtains both theorems as a consequence of a characterization of homogeneous Besov-Lipschitz spaces (defined by iterated difference conditions) of the Hardy-Littlewood type. He uses constructive arguments to show that the Morrey-Campanato spaces agree with the appropriate Besov-Lipschitz space.

We have been informed that our results were simultaneously obtained by DeVore and Sharpley [2] and are used by them in the study of smooth functions. We have also seen, in a preprint, a proof of Theorem 2 by methods even simpler than the ones used here (Jonsson, Sjögren, and Wallin [7]) but our argument includes the additional information which we summarize in Theorem 3.

Finally we note that the corresponding identifications for the non-homogeneous Morrey-Campanato spaces follow trivially from the results for the homogeneous spaces. (These are the spaces $L^q \cap L(\alpha,q,s)$, $\alpha > 0$, with norm $\|\cdot\|_{L^q} + \|\cdot\|_{L(\alpha,q,s)}$.) Note, however, that these results for the non-homogeneous case are fairly straight forward consequences of known results. Thus, if αn is not an integer, one sees that Campanato's characterization of his spaces as Hölder continuous functions (provided they are defined on "nice" bounded domains) extends easily to functions in $L^q(R^n)$. For such functions the principal error term (involving the L^p-norm of the function on the bounded domain) tends to zero as we consider a sequence of balls that expand to R^n . In a similar vein we note that Nagel and Stein have given a characterization of the non-homogeneous Lipschitz spaces and for the Campanato spaces $L(\alpha,\infty,[n\alpha]) \cap L^\infty$

[8; Prop. 3, p.80 and Main Lemma, p.83] which shows that these two classes of spaces agree. As they remark on page 83, the characterization in the Main Lemma would be the same if they had used the spaces $L(\alpha, q, [n\alpha]) \cap L^q$ and so Theorem 2 in the non-homogeneous case follows from their argument. One should also note that in the non-homogeneous case if $k > 0$ or $q < \infty$ then $L(\alpha, q, k)$ is a function space, not a space of equivalence classes $\mathrm{mod}(\mathcal{P}^k)$ since if f and g are both in L^q and $f - g \in \mathcal{P}^k$ then $f - g \equiv 0$.

§2. Proof of Theorem 1. We fix $f \in L(\alpha, q, s)$, $s > n\alpha \geq 0$, $1 \leq q \leq \infty$ and set $\|f\| \equiv \|f\|_{L(\alpha, q, s)}$. For $B = B(x_0, r) = \{|x - x_0| < r\}$ we write $P_B(x) = \sum_{|\nu| \leq s} a_\nu(x_0, r)(x - x_0)^\nu = \sum_{|\nu| \leq s} a_\nu(B)(x - x_0)^\nu$, where P_B is a polynomial (not always unique) in $\mathcal{P}^s$ that minimizes $B_q(f - P)$, $P \in \mathcal{P}^s$. The proof depends on a simple lemma that is essentially due to Campanato [1].

Lemma (2.1). Suppose $f \in L(\alpha, q, s)$, $B \subset B'$ and the radius of B' is at most twice the radius of B. If $|\nu| = s$

$$(2.2) \qquad |a_\nu(B) - a_\nu(B')| \leq C\|f\| \, |B|^{\alpha - \frac{s}{n}}$$

where C is a positive constant that does not depend on f or B.

Proof. Observe that

$$(2.3) \qquad |a_\nu(B) - a_\nu(B')| \leqq C_{s,n,q}|B|^{-\frac{s}{n}} B_q(P_B - P_{B'}) .$$

This is a consequence of two simple observations. The first is that since $a_\nu(B)$ and $a_\nu(B')$ are coefficients of highest order in P_B and $P_{B'}$ they do not depend on the point x_0 about which the polynomials are expanded. Second, for a fixed ball $B = B(x_0, r)$, $P(x) = \sum_{|\nu| \leqq s} c_\nu (x - x_0)^\nu$, we have that $\Sigma |c_\nu| \sim B_q(P)$, and so by homogeneity $|c_\nu| \leqq C|B|^{-s/n} B_q(P)$ if $|\nu| = s$, and so (2.3) follows. Now, $B_q(P_B - P_{B'}) \leqq B_q(f - P_B) + B_q(f - P_{B'})$ $\leqq B_q(f - P_B) + 2^{n/q} B'_q(f - P_{B'}) \leqq \|f\| \, |B|^\alpha + 2^{n/q} \|f\| \, |B'|^\alpha = \|f\| \, |B|^\alpha (1 + 2^{n(1/q + \alpha)})$ and Lemma (2.1) is established.

If we let $B = B(x_0, 2^k)$, $B' = B(x_0, 2^{k+1})$ we obtain $|a_\nu(x_0, 2^k) - a_\nu(x_0, 2^{k+1})| \leqq C\|f\| 2^{k(n\alpha - s)}$. Similarly, if $2^k < r \leqq 2^{k+1}$, $|a_\nu(x_0, r) - a_\nu(x_0, 2^k)| \leqq C\|f\| r^{(n\alpha - s)}$ where $C > 0$ does not depend on f or x_0. Consequently $a_\nu(x_0) = \lim_{r \to \infty} a_\nu(x_0, r)$ exists and there is a $C > 0$, independent of f and x_0 such that $|a_\nu(x_0) - a_\nu(x_0, r)| \leqq C\|f\| r^{(n\alpha - s)}$. If $x_0 \neq y_0$ we let $B = B(x_0, r)$ and $B' = B(y_0, 2r)$ with $r > |y_0 - x_0|$. Then $B \subset B'$ and from Lemma (2.1) $|a_\nu(x_0, r) - a_\nu(y_0, 2r)| \leqq C\|f\| r^{(n\alpha - s)} \to 0$ as $r \to \infty$, and so $a_\nu(x_0) = a_\nu(y_0)$. Thus, $\lim_{r \to \infty} a_\nu(x_0, r) = a_\nu$ exists, is independent of x_0 and

$$(2.4) \qquad |a_\nu(B) - a_\nu| \leqq C\|f\|\,|B|^{\alpha - \frac{s}{n}},$$

where $C > 0$ is a constant independent of f and B.

Let $P(x) = \sum_{|\nu| = s} a_\nu x^\nu$. Note that $P \in \mathcal{P}^s$. Let $Q_B(x) = (P_B(x) - P(x)) - \sum_{|\nu| = s} (a_\nu(B) - a_\nu)(x - x_0)^\nu$. Observe that $Q_B \in \mathcal{P}^{s-1}$ and that $(f(x) - P(x)) - Q_B(x) = (f(x) - P_B(x)) + \sum_{|\nu| = s} (a_\nu(B) - a_\nu)(x - x_0)^\nu$.

Thus,

$$B_q((f - P) - Q_B) \leqq B_q(f - P_B) + \sup_{x \in B} \sum_{|\nu| = s} |a_\nu(B) - a_\nu|\, |x - x_0|^{|\nu|}$$

$$\leqq \|f\|\, |B|^\alpha + C\|f\|\, |B|^{\alpha - s/n} |B|^{s/n} \leqq C\|f\|\, |B|^\alpha .$$

This completes the proof of Theorem 1.

§3. Proof of Theorem 2. We assume that $f \in L(\alpha, 1, s)$, $\alpha > 0$ and $s \geqq [n\alpha]$. We set $\|f\| \equiv \|f\|_{L(\alpha,1,s)}$.

In this section we use a function ψ such that

$$(3.1) \qquad \begin{aligned} &\text{i) } \psi \in C^\infty \text{ and is supported in } B(0,1) \\ &\text{ii) } \int \psi(x)\, x^\nu dx = \begin{cases} 1, & |\nu| = 0 \\ 0, & 0 < |\nu| \leqq s \end{cases} \end{aligned}$$

It will suffice to construct such a ψ for $n = 1$. Fix a function $\varphi \in C^\infty(\mathbb{R})$ that is supported in $[-1/(s+1), 1/(s+1)]$ and $\int \varphi\, dx = 1$.

Let $m_\ell = \int \varphi(x)x^\ell dx$, $\ell = 0,1,\ldots,s$. Let $\varphi_k(x) = 1/k\ \varphi(x/k)$, $k = 1,2,\ldots,s+1$, and observe that $\int \varphi_k(x)\, x^\ell dx = k^\ell m_\ell$. Thus, if we let

$$\psi(x) = \sum_{k=1}^{s+1} c_k\varphi_k(x)\ ,\quad \int \psi(x)x^\ell dx = (\Sigma\, c_k k^\ell)m_\ell\ .$$

Note that the matrix with entries $a_{k\ell} = k^\ell$, $k = 1,\ldots,s+1$; $\ell = 0,1,\ldots,s$ is non-singular so we may find $\{c_k\}$ so that $\Sigma\, c_k = 1$ and $\Sigma\, c_k k^\ell = 0$, $\ell = 1,\ldots,s$. ψ is the required function.

For the proof we will make use of an idea from the atomic H^p theory (the atomic decomposition of certain distributions) but we do not use any detailed knowledge of the atomic theory per se.

<u>Definition</u>. g is a (p,s)-atom on R^n , $0 < p \leq 1$, $s \geq [n(1/p - 1)]$ if g is supported on a ball B , $\sup_{x\in B} |g(x)| \leq |B|^{-1/p}$ and $\int g(x)x^\nu dx = 0$, $|\nu| \leq s$.

It follows that if g is a (p,s)-atom and $f \in L(\alpha,1,s)$ with $\alpha = 1/p - 1$,

$$(3.2)\qquad |\int fg| = |\int_B (f(x) - P_B(x))g(x)dx|$$

$$\leq |B|^{1-\frac{1}{p}} \frac{1}{|B|}\int_B |f(x) - P_B(x)|dx \leq |B|^{1-\frac{1}{p}} |B|^\alpha \|f\| = \|f\|\ .$$

For the rest of this section α and p are fixed with $\alpha = 1/p - 1$.

Lemma (3.3). Let ψ be as in (3.1) and set $\psi_t(x) = t^{-n}\psi(x/t)$, $t > 0$. Then if $0 < p < 1$, $0 < t/2 \leqq \tau \leqq t$, $\psi_t - \psi_\tau = \lambda g$ where g is a (p,s)-atom and $|\lambda| \leqq Ct^{n(1/p\ -1)}$.

Proof. $\psi_t - \psi_\tau$ is supported on $B(0,t)$, $|B(0,t)| \leqq A_n t^n$, and $|\psi_t - \psi_\tau| \leqq (1+2^n)t^{-n}\|\psi\|_{L^\infty} = Ct^{n(1/p\ -1)}|B(0,t)|^{-1/p}$. If $|\nu| \leqq s$, $\int(\psi_t(x) - \psi_\tau(x))x^\nu dx = (t^{|\nu|} - \tau^{|\nu|})\int \psi(x)x^\nu dx$. If $\nu = 0$, $t^{|\nu|} - \tau^{|\nu|} = 1 - 1 = 0$. If $0 < |\nu| \leqq s$, $\int \psi(x)x^\nu dx = 0$ and the conclusion of Lemma (3.3) follows.

Lemma (3.4). If $f \in L(\alpha,1,s)$ and ψ is as in (3.1) then $f*\psi_t(x) \to f(x)$ a.e. as $t \to 0$, $\|f*\psi_t - f\|_{L^\infty} \leqq Ct^{n\alpha}$, and, consequently, f is equal a.e. to a continuous function.

Proof. From (3.2) and (3.3) it follows that if $2^{-k-1} \leqq t < 2^{-k}$ then $\|f*\psi_{2^{-k}} - f*\psi_t\|_{L^\infty} \leqq C\|f\|t^{n\alpha}$, so that (as in §2) $f*\psi_t$ converges uniformly to a continuous function h such that $\|f*\psi_t - h\|_{L^\infty} \leqq C\|f\|t^{n\alpha}$. Since f is locally integrable, $f*\psi_t(x) \to f(x)$ a.e. and the proof of (3.4) is complete.

Remark. The differentiability of ψ did not enter into the proof of (3.4). However, an analogous argument shows that if $|\nu| < n\alpha$ then $D^\nu(f*\psi_t)$ converges uniformly to a continuous function that is $D^\nu f$ in the distribution sense and consequently it is also the pointwise derivative of f. In particular, $f \in C^k$ if $k < n\alpha$.

To prove Theorem 2 we obtain the required estimate with a special choice for $P \in \mathcal{P}^s$; namely, $Q_B^s f$, the unique polynomial in $\mathcal{P}^s$ such that

(3.5) $$\int_B f(x)x^\nu dx = \int_B Q_B^s f(x)x^\nu dx\ , \quad |\nu| \leqq s\ .$$

This is the so-called Gram-Schmidt polynomial. The crucial facts about this polynomial are that

(3.6) $$Q_B^s P = P \quad \forall\ P \in \mathcal{P}^s$$

(since $Q_B^s P$ is the unique polynomial in $\mathcal{P}^s$ that satisfies (3.5)) and that there is a kernel $g_B^s(t;x)$ that is supported on B and

(3.7) $$Q_B^s f(x) = \int_B f(t)g_B^s(t;x)dt\ , \quad \forall\ x \in B\ .$$

By a homogeneity argument,

(3.8) $$|g_B^s(t;x)| \leqq C|B|^{-1}\ ,$$

where $C > 0$ is independent of B and $x \in B$. From (3.6) and (3.7) we have

(3.9) $$\int_B g_B^s(t;x)t^\nu dt = x^\nu\ , \quad 0 \leqq |\nu| \leqq s\ .$$

Now observe that if ψ is as in (3.1)

$$\int \psi_r(x-t)t^{\nu}dt = x^{\nu} \quad , \quad 0 \leqq |\nu| \leqq s \; . \tag{3.10}$$

Fix $B = B(x_0, r)$. It follows from (3.8), (3.9) and (3.10) that $\psi_r(x-t) - g_B^s(t;x) = \lambda_0(x)a_0(t;x)$ where $a_0(t;x)$ is a (p,s)-atom supported on $B(x_0, 2r)$ and $|\lambda_0(x)| \leqq C|B|^{\alpha}$, $x \in B$. Similarly $\psi_{r2^{-k}}(x-t) - \psi_{r2^{-(k-1)}}(x-t) = \lambda_k(x)a_k(t;x)$ where $a_k(t;x)$ is a (p,s)-atom and $|\lambda_k(x)| \leqq C|B|^{\alpha} 2^{-kn\alpha}$ for $x \in B$. Then for a.e. $x \in B$,

$$f(x) - Q_B^s f(x) = \int f(t)(\psi_r(x-t) - g_B^s(t;x))dt$$

$$+ \sum_{k=1}^{\infty} \int f(t)(\psi_{r2^{-k}}(x-t) - \psi_{r2^{-(k-1)}}(x-t))dt$$

$$= \sum_{k=0}^{\infty} \lambda_k(x) \int f(t)a_k(t;x)dt \; .$$

We now use (3.2), so for a.e. $x \in B$

$$|f(x) - Q_B^s f(x)| \leqq \sum_{k=0}^{\infty} |\lambda_k(x)| \, |\int f(t)a_k(t;x)dt|$$

$$\leqq \sum_{k=0}^{\infty} C|B|^{\alpha} 2^{-kn\alpha} \|f\| \leqq C\|f\| \, |B|^{\alpha} \; .$$

This completes the proof of Theorem 2.

<u>The Proof of Theorem 3.</u> It is easy to see that if each component of the gradient of f is in $L(\alpha,\infty,s)$ then f is in $L(\alpha+1/n,\infty,s+1)$ so that the "if" part of the theorem is straight-forward. From the remark following (3.4) we have that $D^\nu f$ is continuous if $|\nu| < n\alpha$, and $D^\nu f$ is the limit (uniformly) of $D^\nu(f*\psi_t) = f*D^\nu\psi_t$. Arguing as above we obtain the estimates, $D^\nu\psi_r(x-t) - D^\nu g_B^s(t;x) = \lambda_0^\nu(x)a_0^\nu(t;x)$, and $D^\nu\psi_{r2^{-k}}(x-t) - D^\nu\psi_{r2^{-(k-1)}}(x-t) = \lambda_k^\nu(x)a_k^\nu(t;x)$ where $a_k^\nu(t;x)$ is a (p,s)-atom and $|\lambda_k^\nu(x)| \leq c^\nu|B|^{\alpha-|\nu|/n}\,2^{-k(n\alpha-|\nu|)}$ provided $|\nu| < n\alpha$, and it follows, as above, that $D^\nu f\in L(\alpha-|\nu|/n,\infty,s-|\nu|)$ and $\|D^\nu f\|_{L(\alpha-|\nu|/n,\infty,s-|\nu|)} \leq c\|f\|$. The other norm estimate is direct and that completes the proof of Theorem 3.

REFERENCES

1. S. Campanato, Proprietà di una famiglia di spazi funzionali, Ann. Scuola Norm. Sup.-Pisa, 18 (1964), 137-160.

2. R. DeVore and R. Sharpley, On the smoothness of functions, to appear.

3. H. Greenwald, On the theory of homogeneous Lipschitz and Campanato spaces, to appear in Pacific Journ. Math.

4. B. Grevholm, On the structure of the spaces $\mathcal{L}_k^{p,\lambda}$, Math. Scand. 26 (1970), 241-254.

5. S. Janson, Generalizations of Lipschitz spaces and an application to Hardy spaces and bounded mean oscillation, Duke Math. J. 47 (1980), 959-982.

6. F. John and L. Nirenberg, On functions of bounded mean oscillation, Comm. Pure Appl. Math. 14 (1961), 415-426.

7. A. Jonsson, P. Sjögren, and H. Wallin, Hardy and Lipschitz spaces on subsets of R^n, University of Umea, Sweeden, Report No. 8, 1981, 38 pp.

8. A. Nagel and E. M. Stein, Lectures on Pseudo-Differential Operators: Regularity Theorems and Applications to Non-Elliptic Problems, Math. Notes No. 24, Princeton University Press, 1979.

9. M. Taibleson and G. Weiss, The molecular characterization of certain Hardy spaces, Astérisque 77 (1980), 68-149.

ON NONISOTROPIC LIPSCHITZ SPACES

by

Paolo M. Soardi

1. Introduction.

Suppose A_t $(t>0)$ is the group of linear transformation of R^N defined by $A_t=\mathrm{diag}(t^{\alpha_1},..,t^{\alpha_N})$, where $\alpha_j\geq 1$ is real $(j=1,..,N)$. It is known that one can associate to any such group of transformations a unique 'norm' function $\rho(x)$ such that $\rho(x-y)$ is a distance in R^N, $\rho(A_t x)=t\rho(x)$ and $\rho(x)=1$ if and only if $|x|=1$ (where $|\cdot|$ is the euclidean norm); see [4]. In this paper we introduce the analogue, with respect to A_t, of the higher order difference operators and define the corresponding Lipschitz spaces. We show that such spaces are equivalent to Campanato-Morrey spaces and study some of their principal properties. We also point out the relationship with the nonisotropic Hardy spaces. Nonisotropic Lipschitz spaces are also studied in [6] and in [11]. These Authors define a general class of nonisotropic Lipschitz spaces of vector-valued functions, by means of convolutions with suitable Borel measures. We confine ourselves to the case of the supremum norm, but we deal with not necessarily bounded tempered distributions (as done in [8] and in [10] in the isotropic case). A basic tool in the treatment of [6] and [11] is Calderon's representation theorem for tempered distributions. Such a theorem was also used in [8] to prove the equivalence between isotropic Lipschitz and Campanato-Morrey spaces. This equivalence can be proved also in our case via Calderon's theorem (as pointed out to us by M. Taibleson). However we preferred to proceed in a perhaps more elementary way, by expressing the moduli of continuity by means of a nonisotropic Peetre's K-functional, which may be of independent interest. On the other hand a discrete analogue of Calderon's method is used in the proofs of theorems 1 and 8 below.

Non isotropic difference operators,Lipschitz spaces and related appr ximation properties of functions of several variables are also studi in [7] from a somewhat different point of view,under the assumption that all the α_j are integers.In this paper we shall have in mind the well known duality between Lipschitz and Hardy spaces([5],[10]).Fina ly,R^N with nonisotropic dilations is a particular instance of a homo geneous group,in the sense of [5].The Campanato spaces defined below coincide with those of Folland and Stein who are,however,mainly inte rested in the case of stratified groups .We refer to [5] for the gen ral theory.

We wish to thank G.Weiss for calling our attention to the problems considered here,and M.Taibleson and G.Weiss for several invaluable con versations on the subject of this paper.

2.Higher order difference operators and nonisotropic Lipschitz space

Suppose A_t is the group of dilations defined in section 1.Through out this paper we use the following notation.We set $\alpha=(\alpha_1,..,\alpha_N)$ and $|\alpha|=\alpha_1+..+\alpha_N$ .We adopt the usual notation for multiindex;in particular we write $|\beta|$ also to denote the lenght of the multiindex β. We let $\alpha\cdot\beta=\alpha_1\beta_1+..+\alpha_N\beta_N$ and denote by F the set of all real numbers k such that $k=n_1\alpha_1+..+n_N\alpha_N$, where n_j ranges over all nonnegative integers.Nonisotropic Lipschitz,Campanato-Morrey and Hardy spaces will be indexed by the elements of F. If all the α_j are integers , then the elements of F are nonnegative integers.

For every r>0 and $x\in R^N$ we denote by B(x,r) the nonisotropic ball $B(x,r) = \{y\in R^N:\ \rho(x-y)\le r\}$. If f is a function on R^N we write $f_r(x) = r^{-|\alpha|}f(A_r^{-1}x)$. We notice that

$$\int_{R^N} f_r(x)\,dx = \int_{R^N} f(x)\,dx \quad \text{and} \quad D^\beta f_r(x) = r^{-\alpha\cdot\beta}(D^\beta f)_r(x) .$$

Finally C will denote a constant which may vary from line to line.

Lemma1. Suppose $n\geq 2$ is an integer and $0<a_1<\dots<a_n$, $0<x_1<\dots<x_{n-1}$ are real numbers. Then there exists exactly one n-tuple $c_1,\dots,c_n$ such that

(1) $\sum_{j=1}^{n} c_j = 1$; (2) $\sum_{j=1}^{n} c_j a_j^{x_k} = 0$ $(k=1,..,n-1)$

(3) $\sum_{j=1}^{n} c_j a_j^{x} \neq 0$ if $x\neq x_k$ $(k=1,..,n-1)$.

Proof. Since every nontrivial linear combination of n distinct exponentialshas at most n-1 zeros ,the proof follows from elementary linear algebra.

Suppose now that k belongs to the set F and denote by $\bar{n}(k)-1$ the number of distinct values smaller than k of $\alpha\cdot\beta$,as β ranges over the set of all multiindex $(\beta\neq 0)$. Taking Lemma1 into account, let ,for every $n\geq\bar{n}(k)$, $c_0=-1, c_1,\dots,c_n$ be a n+1-tuple satisfying

$$(4)\quad \begin{cases} \sum_{j=0}^{n} c_j = 0 \\ \sum_{j=0}^{n} c_j j^{\alpha\cdot\beta} = 0 \quad \text{if} \quad 0<\alpha\cdot\beta<k \\ \sum_{j=0}^{n} c_j j^{\alpha\cdot\beta} \neq 0 \quad \text{if } \alpha\cdot\beta\geq k \end{cases}$$

(Remark that, if $n = \bar{n}(k)$, then such a n+1-tuple is uniquely determined).

To any such a n+1-tuple we associate a measure $\mu_h = \sum_{j=0}^{n} c_j \delta_{-A_j h}$, where $h \in R^N$ and δ_y denotes the Dirac measure at y.

Definition. For every $f \in L^1_{loc}(R^N)$ and every μ_h as above we set :

$$(5) \qquad \Delta(\mu_h) f(x) = f * \mu_h(x) = \sum_{j=0}^{n} c_j f(x + A_j h)$$

$x, h \in R^N$.

In the sequel we will write Δ_h^k instead of $\Delta(\mu_h)$. The arbitrariety in the choice of n and the coefficients c_j is illusory, as far as Lipschitz spaces are concerned. In fact such spaces will be shown to be equivalent to Campanato-Morrey spaces whose definition depends only on k.

If $\alpha_1 = \ldots = \alpha_N = 1$, then $\bar{n}(k) = k$, so that, choosing $n=k$, we obtain

$$(-1)^{k+1} \Delta_h^k f(x) = \sum_{j=0}^{k} (-1)^{k+j} C(k,j) f(x+jh)$$

(where C(k,j) is the binomial coefficient).

Hence, up to the sign, we have the usual difference operators.

If all the α_j's are integers, then $\bar{n}(k) \le k$. Choosing $n=k$, we see that the numbers $c_j = (-1)^{j+1} C(k,j)$ satisfy (4). In this case

(5) gives, up to the sign, the difference operators defined by Torchinsky [11] .

Definition. For every positive k in F we denote by Π^k the linear space of all polynomials $P = \sum_\beta a_\beta x^\beta$ such that $a_\beta = 0$ if $\alpha\cdot\beta \geq k$.

Lemma2. Suppose f is locally integrable on R^N. Then , for every k, $f \in \Pi^k$ if and only if $\Delta_h^k f = 0$ for every $h \in R^N$.

Proof. Suppose β is such that $\alpha\cdot\beta<k$. Then

$$\Delta_h^k(x^\beta) = \sum_\xi C(\beta,\xi)x^{\beta-\xi}h^\xi \sum_{j=0}^{n} c_j j^{\alpha\cdot\xi}$$

where ξ ranges over all multiindex smaller or equal to β . Hence the first part of the lemma follows from (4) . Conversely, suppose $\Delta_h^k f = 0$ for all h . Let g be an infinitely differentiable function with compact support. Then , for every h :

$$(6) \qquad 0 = \langle \Delta_h^k f , g\rangle = \langle f , \Delta_{-h}^k g\rangle \qquad .$$

Set $h = A_t v$, with $\rho(v) = 1$. Denote by $k_1<k_2<\ldots$ the values of $\alpha\cdot\beta$ not smaller than k . Set $m_1=[k_1]+1$. If $|\beta|\geq m_1$ then $\alpha\cdot\beta>k$. By (4) and Taylor's theorem :

$$(7)\qquad \Delta_h^k g(x) = \sum_{\substack{|\beta|<m_1 \\ \alpha\cdot\beta\geq k_1}} t^{\alpha\cdot\beta}\, \gamma_{k,\beta} v^{\beta} D^{\beta} g(x) \quad + $$

$$+ \sum_{|\beta|=m_1} t^{\alpha\cdot\beta}\, v^{\beta} G_{\beta}(x,h)$$

where $\gamma_{k,\beta} = (\beta!)^{-1} \sum_{j=0}^{n} c_j j^{\alpha\cdot\beta}$ and $G_\beta(x,h)$ is a bounded function of x and h. Remark that, by (4), $\gamma_{k,\beta} \neq 0$.

We have from (6) and (7)

$$(8)\qquad 0 = \sum_{\substack{|\beta|<m_1 \\ \alpha\cdot\beta=k_1}} t^{k_1}\, \gamma_{k,\beta} v^{\beta} \langle f, D^{\beta} g\rangle \quad + \text{ higher order terms .}$$

Dividing both members of (8) by t^{k_1} and letting t tend to zero we get

$$0 = \sum_{\substack{|\beta|<m_1 \\ \alpha\cdot\beta=k_1}} \gamma_{k,\beta} v^{\beta} \langle f, D^{\beta} g\rangle$$ for all v and g, i.e. $D^{\beta} f = 0$ for all β such that $\alpha\cdot\beta = k_1$. Now, repeating the argument above with k_n and $m_n = [k_n]+1$ in place of k_1 and m_1 we obtain $D^{\beta} f = 0$ for all β such that $\alpha\cdot\beta \geq k$. This implies that f belongs to Π^k.

Definition. Let f be a locally bounded function on R^N. Suppose $\delta>0$ and $k>\delta, k\in F$. We say that f belongs to $\mathrm{Lip}(\delta,k)$ if

$$(9)\qquad \sup_{h\neq 0}\frac{\|\Delta_h^k f\|_\infty}{\rho(h)^\delta} = \|f\|_{Lip(\delta,k)} < \infty$$

By Lemma 2,(9) defines a norm on $Lip(\delta,k)$ if we identify two functions f and g such that f-g belongs to Π^k . We shall see later that $Lip(\delta,k)$ is a Banach space for the norm (9) . We did not stress in our notation the dependence of the space $Lip(\delta,k)$ upon the choice of the coefficients c_j for the reasons explained above . In any case , no confusion can arise .

Theorem1 . If $f\in Lip(\delta,k)$ then $D^\beta f$ is continuous for $\alpha\cdot\beta<\delta$.
Proof. Let g denote an infinitely differentiable function whose support is contained in the ball $B = B(0,1)$ and such that $\int_B g(x)dx = 1$. Set $p^r = \sum_{j=1}^n c_j g_{j2^{-r}}$, $P^r = p^r - p^{r-1}$, $Q^r = p^r + p^{r-1}$. By (4) and homogeneity properties we have $\int_{R^N} p^r(x)dx = 1$, so that

$$(10)\qquad D^\beta f = f*p^o*D^\beta p^o + \sum_{j=1}^\infty f*P^j*D^\beta Q^j$$

in the sense of distributions . Since every term on the right of (10) is continuous , it will suffice to prove that the series converges absolutely in $L^\infty(R^N)$.
First we remark that

$$f*P^j(x) = \int_{R^N} g(-y)\Delta^k_{A_{2^{-j}}y} f(x)\,dy -$$

$$- \int_{R^N} g(-y)\Delta^k_{A_{2^{-j+1}}y} f(x)\,dy$$

so that

$$(11) \qquad ||f*P^j||_\infty \leq C||f||_{Lip(\delta,k)} 2^{-j\delta} \quad .$$

Now , for for every compactly supported infinitely differentiable function ϕ we have

$$|\int_{R^N} \phi(x)\, f*P^j*D^\beta Q^j(x)dx| = |\int_{R^N} \tilde{\phi}* D^\beta Q^j(x) f*P^j(x)dx|$$

$$\leq C\,||f||_{Lip(\delta,k)} 2^{-j\delta}||\,\phi\,||_1||D^\beta Q^j||_1$$

by (11) (here $\tilde{\phi}(x) = \phi(-x)$). Since $||D^\beta Q^j||_1 = C2^{j\alpha\cdot\beta}$ for all j , we obtain $\sum_{j=1}^\infty ||f*P^j* D^\beta Q^j||_\infty \leq C||f||_{Lip(\delta,k)}$ and this concludes the proof of the theorem .

We shall see later that $D^\beta f$ belongs to a suitable Lipschitz space.

3.Equivalence between nonisotropic Lipschitz and Campanato spaces.

Suppose γ is a real positive number and $k>|\alpha|\gamma$, $k\in F$.

The (nonisotropic) Campanato-Morrey space $L(\gamma,\infty,k)$ is defined as the space of all (classes of equivalence mod.Π^k of) locally bounded functions f such that

$$(12) \qquad \sup_B m(B)^{-\gamma} \inf_{P\in\Pi^k} ||\,f-P||_{L^\infty(B)} = ||f||_{L(\gamma,\infty,k)} < \infty \,.$$

where B ranges over all nonisotropic balls in R^N and m denotes the Lebesgue measure . Remark that $m(B) = C_N r^{|\alpha|}$ if r is the (nonisotropic) radius of B .

Spaces similar to $L(\gamma,\infty,k)$ were studied in [2] and [9] .

As in the isotropic case the norm (12) can be realized by means of Graham-Schmidt orthonormalizations . Namely , for all ball B denote by $P_B(f)$ the unique polynomial in Π^k such that $\int_B (P_B(f)(x)-f(x))x^\beta dx = 0$ for all β such that $\alpha\cdot\beta<k$

Theorem2 . The norm defined by (12) is equivalent to

$$\sup_B m(B)^{-\gamma} \|f-P_B(f)\|_{L^\infty(B)} \quad . \tag{13}$$

The proof of this theorem is identical with the proof of Lemma 8.2 in [10] and we shall omit it .

In the following we will use for $P_B(f)$ the notation :

$$P_B(f) = \sum_{\alpha\cdot\beta<k} a_\beta(B)(x-x_o)^\beta \tag{14}$$

where x_o is the center of B .

Theorem3 . $L(\gamma,\infty,k)$ is a Banach space .

Proof.Suppose f_n is a Cauchy sequence in $L(\gamma,\infty,k)$.By Theorem 2 and the linearity of P_B we have that $F_{B,n}= f_n-P_B(f_n)$ is a Cauchy sequence in $L^\infty(B)$ for all ball B .Hence there is g_B belonging to $L^\infty(B)$ such that $\|F_{B,n}-g_B\|_{L^\infty(B)} \to 0$ as $n \to \infty$.

If $B_1 \subseteq B_2$, $F_{B_i,n} \to g_{B_i}$ on B_1, so that $P_{B_1}(f_n)-P_{B_2}(f_n) \to g_{B_2}-g_{B_1}$ uniformly on B_1. Thus $g_{B_2}-g_{B_1}$ coincides on B_1 with a polynomial of Π^k . Let $B_m = B(0,m)$, $m = 1,2\ldots$, and define g on R^N in the following way : $g(x) = g_{B_1}(x)$ if $x \in B_1$, while $g(x) = g_{B_m}(x) + \sum_{j=2}^{m} P_j(x)$ (where P_j denotes the restriction of $g_{B_{j-1}} - g_{B_j}$ to B_{j-1}) if $x \in B_m$. Now , if B is any ball and m is such that $B \subseteq B_m$, $g = g_{B_m} + P$, where P is the above sum of polynomials , on B . Moreover $g_B - g_{B_m} = Q \in \Pi^k$ on B . Hence

$$||g-P+Q||_{L^\infty(B)} = ||g_B||_{L^\infty(B)} = \lim_n ||F_{B,n}||_{L^\infty(B)}$$
$$\leq \sup ||f_n||_{L(\gamma,\infty,k)}\, m(B)^\gamma \quad .$$

Therefore $g \in L(\gamma,\infty,k)$. Analogously , fix $\varepsilon>0$. If i and j are large enough we have for every ball B :

$$\varepsilon \geq ||f_j-f_i||_{L(\gamma,\infty,k)} \geq m(B)^{-\gamma}\, ||F_{B,j} - F_{B,i}||_{L^\infty(B)} \quad .$$

Letting j tend to infinity we get

$$\varepsilon \geq m(B)^{-\gamma}\, ||g_B-F_{B,i}||_{L^\infty(B)} = m(B)^{-\gamma}\, ||g-P+Q-F_{B,i}||_{L^\infty(B)} \, .$$

By (12) f_i converges to g in $L(\gamma,\infty,k)$ as i tends to infinity .

Our next purpose is to prove an equivalence theorem between nonisotropic Lipschitz and Campanato-Morrey spaces. To do this we need an expression of the nonisotropic moduli of continuity by means of a nonisotropic Peetre's K functional.

Definition. Suppose f is a continuous function on R^N. For every $k \in F$ and $t>0$ the (possibly infinite) k-th nonisotropic modulus of continuity of f is defined as

$$\omega_k(t,f) = \sup \|\Delta_h^k f\|_\infty \tag{15}$$

where the supremum in (15) is over all h such that $\rho(h) \leq t$. Theorem 4 below shows that the quantity defined in (15) depends only on k and not on the measure used in (5) to define the difference operator.

In the following we denote by k_1 the smallest integer greater or equal to $k \in F$. We denote also by C^{k_1} the linear space of all continuous functions k_1 times continuously differentiable.

Theorem 4 . Suppose $k \in F$. There exists positive constants C_1 and C_2 such that, for everycontinuous function f one has , for all $t>0$

$$C_1 \omega_k(t,f) \leq \inf_{g \in C^{k_1}} \Big(\|f-g\|_\infty + \sum_{\substack{\alpha \cdot \beta \geq k \\ |\beta| \leq k_1}} t^{\alpha \cdot \beta} \|D^\beta g\|_\infty \Big) \leq \tag{16}$$

$$\leq C_2 \omega_k(t,f)$$

Proof. On one side we have , for all $g \in C^{k_1}$

$$\omega_k(t,f) \leq \omega_k(t,g) + \omega_k(t,f-g) \quad .$$

By (7) (with m_1 in the role of k_1 and k_1 in the role of k) we get immediately the left inequality in (16) .

On the other side we proceed as follows. Let $\beta_1,..,\beta_s$ be the multiindex β such that $\alpha\cdot\beta\geq k$, $|\beta|\leq k_1$. Let ϕ be an infinitely differentiable function such that $\int_{R^N}\phi(x)\,dx = 1$ and $\operatorname{supp}\phi\subseteq B(0,z)$, where $z\leq k^{-(s+1)}$.

Set $P^o=\phi$, and, for $i=1,..,s$, $P^i=b_i\sum_{j_i=1}^n c_{j_i} j_i^{\alpha\cdot\beta_i} P^{i-1}_{j_i}$, where n is the same number which appears in the definition (5) and $b_i=(\sum_{j_i=1}^n c_{j_i} j_i^{\alpha\cdot\beta_i})^{-1}$. This is possible by the third equation of (4). Set also $Q=\sum_{j_{s+1}=1}^n c_{j_{s+1}} P^s_{j_{s+1}}$. Remark that, for $i=1,..,s$, one ha

$$Q=B_i\sum_{j_{s+1}=1}^n..\sum_{j_i=1}^n c_{j_i}..c_{j_{s+1}} j_i^{\alpha\cdot\beta_i}..j_s^{\alpha\cdot\beta_s} P^{i-1}_{j_i..j_{s+1}} \tag{17}$$

where $B_i=b_i..b_s$. Moreover $\int_{R^N}Q_t(x)\,dx=\int_{R^N}P^i_t(x)\,dx=1$ and $\operatorname{supp}Q_t\subseteq B(0,t)$, $\operatorname{supp}P^i_t\subseteq B(0,t)$ for every $t>0$.

Now we let $g_t=Q_t*f$. Then

$$(Q_t*f-f)(x)=\int_{B(0,t)}f(x-y)Q_t(y)dy - f(x)\int_{B(0,t)}P^s_t(y)dy$$

$$=\int_{B(0,t)}P^s_t(y)\Delta^k_{-y}f(x)dy \qquad \text{whence}$$

$$||g_t-f||_\infty\leq C\omega_k(t,f) \tag{18}$$

Now $D^{\beta_i}g_t=D^{\beta_i}Q_t*f$, and by (17)

$$D^{\beta_i}Q_t=t^{-\alpha\cdot\beta_i}B_i\sum_{j_{s+1}=1}^n..\sum_{j_{i+1}=1}^n c_{j_{s+1}}..c_{j_{i+1}} j^{\alpha\cdot\beta_s}..j^{\alpha\cdot\beta_{i+1}} \times$$

$$\times \sum_{j_i=1}^{n} c_{j_i} (D^{\beta_i} P^{i-1}_{j_{i+1}\cdots j_{s+1}})_{tj_i} \quad .$$

Since $\int_{R^N} D^{\beta_i} P^{i-1}(x)\,dx = 0$, arguing as before , we get

$$t^{\alpha\cdot\beta_i} ||D^{\beta_i} g_t||_\infty \leq C \sum_{j_{i+1},\ldots,j_{s+1}=1}^{n} \int_{B(0,1)} |D^{\beta_i} P^{i-1}_{j_{s+1}\cdots j_{i+1}}(y)|\,||\Delta^k_{A_t y} f||_\infty \,dy .$$

Hence we get

$$(19) \qquad t^{\alpha\cdot\beta_i} ||D^{\beta_i} g_t||_\infty \leq C\omega_k(t,f) \quad .$$

Therefore the right inequality in (16) follows from (18) and (19) .

Remark . If $f\in \mathrm{Lip}(\delta,k)$ then $\omega_k(t,f) < \infty$ for all positive t.Denote by E_o the space of all continuous bounded functions with the supremum norm and by E_1 the space $C_b^{k_1}/\Pi^k$,where $C_b^{k_1}$ denotes the subspace of C^{k_1} consisting of those functions whose derivatives of order β $(\alpha\cdot\beta\geq k$, $|\beta|\leq k_1)$ are bounded.We put on E_1 the norm $\sum_{\substack{\alpha\cdot\beta\geq k\\ |\beta|\leq k_1}} ||D^\beta\cdot||_\infty$.

It follows immediately from (16) that , in the isotropic case,Lip(δ,k) is isomorphic to $[E_o,E_1]_{\delta/k,\infty}$,since in this case $k=k_1$ and the second member of (16) coincides with Peetre's K-functional.In the nonisotropic case we can think of the second member of (16) as of a nonisotropic K-functional . Namely, it easy to see that if T is alinear operator defined on , say , C (the space of all continuous functions on R^N), with values in C, and if T satisfies the following conditions

i) $\|Ta\|_\infty \leq M_o \|a\|_\infty$ for all continuous bounded a

ii) $$\sum_{\substack{\alpha\cdot\beta=s \\ |\beta|\leq k_1}} \|D^\beta(Ta)\|_\infty \leq M_1 \sum_{\substack{\alpha\cdot\beta=s \\ |\beta|\leq k_1}} \|D^\beta a\|_\infty \quad \text{for all } a\in C_b^{k_1}$$

and for all $s\in F$, $k \leq s \leq \max_{|\beta|\leq k_1} \alpha\cdot\beta$

then $\|Tf\|_{Lip(\delta,k)} \leq C\max(M_o,M_1)\|f\|_{Lip(\delta,k)}$ for all f in $Lip(\delta,k)$ (compare with [12,I.1.3]).

It is also fairly clear that a result similar to theorem 4 holds if the supremum norm is replaced by the L^p norm.

Theorem 5. For every $\gamma>0$ and every $k\in F$,such that $k>|\alpha|\gamma$, $Lip(|\alpha|\gamma,k)$ is isomorphic to $L(\gamma,\infty,k)$.

Proof.First suppose $f\in L(\gamma,\infty,k)$. By lemma 2 we have , for all $x,h\in R^N$ and for all $P\in\Pi^k$

$$|\Delta_h^k f(x)| = |\Delta_h^k(f-P)(x)| \leq C \|f-P\|_{L^\infty(B)}$$

where $B = B(x, nh)$. Hence $||\Delta_h^k f||_\infty \leq C||f||_{L(\gamma,\infty,k)}\ \rho(h)^{|\alpha|\gamma}$ and $f \in Lip(|\alpha|\gamma,k)$.

Conversely let f belong to $Lip(|\alpha|\gamma,k)$. Suppose k_1 as in (16).

By theorem 4 there exists g belonging to C^{k_1} such that for every ball $B = B(x,t)$

$$(20)\qquad \inf_{P\in\Pi^k} ||f-P||_{L^\infty(B)} \leq ||f-g||_{L^\infty(B)} + \sup_{y\in B} |g(y) -$$

$$- \sum_{\substack{|\beta|\leq k_1\\ \alpha\cdot\beta<k}} (\beta!)^{-1} D^\beta g(x)\ (y-x)^\beta| \leq ||f-g||_{L^\infty(B)} +$$

$$+ C\sum_{\substack{|\beta|\leq k_1\\ \alpha\cdot\beta\geq k}} t^{\alpha\cdot\beta}\ ||D^\beta g||_{L^\infty(B)} \leq \qquad C\omega_k(t,f) \leq C||f||_{Lip(|\alpha|\gamma,k)}\, t^{|\alpha|\gamma}$$

which completes the proof of the theorem.

Corollary1 . $Lip(\delta,k)$ is a Banach space .

Corollary2 . If $0<\delta<k_1\leq k_2$ then , then for all locally bounded function f

$$\sup_{h\neq 0} \frac{||\Delta_h^{k_2} f||_\infty}{\rho(h)^\delta} \leq C \sup_{h\neq 0} \frac{||\Delta_h^{k_1} f||_\infty}{\rho(h)^\delta}$$

4. Further properties of nonisotropic Campanato-Morrey spaces.

In this section we extend to the nonisotropic case some of the results of [10,sect.8] for isotropic Campanato-Morrey spaces . The isotropic versionsof Lemma 3 and Theorem 6 below were communicated to us by G. Weiss.

Lemma 3. Suppose $B = B(x_o, r) \subseteq B' = B'(y_o, r')$, with $r' \leq 2r$. Then, if $f \in L(\gamma, \infty, k)$ and $a_\beta(B)$, $a_\beta(B')$ are as in (14), we have :

$$(21) \qquad |a_\beta(B) - a_\beta(B')| \leq C||f||_{L(\gamma,\infty,k)} m(B)^{\gamma - |\alpha|^{-1}\alpha\cdot\beta}$$

for every β such that $k' = \alpha\cdot\beta$, where k' is the antecedent of k in F. If $x_o = y_o$, then (21) holds for every β.

Proof. Since all norms are equivalent on a finite dimennsional space, there exists a constant C such that for all $P \in \Pi^k$, $P(x) = \sum(\beta!)^{-1} D^\beta P(x_o)(x - x_o)^\beta$, one has

$$(22) \qquad (\beta!)^{-1} r^{\alpha\cdot\beta} D^\beta P(x_o) \leq \sum_{\alpha\cdot\beta<k} r^{\alpha\cdot\beta}(\beta!)^{-1} D^\beta P(x_o)$$

$$\leq C||P||_{L^\infty(B)} \quad .$$

By a homogeneity argument C does not depend on the ball B. Suppose $\alpha\cdot\beta = k'$ or $x_o = y_o$. Writing (22) for $P_B(f)$ and $P_{B'}(f)$ we get

$$|a_\beta(B) - a_\beta(B')| \leq Cr^{-\alpha\cdot\beta}||P_B(f) - P_{B'}(f)||_{L^\infty(B)}$$

$$\leq Cr^{-\alpha\cdot\beta}\left(||P_B(f) - f||_{L^\infty(B)} + \right.$$

$$\left. + ||P_{B'}(f) - f||_{L^\infty(B')}\right)$$

$$\leq C||f||_{L(\gamma,\infty,k)}\, r^{|\alpha|\gamma - \alpha\cdot\beta}$$

Theorem 6. Suppose $k'>|\alpha|\gamma$ and k' antecedent of k in F. Then there is a constant C such that for every $f\in L(\gamma,\infty,k)$ there exists $P\in\Pi^k$ such that

$$\|f-P\|_{L(\gamma,\infty,k')} \le C\|f\|_{L(\gamma,\infty,k)} \tag{23}$$

Proof. By Lemma 3, if $B = B(x_o,r)$, $\lim_{r\to\infty} a_\beta(B)$ exists and is independent of x_o for all multiindex β such that $\alpha\cdot\beta=k'$. Call a_β this limit. Set $P(x) = \sum_{\alpha\cdot\beta=k'} a_\beta x^\beta$ and $Q_B(x) = P_B(f) - P(x) - \sum_{\alpha\cdot\beta=k'} (a_\beta(B) - a_\beta)(x-x_o)^\beta$. Then $Q_B\in\Pi^{k'}$ and

$$\|f-P-Q_B\|_{L^\infty(B)} \le \|f-P_B(f) + \sum_{\alpha\cdot\beta=k'} (a_\beta(B)-a_\beta)(x-x_o)^\beta\|_{L^\infty(B)} \tag{24}$$

The thesis follows at once from (24).

We notice that the Authors of [8] give another proof of Theorem 6 in the isotropic case using Calderon's Theorem.

Theorem 7. Suppose $f\in L(\gamma,\infty,k)$, where k is the smallest element of F greater than $|\alpha|\gamma$. Then, as $x\to\infty$

$$f(x) = O(\rho(x)^{|\alpha|\gamma}) \qquad \text{if } |\alpha|\gamma\in F \tag{25}$$

$$f(x) = O(\rho(x)^{|\alpha|\gamma}\log\rho(x)) \qquad \text{if } |\alpha|\gamma\in F \tag{26}$$

Proof. For every (large) x set $r = \rho(x)$ and $B = B(O,r)$. Then

$$\begin{aligned} |f(x)| &\le \|f-P_B(f)\|_{L^\infty(B)} + \|P_B(f)\|_{L^\infty(B)} \\ &\le C\|f\|_{L(\gamma,\infty,k)} r^{|\alpha|\gamma} + \sup_\beta |a_\beta(B)| r^{\alpha\cdot\beta} \end{aligned} \tag{27}$$

By Lemma 3

$$|a_\beta(B)| \le |a_\beta(B(0,1))| + \sum_j |a_\beta(B(0,e^j) - a_\beta(B(0,e^{j-1}))|$$

$$\le |a_\beta(B(0,1))| + C\sum_j e^{j(|\alpha|\gamma-\alpha\cdot\beta)}$$

where $j=1,..,O(\log r)$. From this , (27),and the assumptions on k we obtain (25) and (26) .

Theorem 7 is the nonisotropic version of Proposition 8 of [10].

Our next theorem shows that derivatives map continuously Lipschitz spaces into Lipschitz spaces.

Theorem 8. Suppose $f\in Lip(\delta,k)$, where k is the smallest element of F larger than δ. Then for all multiindex β such that $\alpha\cdot\beta<\delta$ and for all $s\in F$ such that $s>\delta-\alpha\cdot\beta$,$D^\beta f\in Lip(\delta-\alpha\cdot\beta,s)$ and

$$(28) \qquad ||D^\beta f||_{Lip(\delta-\alpha\cdot\beta,s)} \le C||f||_{Lip(\delta,k)} \quad .$$

Proof . Let g,p^j,P^j,Q^j have the same meaning as in the proof of Theorem 1. This time, however, we let j range over all the relative integers. First we claim that for every fixed $h\in R^N$

$$(29) \qquad \Delta_h^s D^\beta f = \sum_{-\infty}^{+\infty} f*p^j*\Delta_h^s D^\beta Q^j$$

in the sense of distributions. Set $v(x)=p^o*p^o(x)$. It is immediately seen that it suffices to prove that $<f,v_j*\Delta_h^s D^\beta\psi> \to 0$ as $j\to-\infty$ for every infinitely differentiable compactly supported function ψ (here we made $v_j = v_{2^{-j}}$).

Since, by Lemma 2, $\int_{R^N} \Delta_h^s D^\beta \psi(y)\, y^\xi dy = 0$ for all multiindex ξ such that $\alpha\cdot(\xi-\beta)<s$, an application of Taylor's theorem gives

$$<f,v_j*\Delta_h^s D^\beta\psi> = \int_{R^N}\int_{R^N} f(A_{2^{-j}}x)R(x,A_{2^j}y)\Delta_h^s D^\beta\psi(y)\, dydx$$

where the remainder term satisfies (since the support of ψ is compact) an inequality of the type

$$|R(x,A_{2^j}y)| \leq 2^{j(s+\alpha\cdot\beta)}G(x)$$

with G a suitable bounded function which decreases rapidly at infinity. Hence we get from Theorem 7, for a suitable (small) positive ε

$$|<f,v_j*\Delta_h^s D^\beta\psi>| \leq 2^{j(s+\alpha\cdot\beta)}2^{-j(\delta+\varepsilon)}\iint_{R^{2N}} \rho(x)^{\delta+\varepsilon}G(x)|\Delta_h^s D^\beta\psi(y)|dydx$$

$$\leq C\ 2^{j(s+\alpha\cdot\beta-\delta-\varepsilon)} \to 0 \text{ as } j\to-\infty .$$

thus proving our claim.

Next we have by Taylor's theorem applied to Q^j

$$\sum_{-\infty}^{+\infty}||f*P^j||_\infty||\Delta_h^s D^\beta Q^j||_1 \leq C||f||_{Lip(\delta,k)}\sum_{-\infty}^{+\infty}2^{-j\delta}\min(||D^\beta Q^j||_1, \sum_{\substack{\alpha\cdot\xi\geq s\\ |\xi|\leq s_1}}\rho(h)^{\alpha\cdot\xi}\times$$

$$\times ||D^{\beta+\xi}Q^j||_1)$$

$$\leq C||f||_{Lip(\delta,k)}\sum_{-\infty}^{+\infty}2^{-j\delta}\min(2^{j\alpha\cdot\beta}, \sum_{\substack{\alpha\cdot\xi\geq s\\ |\xi|\leq s_1}}\rho(h)^{\alpha\cdot\xi}\, 2^{\alpha j(\beta+\xi)}) ,$$

where s_1 is the smallest integer not smaller than s.
For every h we choose an integer r in such a way that $2^{-r} \leq \rho(h) < 2^{-r+}$
Then the series above is majorized by

$$C||f||_{Lip(\delta,k)} \left(\sum_{\xi} \rho(h)^{\alpha\cdot\xi} \sum_{-\infty}^{r} 2^{-j(\delta-\alpha(\beta+\xi)} + \sum_{r}^{+\infty} 2^{-j(\delta-\alpha\cdot\beta)} \right)$$

which is in turn majorized by $C||f||_{Lip(\delta,k)} \, \rho(h)^{\delta-\alpha\cdot\beta}$.
This ,together with (29), proves (28) and the theorem.

Remark . It is not difficult to show that , in analogy with the isotropic case , if the derivatives of f are lipschitzian , then also f must belong to a suitable Lipschitz spaces .
Namely , we have

$$\Delta_h^k f(x) = \sum_{m=1}^{N} \alpha_m h_m \int_0^1 \sum_{j=1}^{n} c_j \, j^{\alpha_m} D_{x_m} f(x+A_{ju}h) u^{\alpha_m - 1} \, du \qquad .$$

Hence , remembering (4) , arguing as we did to establish the left hand inequality in (16) and then using the right hand inequality in (16) we get the following inequality :

$$||\Delta_h^k f||_\infty \leq C \sum_{m=1}^{N} \rho(h)^{\alpha_m} \omega_{k-\alpha_m}(\rho(h), D_{x_m} f)$$

Therefore we have the following corollary

Corollary 3. Suppose $k>\delta>0$, and suppose that k is the smallest element of F which satisfies this inequality. Let moreover $\delta-\alpha_m>0$ for $m=1,\ldots,N$. Then $f\in Lip(\delta,k)$ if and only if $D_{x_m}f$ belongs to $Lip(\delta-\alpha_m,k-\alpha_m)$ for $m=1,\ldots,N$.

5. Nonisotropic Hardy spaces.

In this section we study the relationship between nonisotropic Hardy and Campanato-Morrey spaces . In particular we show that the duality results hold also in the nonisotropic case . On account of the results of the preceding sections the proofs of the theorems below can be obtained by a mere translation of those in [3] and [10]

Suppose $0<p\leq 1\leq q\leq\infty, p<q$, and let $s\in F$, $s>|\alpha|(p^{-1}-1)$. A (p,q,s) atom is a function in $L^q(R^N)$ supported in a (nonisotropic)ball B and satisfying

(30) $$m(B)^{1/p-1/q}\left(\int_B |a(x)|^q\,dx\right)^{1/q}\leq 1$$

(31) $$\int_B a(x)x^\beta dx = 0 \qquad \text{if } \alpha\cdot\beta<s \quad .$$

(In the isotropic case these are usually called $(p,q,s-1)$-atoms)

Definition. the nonisotropic $H^{p,q,s}(R^N)$ space is the space of all tempered distributions g on R^N such that

(32) $$g = \sum_1^\infty \mu_n a_n$$

where a_n is a (p,q,s)-atom and

(33) $$||g||_{p,q,s} = \inf \, (\sum|\mu_n|^p)^{1/p} < \infty$$

the infimum in (33) being taken over all the representations (32) of g. As in the isotropic case, it is possible to show that nonisotropic Hardy spaces depend essentially only on the index p.

Theorem 9 . Suppose p,q and s are related as in the definition of (p,q,s)-atoms. Let $\bar{s}$ be the smallest element of F such that $\bar{s} > |\alpha|(p^{-1}-1)$. Then $H^{p,q,s}(R^N)$ is isomorphic to $H^{p,\infty,\bar{s}}(R^N)$.

Proof. The dependence on the index q can be disposed of by the same arguments used to prove theorem A of [3]. One has only to subtract appropriate polynomials instead of constants . Analogously we see that Hardy spaces do not depend on the index s by the same arguments used in [10, p.80-83] , which can be obviously adapted to the nonisotropic case.

Remark. Theorem 9 shows that the nonisotropic Hardy spaces defined above coincide with those defined by Calderon'[1], who proved for such spaces a maximal function characterization.

Definition. Suppose $1\leq q \leq\infty, \gamma<0, s*|\alpha|\gamma$. The nonisotropic Campanato-Morrey space $L(\gamma,q,s)$ is defined as the space of all (classes of equivqlence of) locally q-integrable functions f such that

$$\sup_{B} m(B)^{-\gamma} \inf_{P\in\Pi^s} \left(\int_{R^N} |f(x)-P(x)|^q \, m(B)^{-1} \, dx\right)^{1/q} < \infty \tag{34}$$

where B varies over the set of all nonisotropic balls in R^N. The quantity in (34) is a norm on $L(\gamma,q,s)$ and will be denoted by $||f||_{L(\gamma,q,s)}$.

It is not difficult to prove that $L(\gamma,q,s)$ is continuously imbedded in $L(\gamma,q',s)$ for $q'<q$ .Moreover $L(\gamma,q,s)$ is the dual of $H^{p,q,s}(R^N)$ provided that $p<1$ and $\gamma = |\alpha|(p^{-1}-1)$ .More precisely for every continous linear functional L on $H^{p,q,s}$ there is f belonging to $L(\gamma,q,s)$ such that , for every g of the form (30) one has $|L(g)| = |\lim_j \sum_{n=1}^{j} \mu_n \int_{R^N} f(x)a_n(x)\,dx| \le ||f||_{L(\gamma,q,s)} \times ||g||_{H^{p,q,s}}$.

The proof of the latter fact can be obtained using the arguments of theore in [3] .

Theorem 10. For every $q\geq 1$,$\gamma>0$, and $s>|\alpha|\gamma$, $L(\gamma,q,s)$ is isomorphic to $L(\gamma,\infty,s)$.

Proof. First of all we can repeat the arguments of [10,Theorem8.6] to prove that for all $p<1$ $||\Delta_h^s\delta||_{H^{p,\infty,s}} \leq C\rho(h)^{|\alpha|(1/p - 1)}$ for all h. Then using the duality result above one gets for g belonging to $L(\gamma,q,s)$: $||\Delta_h^s g||_\infty \leq C||g||_{L(\gamma,q,s)} \rho(h)^{|\alpha|\gamma}$ .Hence the theorem follows from Theorem 5 .

References

[1] A.P.Calderon,An atomic decomposition of distributions in parabolic H^p spaces,Advances in Math.,25 (1977),216-225.

[2] S;Campanato,Proprietà di una famiglia di spazi funzionali,Ann. Scuola Norm; Sup. Pisa,18 (1964) ,137-160.

[3] R.R.Coifman and G.Weiss,Extension of Hardy spaces and their use in analysis,Bull.A.M.S. ,83 (1977),569-645.

[4] E.B. Fabes and N.M.Rivière,Singular integrals with mixed homogeneity, Studia Math. ,27 (1966),19-38.

[5] G.B.Folland and E.M.Stein,Hardy spaces on homogeneous groups, Princeton 1982.

[6] N.H.Heideman,Duality and fractional integration in Lipschitz spaces,Studia Math. 50 (1974),65-85.

[7] I.R.Inglis,Difference operators and nonisotropic Lipschitz spaces ,Preprint.

[8] S.Janson and M.Taibleson,I teoremi di rappresentazione di Calderon,Atti Accad; Scienze Torino,to appear.

[9] C.B.Morrey,Second order elliptic systems of differential equations,Ann. of Math. Studies ,33(1954).

[10] M.H.Taibleson and G.Weiss,The molecular characterization of certain Hardy spaces , Asterisque ,77 (1980),67_149.

[11] A.Torchinsky,Singular integrals in the space (B,X),Studia Math. ,48 (1973),165-189.

[12] H.Triebel,Interpolation theory,function spaces,differential operators,Amsterdam ,New York,Oxford , 1978.

Address of the Author : Istituto Matematico , Università di Milano , via Saldini 50 , 20133 Milano , Italy .

LIPSCHITZ SPACES ON COMPACT RANK ONE SYMMETRIC SPACES

by Leonardo Colzani *)

Let X be a compact rank one symmetric space . A natural definition of the Lipschitz space $\Lambda(\alpha,p,X)$ $(\alpha>0\ ,\ 1\leq p\leq+\infty)$, intrinsic to the homogeneous stucture of $X = G/K$ is the following :

A function f in $L^p(X)$ belongs to the space $\Lambda(\alpha,p,X)$ if for every D in the Lie algebra g of G , with $\|D\| \leq 1$, we have

$$\operatorname{Sup.}_{t\in R} |t|^{-\alpha} \left\| \sum_{j=0}^{n} (-)^{n-j}\binom{n}{j} f(\exp(tD)\cdot x) \right\|_p \leq c \quad ,$$

where n is an integer greater than α .

It is known that this definition is equivalent to : f belongs to the intermediate space (in the real method of interpolation) $(L^p(X),L^p_n(X))_{\alpha/n}$, where $L^p_n(X)$ is the space of all functions with derivatives up to the order n in $L^p(X)$ (see [5]) . However many other approaches to Lipschitz spaces are known ; in particular when X = T , the one dimensional torus , G.H.Hardy and J.E.Littlewood , and A.Zygmund for integer values of α , proved that a function f is in $\Lambda(\alpha,p,T)$ if and only if the Poisson integral of f satisfies the estimates :

$$\operatorname{Sup.}_{0\leq r<1} (1-r)^{n-\alpha} \left\| \frac{d^n}{dx^n} \left(\sum_{-\infty}^{+\infty} r^{j} \hat{f}(j) \exp(ijx) \right) \right\|_p \leq c \quad ,$$

where again $n>\alpha$ (see [8] , chapter VII , for the case $p=+\infty$, and for further references). This approach to Lipschitz spaces has been developed in the euclidean space R^N by M.H.Taibleson ([7] , or also [6] , chapter V), and for the sphere S^N by H.C.Greenwald ([4]) .

---

*) Research supported by the Consiglio Nazionale delle Ricerche

In this paper we shall study Lipschitz spaces by means of "harmonic functions" on any compact rank one symmetric space . In particular we shall prove the equivalence between the definition of Lipschitz spaces in terms of interpolation spaces in the real method of interpolation and the one in terms of harmonic functions , and also we shall study the imbedding between different Lipschitz spaces and the boundedness of fractional integral operators . The main tool for this study is given by precise estimates of the derivatives of the Poisson kernel associated to any compact rank one symmetric space . The paper is divided into three sections . The first one contains definitions and introductory material . In the second we shall study in some details the Poisson kernel associated to any compact rank one symmetric space . We do not know if the contents of this section are completely new , but in any case they have also some independent interest . The third section is properly devoted to the study of the Lipschitz spaces $\Lambda(\alpha,p,X)$.

1 - DEFINITIONS AND INTRODUCTORY MATERIAL :

The complete list of compact symmetric spaces of rank one is the following :

i) The sphere $S^N = SO(N+1)/SO(N)$
$N = 1,2,3,\ldots$

ii) The real projective space $P^N(R) = SO(N+1)/O(N)$
$N = 2,3,4,\ldots$

iii) The complex projective space $P^N(C) = SU(L+1)/S(U(L)\times U(1))$
$L = N/2$, $N = 4,6,8,\ldots$

iv) The quaternionic projective space $P^N(H) = Sp(L+1)/Sp(L) \times Sp(1)$

$L = N/4$, $N = 8,12,16,\ldots$

v) The Cayley projective plane $P^{16}(\text{Cayley}) = F_{4(-52)}/SO(9)$

N denotes the real dimension of any one of these spaces . Denote by $X = G/K$ any one of these spaces . If $x \in X$ and $\xi \in G$, denote by $\xi \cdot x$ the action of G on X . G acts as a group of isometries of X , and K is the stabilizer of some fixed element of X . The Haar measure of G induces an invariant measure on X . Denote by dx the element of this invariant measure , normalized so that the total measure of X is 1 , and denote by $L^p(X)$ the Banach space of Borel measurable functions on X with norm

$$|| f ||_p = (\int_X |f(x)|^p \, dx)^{1/p}$$

if $1 \leq p < +\infty$, and

$$|| f ||_\infty = \operatorname{Ess.Sup.}_{x \in X} |f(x)| \quad .$$

Let g be the Lie algebra of G , and let exp be the exponential map from g onto G . If $D \in g$ and if f is a differentiable function on X , we shall put

$$Df(x) = \frac{d}{dt} f(\exp(tD)\cdot x)_{t=0} \quad .$$

More in general it is possible to define Df(x) for any left invariant differential operator D on G , i.e. for any D in the universal enveloping algebra U(G) . Let $\{D_i\}$ be a basis of g . If $D = D_{i_1} D_{i_2} \ldots D_{i_k}$, we shall say that h = Df (in the weak sense) , if for every indefinitely differentiable function ϕ

$$\int_X f(x)D\phi(x)dx = (-)^k\int_X h(x)\phi(x)dx \quad .$$

For any nonnegative integer n the Sobolev space $L^p_n(X)$ is the set of all functions $f\varepsilon L^p(X)$ with all weak derivatives Df of order at most n in $L^p(X)$. A norm in $L^p_n(X)$ is given by :

$$|| f ||_{L^p_n} = || f ||_p + \sum_D || Df ||_p \quad ,$$

where the sum is taken over all $D = D_{i_1} D_{i_2} \ldots D_{i_k}$, with $k \leq n$.
Another space we shall consider in the sequel is the space of distributions , which is the space of all continuous linear functionals on the space of indefinitely differentiable functions .
If B_0 and B_1 are two Banach spaces , with B_1 continuously imbedded into B_0 , the intermediate space in the real method of interpolation $(B_0,B_1)_\theta$, $0<\theta<1$, is the set of all f in B_0 for which there exists a constant A such that for every $t\varepsilon(0,1]$ there exist $f_0\varepsilon B_0$ and $f_1\varepsilon B_1$, with $f_0+f_1=f$, $|| f_0 ||_{B_0} \leq At^\theta$, $|| f_1 ||_{B_1} \leq At^{\theta-1}$. The smallest A above defines a norm in $(B_0,B_1)_\theta$.
We shall be interested in particular in the intermediate spaces $(L^p_s(X),L^p_n(X))_\theta$.
The Laplace-Beltrami operator Δ of λ has eigenvalues $\{j(j+a+b+1)\}$, where j ranges over all nonnegative integers if $X = S^N$, $P^N(C)$, $P^N(H)$, or P^{16}(Cayley) , and j ranges over all even nonnegative integers if $X = P^N(R)$, and

i) if $X = S^N$ $a=(N-2)/2$ $b=(N-2)/2$

ii) if $X = P^N(R)$ $a=(N-2)/2$ $b=(N-2)/2$

iii) if $X = P^N(C)$ $a=(N-2)/2$ $b=0$

iv) if $X = P^N(H)$ $a=(N-2)/2$ $b=1$

v) if $X = P^{16}(\text{Cayley})$ $a=7$ $b=3$

Denote by $Exp_x(tU)$, $t \in R$, $U \in S^{N-1}$, the exponential map from the tangent space of X at the point x onto X . The curve $t \longrightarrow Exp_x(tU)$ is a geodesic in X . Since all geodesics in a given space are closed and have the same length , without loss of generality we shall suppose this length to be 2π . Then the zonal spherical functions $\{Z_x^{(j)}(Exp_x(tU))\}$ are , up to constants , the Jacobi polynomials $\{P_j^{(a,b)}(\cos t)\}$, where a and b are given by the above list , and

$$P_j^{(a,b)}(z) = \sum_{k=0}^{j} \binom{j+a}{j-k}\binom{j+b}{k}\left(\frac{z-1}{2}\right)^k\left(\frac{z+1}{2}\right)^{j-k}$$

(for all of this see e.g. [3] , and [2] , Vol.2 , chapter X) .

Denote by H_j the eigenspace corresponding to the eigenvalue $j(j+a+b+1)$, and denote by H_jf the projection of f on H_j . If $0 \leq r < 1$, the Poisson integral of f is defined by :

$$Pf(r,x) = \sum_{j=0}^{\infty} r^j \, H_jf(x) \quad ,$$

or , equivalently , by :

$$Pf(r,x) = \int_X P(r,x,y)f(y)dy \quad ,$$

where

$$P(r,x,\mathrm{Exp}_x(tU)) = \sum_{j=0}^{\infty} r^j \| P_j^{(a,b)} \|_2^{-2} P_j^{(a,b)}(1) P_j^{(a,b)}(\cos t)$$

(and only even values of j are allowed if $X = P^N(R)$) . We shall call $P(r,x,y)$ the Poisson kernel of X .

Suppose that f is an integrable function , or , more generally , a distribution , and suppose also that $0\leq r<1$. Then , since $\| H_j f \|_\infty$ is at most of polynomial growth in j , the series $\sum_{j=0}^{\infty} r^j H_j f(x)$ is absolutely convergent , and define an indefinitely differentiable function .

DEFINITION : Let $1\leq p\leq +\infty$, $\alpha>0$, and let m be the smallest integer greater than α . The space $\Lambda(\alpha,p,X)$ is the set of all functions f in $L^p(X)$ with norm

$$\| f \|_{\Lambda(\alpha,p)} = \| f \|_p + \sup_{0\leq r<1} (1-r)^{m-\alpha} \| Pf(r,\cdot) \|_{L^p_m} .$$

This paper is mainly devoted to the study of the Lipschitz spaces $\Lambda(\alpha,p,X)$. We want to remark here that our spaces $\Lambda(\alpha,p,X)$ correspond to the spaces $\Lambda(\alpha,p,\infty,X)$ studied by M.H.Taibleson ([7]) and H.C.Greenwald ([4]). Following these two authors it is possible to define also the spaces $\Lambda(\alpha,p,q,X)$ of all functions f in $L^p(X)$ with norm

$$\| f \|_{\Lambda(\alpha,p,q)} = \| f \|_p + \left(\int_0^1 |(1-r)^{m-\alpha} \| Pf(r,\cdot) \|_{L^p_m}|^q \frac{dr}{1-r}\right)^{1/q}$$

$(m>\alpha)$, but we restrict our attention only to the case $q=+\infty$ just to simplify our exposition .

2 - THE POISSON KERNEL :

Let X be the N-dimensional sphere S^N . It is well known that the Poisson kernel P(r,x,y) can be written explicitely as

$$P(r,x,Exp_x(tU)) = \frac{1 - r^2}{(1-2r\cos t + r^2)^{(N+1)/2}}$$

Since $P_j^{(a,b)}(-z) = (-)^j P_j^{(b,a)}(z)$, if a=b we have

$$\sum_{j=0}^{\infty} r^{2j} \| P_{2j}^{(a,b)} \|_2^{-2} P_{2j}^{(a,b)}(1) P_{2j}^{(a,b)}(\cos t)$$

$$= \frac{1}{2}\Big(\sum_{j=0}^{\infty} r^j \| P_j^{(a,b)} \|_2^{-2} P_j^{(a,b)}(1) P_j^{(a,b)}(\cos t) +$$

$$+ \sum_{j=0}^{\infty} r^j \| P_j^{(a,b)} \|_2^{-2} P_j^{(a,b)}(1) P_j^{(a,b)}(-\cos t) \Big) .$$

Thus if $X = P^N(R)$, the N-dimensional real projective space ,

$$P(r,x,Exp_x(tU)) = \frac{1}{2}\Big(\frac{1 - r^2}{(1-2r\cos t + r^2)^{(N+1)/2}} + \frac{1 - r^2}{(1+2r\cos t + r^2)^{(N+1)/2}}\Big) .$$

For the other projective spaces in order to write explicitely the Poisson kernel we have to make use of more complicate expressions .

Let $\{P_j^{(a,b)}\}$, a,b>-1 , be the system of Jacobi polynomials , and let

$$h_j^{(a,b)} = \int_{-1}^{1} |P_j^{(a,b)}(z)|^2 (1-z)^a (1+z)^b dz \quad ;$$

if $u = (\sin\phi)(\sin\psi)$, $v = (\cos\phi)(\cos\psi)$, $k = \frac{1}{2}(r^{1/2} + r^{-1/2})$, the following formula holds :

$$\sum_{j=0}^{\infty} r^j \, (h_j^{(a,b)})^{-1} \, P_j^{(a,b)}(\cos 2\phi) \, P_j^{(a,b)}(\cos 2\psi)$$

$$= \frac{\Gamma(a+b+2)}{2^{a+b+1}\,\Gamma(a+1)\Gamma(b+1)} \, \frac{(1-r)}{(1+r)^{a+b+2}} \, \left(\frac{k}{u+v+k}\right)^{a+b+2}$$

$$\times \; F_2(a+b+2;a+1/2,b+1/2;2a+1,2b+1;\frac{2u}{u+v+k},\frac{2v}{u+v+k}) \quad ,$$

where F_2 is the Appell's hypergeometric function of two variables of the second type . This is proved in [1] . Since

$$F_2(\alpha;\beta',\beta;\gamma',\gamma;w,z) = \sum_{n,m=0}^{\infty} \frac{(\alpha)_{m+n}(\beta')_m(\beta)_n}{m!n!(\gamma')_m(\gamma)_n} \, w^m z^n$$

(here $(\alpha)_0 = 1$, and $(\alpha)_n = \alpha(\alpha+1)(\alpha+2)\dots(\alpha+n-1))$, if $w = 0$ $F_2(\alpha;\beta',\beta;\gamma',\gamma;w,z)$ reduces to

$$\sum_{n=0}^{\infty} \frac{(\alpha)_n(\beta)_n}{n!(\gamma)_n} \, z^n = F(\alpha,\beta;\gamma;z) \quad ,$$

the classical hypergeometric series (see for example [2], vol.1 , chapters II and V , for the definitions of F and F_2) . We deduce from the above considerations that if X is not the real projective space ,then

$$P(r,x,\mathrm{Exp}_x(tU))$$

$$= c \, \frac{(1-r)}{(1+r)^{a+b+2}} \left(\frac{k}{k+\cos(t/2)}\right)^{a+b+2} F(a+b+2,b+1/2;2b+1;\frac{2\cos(t/2)}{k+\cos(t/2)}) \quad ,$$

where c is a constant depending only on X , $k = \frac{1}{2}(r^{1/2} + r^{-1/2})$, and , again F is the classical hypergeometric series . Note that $r \neq 1$ implies $(2\cos(t/2))/(k+\cos(t/2)) < 1$, and the hypergeometric series converges .

PROPOSITION 1 : <u>Let</u> X <u>be a compact rank one symmetric space of dimension</u> N , <u>and let</u> P(r,x,y) <u>be the associated Poisson kernel . Then</u> P(r,x,y) <u>is a positive approximation of the identity , and , if</u> $|t| \leq \pi$,

i) $$|P(r,x,\mathrm{Exp}_x(tU))| \leq c(1-r)((1-r)+|t|)^{-N-1}$$

ii) $$\left|\frac{\partial^i}{\partial r^i}\frac{\partial^j}{\partial t^j} P(r,x,\mathrm{Exp}_x(tU))\right| \leq c((1-r)+|t|)^{-N-i-j} \quad .$$

Proof : If X is the sphere S^N or the real projective space $P^N(R)$, the proposition easily follows by explicit computations . To deal with the other cases we need estimates of the derivatives of an hypergeometric series . If $\gamma>\beta>0$ the hypergeometric function $F(\alpha,\beta;\gamma;z)$ can be expressed in terms of the integral of an elementary function by means of Euler's formula :

$$F(\alpha,\beta;\gamma;z) = \frac{\Gamma(\gamma)}{\Gamma(\beta)\Gamma(\gamma-\beta)} \int_0^1 w^{\beta-1}(1-w)^{\gamma-\beta-1}(1-wz)^{-\alpha}\,dw$$

([2] , vol.1 , chapter II) . Hence , if $0\leq z<1$,

$$|F(\alpha,\beta;\gamma;z)|$$

$$\leq c\Big(\int_0^z w^{\beta-1}(1-w)^{\gamma-\beta-\alpha-1}\,dw + (1-z)^{-\alpha} \int_z^1 w^{\beta-1}(1-w)^{\gamma-\beta-1}\,dw \Big)$$

$$\leq c(1+(1-z)^{\gamma-\beta-\alpha}) \quad .$$

Using these estimates we obtain :

$$|P(r,x,\mathrm{Exp}_x(tU))|$$

$$= \left| c \frac{(1-r)}{(1+r)^{a+b+2}} \left(\frac{k}{k+\cos(t/2)}\right)^{a+b+2} F(a+b+2, b+1/2; 2b+1; \frac{2\cos(t/2)}{k+\cos(t/2)}) \right|$$

$$\leq c(1-r)\left(1+\left(1-\frac{2\cos(t/2)}{k+\cos(t/2)}\right)^{-a-3/2}\right) \leq c(1-r)(k-\cos(t/2))^{-a-3/2} .$$

Since $-a-3/2 = -(N+1)/2$, and $\frac{1}{2}(r^{1/2}+r^{-1/2}) - \cos(t/2) = \frac{1}{8}((1-r)^2+t^2) + O((1-r)^3+|t|^3)$, i) is proved . A term by term derivation of the hypergeometric series shows that

$$\frac{d^j}{dz^j} F(\alpha,\beta;\gamma;z) = \frac{(\alpha)_j(\beta)_j}{(\gamma)_j} F(\alpha+j,\beta+j;\gamma+j;z) \quad ;$$

hence ii) can be proved in an analogous way as i) . We omit the details .

[]

Let f be an indefinitely differentiable function , and let $D\in U(G)$. Since the Poisson integral commutes with the action of G , we have

$$DPf(r,x) = PDf(r,x) .$$

This observation and the above proposition imply the following :

COROLLARY 2 : <u>Let</u> $1\leq p\leq+\infty$, $0\leq m\leq n$, <u>and let</u> $0\leq r<1$. <u>Then if</u> f <u>is in</u> $L^p_m(X)$, <u>we have</u>

$$\| Pf(r,\cdot) \|_{L^p_n} \leq c(1-r)^{m-n} \| f \|_{L^p_m} .$$

<u>If , in addition</u> , $p<+\infty$, <u>we also have</u>

$$\| Pf(r,\cdot) - f \|_{L^p_m} \longrightarrow 0 \qquad \text{as } r \longrightarrow 1 .$$

COROLLARY 3 : *If* $1\leq p\leq+\infty$, $0\leq n$, $0\leq r<1$, *and if* f *is in* $L^p(X)$, *then*

$$\left\| \frac{d^n}{dr^n} Pf(r,\cdot) \right\|_p \leq c(1-r)^{-n} \| f \|_p .$$

The proof of these corollaries is left to the reader . Let f be a distribution $0\leq r<1$, $x\in X$. Since $P(r,x,y)$ as function of y is indefinitely differentiable , it makes sense to consider the Poisson integral of f , and $Pf(r,x)$ is an indefinitely differentiable function of r and x . Moreover we have the following :

PROPOSITION 4 : *Let* f *be a distribution on* X , m *be a positive integer* $1\leq p\leq+\infty$, *and* $\varepsilon>0$. *Then if there exists a constant* A *such that*

$$\left\| \frac{d^m}{dr^m} Pf(r,\cdot) \right\|_p \leq A(1-r)^{\varepsilon-m} ,$$

the distribution f *can be identified with a function in* $L^p(X)$, *and*

$$\| f \|_p \leq c\Big(A + \sum_{j=0}^{m-1} \| H_j f \|_p\Big) .$$

Proof : By Taylor's formula we have

$$Pf(r,x) = \sum_{j=0}^{m-1} r^j H_j f(x) + \int_0^r \frac{d^m}{du^m} Pf(u,x) \frac{(r-u)^{m-1}}{(m-1)!} du$$

Let $0\leq r_2\leq r_1<1$. Since

$$\| Pf(r_1,\cdot) - Pf(r_2,\cdot) \|_p \leq \sum_{j=0}^{m-1} (r_1^j - r_2^j) \| H_j f \|_p +$$

$$+ A\int_0^{r_2} (1-u)^{\varepsilon-m} \frac{(r_1-u)^{m-1} - (r_2-u)^{m-1}}{(m-1)!} du + A\int_{r_2}^{r_1} (1-u)^{\varepsilon-m} \frac{(r_1-u)^{m-1}}{(m-1)!} du ,$$

$Pf(r,x)$ converges in the L^p-norm as $r \longrightarrow 1$, and the proposition follows . Note that when $p=+\infty$, f , as uniform limit of continuous functions , can be identified with a continuous function .

[]

3 - LIPSCHITZ SPACES :

PROPOSITION 5 : Let $\alpha>0$, m be the smallest integer greater than α , and let n be an integer greater or equal to m . Then the three norms

$$\|f\|_{\Lambda(\alpha,p)} = \|f\|_p + \sup_{0\le r<1} (1-r)^{m-\alpha} \|Pf(r,\cdot)\|_{L^p_m} ,$$

$$\|f\|_{\Lambda(\alpha,p,n)} = \|f\|_p + \sup_{0\le r<1} (1-r)^{n-\alpha} \|Pf(r,\cdot)\|_{L^p_n} ,$$

$$\|f\|^*_{\Lambda(\alpha,p,n)} = \|f\|_p + \sup_{0\le r<1} (1-r)^{n-\alpha} \left\|\frac{d^n}{dr^n} Pf(r,\cdot)\right\|_p ,$$

are equivalent .

Proof : The proof is divided into four steps , and is similar to the proof given in [6] , chapter V , for the euclidean space R^N (see also [4] for a discussion of other equivalent norms in the spaces $\Lambda(\alpha,p,q,S^N)$) .

1) $\|f\|_{\Lambda(\alpha,p,n+1)} \le c\|f\|_{\Lambda(\alpha,p,n)}$.

By the semigroup property of the Poisson integral we have

$$Pf(r,x) = P(Pf(\sqrt{r},\cdot))(\sqrt{r},x) \quad ;$$

hence , by corollary 2 ,

$$\| Pf(r,\cdot) \|_{L^p_{n+1}} \leq c(1-\sqrt{r})^{-1} \| Pf(\sqrt{r},\cdot) \|_{L^p_n}$$

$$\leq c(1-\sqrt{r})^{\alpha-n-1} \| f \|_{\Lambda(\alpha,p,n)}$$

$$\leq c(1-r)^{\alpha-n-1} \| f \|_{\Lambda(\alpha,p,n)}$$

2) $\| f \|_{\Lambda(\alpha,p,n)} \leq c \| f \|^*_{\Lambda(\alpha,p,n)}$.

Let D be a left invariant differential operator of order less or equal to

Since

$$\frac{d^n}{dr^n} DPf(r,x) = D \frac{d^n}{dr^n} Pf(r,x) \quad ,$$

applying again corollary 2 and the semigroup property of the Poisson integral we obtain

$$\| \frac{d^n}{dr^n} DPf(r,\cdot) \|_p \leq c(1-\sqrt{r})^{-n} \| \frac{d^n}{d\sqrt{r}^n} Pf(\sqrt{r},\cdot) \|_p$$

$$\leq c(1-r)^{\alpha-2n} \| f \|^*_{\Lambda(\alpha,p,n)} \quad .$$

Thus

$$\| \frac{d^{n-1}}{dr^{n-1}} DPf(r,\cdot) \|_p = \| \frac{d^{n-1}}{du^{n-1}} DPf(u,\cdot)_{u=0} + \int_0^r \frac{d^n}{du^n} DPf(u,\cdot)\, du \|_p$$

$$\leq c(1 + \int_0^r (1-u)^{\alpha-2n} du) \| f \|^*_{\Lambda(\alpha,p,n)}$$

$$\leq c(1-r)^{\alpha-2n+1} \| f \|^*_{\Lambda(\alpha,p,n)} \quad ,$$

and iterating

$$\| DPf(r,\cdot) \|_p \le c(1-r)^{\alpha-n} \| f \|^*_{\Lambda(\alpha,p,n)} .$$

3) $\| f \|^*_{\Lambda(\alpha,p,n)} \le c \| f \|^*_{\Lambda(\alpha,p,n+1)} \le c^2 \| f \|^*_{\Lambda(\alpha,p,n)}$.
The inequality $\| f \|^*_{\Lambda(\alpha,p,n+1)} \le c \| f \|^*_{\Lambda(\alpha,p,n)}$ follows as in 1) by the semigroup property of the Poisson integral and corollary 3) . The converse inequality $\| f \|^*_{\Lambda(\alpha,p,n)} \le c \| f \|^*_{\Lambda(\alpha,p,n+1)}$ follows by the formula

$$\| \frac{d^n}{dr^n} Pf(r,\cdot) \|_p \le \| \frac{d^n}{du^n} Pf(u,\cdot)_{u=0} \|_p + \int_0^r \| \frac{d^{n+1}}{du^{n+1}} Pf(u,\cdot) \|_p$$

4) If n is even , then $\| f \|^*_{\Lambda(\alpha,p,n)} \le c \| f \|_{\Lambda(\alpha,p,n)}$.
Let Δ be the Laplace-Beltrami operator of X . We have

$$\| \Delta^{n/2} Pf(r,\cdot) \|_p \le c(1-r)^{\alpha-n} \| f \|_{\Lambda(\alpha,p,n)} .$$

The idea is to estimate the difference between

$$\Delta^{n/2} Pf(r,x) = \sum_{j=0}^{\infty} r^j (j(j+a+b+1))^{n/2} H_j f(x) ,$$

and

$$r^n \frac{d^n}{dr^n} Pf(r,x) = \sum_{j=0}^{\infty} r^j (j(j-1)\ldots(j-n+1)) H_j f(x) ,$$

and to do this it is enough to apply n times the following :

LEMMA 6 : Let $\{h_j\}$ be a sequence of functions , with $\{\| h_j \|_p\}$ slowly increasing , and let γ,δ, and η be real numbers , with $\eta<0$. Suppose that for $0\le r<1$ we have :

$$\| \sum_{j=0}^{\infty} r^j \, (j+\gamma) \, h_j(\cdot) \|_p \leq (1-r)^{\eta} \quad .$$

Then we also have :

$$\| \sum_{j=0}^{\infty} r^j \, (j+\delta) \, h_j(\cdot) \|_p \leq c(1-r)^{\eta} \quad .$$

Proof : Since

$$\sum_{j=0}^{\infty} r^j \, (j+\gamma) \, h_j(x) \; - \; \sum_{j=0}^{\infty} r^j \, (j+\delta) \, h_j(x) \; = \; (\gamma-\delta) \sum_{j=0}^{\infty} r^j \, h_j(x) \quad ,$$

it is enough to estimate $\| \sum_{j=0}^{\infty} r^j \, h(\cdot) \|_p$. If $j_0 > \gamma$ we have

$$\sum_{j=j_0}^{\infty} r^j \, h_j(x) \; = \; r^{-\gamma} \int_0^r \Big(\sum_{j=j_0}^{\infty} u^j \, (j+\gamma) \, h_j(x) \Big) \, u^{\gamma-1} \, du \quad .$$

Hence

$$\begin{aligned} \| \sum_{j=0}^{\infty} r^j \, h_j(\cdot) \|_p &\leq c \int_0^r (1-u)^{\eta} \, du \\ &\leq c && \text{if } -1 < \eta \\ &\leq c \log 1/(1-r) && \text{if } \eta = -1 \\ &\leq c(1-r)^{\eta+1} && \text{if } \eta < -1 \quad . \end{aligned}$$

[]

Proposition 5 has the following corollary :

COROLLARY 7 : Let $\alpha>0$, and let s be an integer smaller than α . Then if f is in $\Lambda(\alpha,p,X)$, f belongs also to $L_s^p(X)$. More precisely , if D is a left *invariant differential operator of order* s , *then* Df *belongs to* $\Lambda(\alpha-s,p,X)$, *and* $\| Df \|_{\Lambda(\alpha-s,p)} \le c \| f \|_{\Lambda(\alpha,p)}$. *When* $p=+\infty$, *then* Df *can be identified with a continuous function* .

Proof : Let $n>\alpha$; then

$$\| \frac{d^n}{dr^n} DPf(r,\cdot) \|_p \le c(1-r)^{\alpha-s-n} \| f \|_{\Lambda(\alpha,p)} .$$

The existence of $Df(x) = \lim_{r\to 1} DPf(r,x)$ in the L^p-norm is a consequence of this inequality and proposition 4 .

[]

PROPOSITION 8 : *Let* $\alpha>0$, *and let* s *and* n *be two integers , with* $0\le s<\alpha<n$. *Then*

$$\Lambda(\alpha,p,X) = (L^p_s(X), L^p_n(X))_{(\alpha-s)/(n-s)} ,$$

moreover the corresponding norms are equivalent .

Proof : Let f be in $(L^p_s(X), L^p_n(X))_{(\alpha-s)/(n-s)}$ and for simplicity suppose that the norm of f is 1 . Then for every $t\in(0,1]$ there exist f_0 and f_1 , with $f_0+f_1 = f$, $\| f_0 \|_{L^p_s} \le t^{(\alpha-s)/(n-s)}$, $\| f_1 \|_{L^p_n} \le t^{(\alpha-n)/(n-s)}$. Take $t = (1-r)^{n-s}$. Then , by corollary 2 ,

$$\| Pf(r,\cdot) \|_{L^p_n} \le c(1-r)^{s-n} \| f_0 \|_{L^p_s} + \| f_1 \|_{L^p_n}$$

$$\le c(1-r)^{\alpha-n} .$$

Since also $|| f ||_p \leq || f ||_{L^p_s} \leq c$, we have that f is in $\Lambda(\alpha,p,X)$, and $|| f ||_{\Lambda(\alpha,p)} \leq c$.

Suppose now that f is in $\Lambda(\alpha,p,X)$ and again suppose f of norm 1 . Let $t\varepsilon(0,1)$ and take r and u so that $(1-r)^{n-s} = t$ and $0\leq r\leq u<1$. Then we can write

$$f(x) = \left(f(x) - \sum_{j=0}^{n-1} \frac{d^j}{dr^j} Pf(r,x) \frac{(u-r)^j}{j!} \right) + \left(\sum_{j=0}^{n-1} \frac{d^j}{dr^j} Pf(r,x) \frac{(u-r)^j}{j!} \right)$$

$$= f_0(x) + f_1(x) \quad .$$

Since , by corollary 7 , f is in $L^p_s(X)$ (and the derivatives of f are continuous if $p=+\infty$) , we can choose u so close to 1 that

$$|| f - Pf(u,\cdot) ||_{L^p_s} \leq t^{(\alpha-s)/(n-s)} \quad .$$

Moreover , again by corollary 7 ,

$$|| \frac{d^n}{dr^n} Pf(r,\cdot) ||_{L^p_s} \leq c(1-r)^{\alpha-s-n} \quad ,$$

and then

$$|| Pf(u,\cdot) - \sum_{j=0}^{n-1} \frac{d^j}{dr^j} Pf(r,\cdot) \frac{(u-r)^j}{j!} ||_{L^p_s}$$

$$= || \int_r^u \frac{d^n}{dh^n} Pf(h,\cdot) \frac{(u-h)^{n-1}}{(n-1)!} dh ||_{L^p_s}$$

$$\leq c\int_r^u (1-h)^{\alpha-s-n} (u-h)^{n-1} dh \leq c(1-r)^{\alpha-s} = ct^{(\alpha-s)/(n-s)} \quad .$$

Hence $|| f ||_{L^p_s} \leq ct^{(\alpha-s)/(n-s)}$.

$$\| f_1 \|_{L^p_n} \le \sum_{j=0}^{n-1} \| \frac{d^j}{dr^j} Pf(r,\cdot) \|_{L^p_n} \frac{(u-r)^j}{j!}$$

$$\le c \sum_{j=0}^{n-1} (u-r)^j (1-\sqrt{r})^{-j} \| Pf(\sqrt{r},\cdot) \|_{L^p_n}$$

$$\le c(1-r)^{\alpha-n} = ct^{(\alpha-n)/(n-s)} .$$

Concluding , f is in the interpolation space $(L^p_s(X), L^p_n(X))_{(\alpha-s)/(n-s)}$, and the norm of f in this space is controlled by the $\Lambda(\alpha,p)$-norm .

[]

If X is the euclidean space R^N (or also the N-dimensional torus T^N) , $1 \le p \le q \le +\infty$, $\alpha - N/p = \beta - N/q$, and if the function f is in $\Lambda(\alpha,p,X)$, then f belongs also to the space $\Lambda(\beta,q,X)$ (see [6] , or [7]) . An analogous imbedding theorem holds when X is a compact rank one symmetric space .

PROPOSITION 9 : Suppose that $\alpha \ge \beta > 0$, $1 \le p \le q \le +\infty$, and $\alpha - N/p = \beta - N/q$. Then we have a continuous inclusion of the space $\Lambda(\alpha,p,X)$ into $\Lambda(\beta,q,X)$.

Proof : By Young's inequality we have

$$\| r^n \frac{d^n}{dr^n} Pf(r,\cdot) \|_q \le \| P(\sqrt{r},x,\cdot) \|_s \| (\sqrt{r})^n \frac{d^n}{d\sqrt{r}^n} Pf(\sqrt{r},\cdot) \|_p ,$$

where $1/p + 1/s = 1 + 1/q$. Using proposition 1 it is immediate to verify that $\| P(\sqrt{r},x,\cdot) \|_s \le c(1-\sqrt{r})^{N(1/s-1)}$. Thus , if $n>\alpha$,

$$\| \frac{d^n}{dr^n} Pf(r,\cdot) \|_q \le c(1-\sqrt{r})^{N(1/s-1)+\alpha-n} \| f \|_{\Lambda(\alpha,p)}$$

$$\le c(1-r)^{\alpha+N(1/q-1/p)-n} \| f \|_{\Lambda(\alpha,p)} .$$

Since $\alpha+N(1/q-1/p) = \beta > 0$, by proposition 4 the function f is in $L^q(X)$, and the proposition follows .

[]

REMARK : To see that this proposition is sharp , simply test it on the Poisson kernel . Indeed an easy calculation using proposition 1 shows that $\| P(r,x,\cdot) \|_{\Lambda(\alpha,p)} \sim (1-r)^{N(1/p-1)-\alpha}$.

We want to conclude this section with the remark that the Poisson integral characterization of Lipschitz spaces is also usefull to derive easily some of the properties of fractional integrals on X . Let $f = \sum_{j=0}^{\infty} H_j f$ be a distribution on X . The fractional integral $I^\beta f$ is defined by :

$$I^\beta f = \sum_{j=0}^{\infty} (1+j)^{-\beta} H_j f .$$

Note that this series converges at least in the topology of distributions .

PROPOSITION 10 : <u>Let</u> $\alpha>0$, <u>and</u> $\alpha+\beta>0$. <u>The fractional integral</u> I^β <u>is an isomorphism of</u> $\Lambda(\alpha,p,X)$ <u>onto</u> $\Lambda(\alpha+\beta,p,X)$.

Proof : Obviously we can restrict our attention only to the case $\beta>0$. A term by term integration shows that

$$PI^\beta f(r,x) = \sum_{j=0}^{\infty} r^j (1+j)^{-\beta} H_j f(x) = \Gamma(\beta)^{-1} \int_0^1 (\log 1/t)^{\beta-1} Pf(rt,x)\, dt .$$

Thus , if $n>\alpha+\beta$,

$$\| PI^\beta f(r,\cdot) \|_{L^p_n} \leq \Gamma(\beta)^{-1} \int_0^1 (\log 1/t)^{\beta-1} \| Pf(rt,\cdot) \|_{L^p_n} dt$$

$$\leq c\| f \|_{\Lambda(\alpha,p)} \int_0^1 (\log 1/t)^{\beta-1} (1-rt)^{\alpha-n} dt$$

$$\leq c(1-r)^{\alpha+\beta-n} \| f \|_{\Lambda(\alpha,p)} .$$

Similar computations also show that $I^\beta f \in L^p(X)$ whenever f is in $L^p(X)$, and $\| I^\beta f \|_p \leq \| f \|_p$. Hence if f is in $\Lambda(\alpha,p,X)$, $I^\beta f$ is in $\Lambda(\alpha+\beta,p,X)$, and $\| I^\beta f \|_{\Lambda(\alpha+\beta,p)} \leq c\| f \|_{\Lambda(\alpha,p)}$.

To prove that the map I^β is onto it is enough to prove the continuity of the map $I^{-\beta}$ from $\Lambda(\alpha+\beta,p,X)$ into $\Lambda(\alpha,p,X)$. Suppose first that $\beta=s$ is an integer , and let m be an integer greater than α . If f is in $\Lambda(\alpha+s,p,X)$,

$$\| I^{-s}f \|_{\Lambda(\alpha,p)} \leq c(\| I^{-s}f \|_p + \sup_{0\leq r<1} (1-r)^{m-\alpha} \| \frac{d^m}{dr^m} PI^{-s}f(r,\cdot) \|_p)$$

Since

$$\| \frac{d^{m+s}}{dr^{m+s}} Pf(r,\cdot) \|_p$$

$$= \| r^{-m-s} \sum_{j=0}^{\infty} r^j (j(j-1)\ldots(j-m+1)(j-m)\ldots(j-m-s+1)) H_j f \|_p$$

$$\leq c(1-r)^{\alpha-m} \| f \|_{\Lambda(\alpha+s,p)} ,$$

a repeated application of lemma 6 yelds :

$$\| \frac{d^m}{dr^m} PI^{-s}f(r,\cdot) \|_p$$

$$= \| r^{-m} \sum_{j=0}^{\infty} r^j (j(j-1)\ldots(j-m+1))(1+j)^s H_j f \|_p$$

$$\leq c(1-r)^{\alpha-m} \| f \|_{\Lambda(\alpha+s,p)} .$$

To evaluate $|| I^{-s}f ||_p$ simply apply proposition 4 . The continuity of the map $I^{-\beta}$ from $\Lambda(\alpha+\beta,p,X)$ into $\Lambda(\alpha,p,X)$ is then proved if β is an integer . If β is not an integer , let s be an integer greater than β . Then $I^{-\beta}f = I^{-s}(I^{s-\beta}f)$, and since we have already proved that $I^{s-\beta}$ is a continuous map from $\Lambda(\alpha+\beta,p,X)$ into $\Lambda(\alpha+s,p,X)$, we have

$$|| I^{-\beta}f ||_{\Lambda(\alpha,p)} \leq c|| I^{s-\beta}f ||_{\Lambda(\alpha+s,p)} \leq c^2|| f ||_{\Lambda(\alpha+\beta,p)} ,$$

and the proof of the proposition is complete .

[]

REMARK : With small changes in the proof of this proposition it is also possible to prove that if $\beta>0$, I^{β} is a continuous map of $L^p(X)$ into $\Lambda(\beta,p,X)$. However this map is not onto . For example , let p=2 . It is not difficult to check that the function $f(x) = \sum_{j=0}^{\infty} (1+j)^{-N/2-\beta} Z_y^{(j)}(x)$ belongs to $\Lambda(\beta,2,X)$, but $I^{-\beta}f$ is not in $L^2(X)$. Indeed $|| I^{-\beta}f ||_2 = (\sum_{j=0}^{\infty} (1+j)^{-N}|| Z_y^{(j)} ||_2^2)^{1/2} = +\infty$.

Proposition 10 has been proved for the sphere S^N by H.C.Greenwald ([4]) ; however the proof presented here seems more elementary .

R E F E R E N C E S

[1] W.N.Bailey , The generating function of Jacobi polynomials , J.London Math.Soc. , 13 (1938) , 8-12 .

[2] A.Erdelyi , W.Magnus , F.Oberhettinger , F.Tricomi , Higher transcendental functions , vols.I,II,III (Bateman Manuscript Project) , McGraw-Hill , New York , 1955 .

[3] R.Gangolli , Positive definite kernels on homogeneous spaces and certain stochastic processes related to Levy's Brownian motion of several parameters , Ann.Inst.H.Poincaré Sect.B 3 , No.2 (1967) , 121-225

[4] H.C.Greenwald , Lipschitz spaces on the surface of the unit sphere in euclidean n-space , Pacific J.Math. , 50 (1974) , 63-80 .

[5] H.Johnen , Sätze von Jackson-Typ auf Darstellungräumen kompakter , zusammenhängender Liegruppen , Linear operators and approximation (Proc. Conf. , Oberwolfach , 1971) , pp. 254-272 , Internat.Ser.Numer.Math. , vol. 20 , Birkhäuser , Basel , 1972 .

[6] E.M.Stein , Singular integrals and differentiability properties of functions , Princeton Univ.Press , Princeton , 1970 .

[7] M.H.Taibleson , On the theory of Lipschitz spaces of distributions on euclidean n-space , I , J.Math.Mech. , 13 (1964) , 407-480 ; II , (ibid) , 14 (1965) , 821-840 ; III , (ibid) , 15 (1966) , 973-981 .

[8] A.Zygmund , Trigonometric series , Sec.Ed. , Cambridge Univ.Press , Cambridge , 1968 .

UNIVERSITA' DEGLI STUDI DI MILANO
ISTITUTO MATEMATICO "F.ENRIQUES" ,
Via C.SALDINI , 50
20133 MILANO , ITALIA

ON THE SOBOLEV SPACES $W^{k,1}(R^n)^*$

S. POORNIMA

For any non negative integer k and any p, $1 \leq p < \infty$, let $W^{k,p}(R^n)$ denote the Sobolev space $\{f \in L^p(R^n) : \partial^\alpha f \in L^p(R^n), |\alpha| \leq k\}$ where ∂^α is the distributional derivative of order α.

Our interest here is to observe how the space $W^{k,1}(R^n)$ distinguishes itself from the spaces $W^{k,p}(R^n)$ $(1 < p < \infty)$. We are concerned with the following two problems in this light :

(i) the embedding problem ;

(ii) the multiplier problem.

The main tool for the solution of these problems on $W^{k,p}$ for $1 < p < \infty$ is the existence of an isomorphism between $L^p(R^n)$ and $W^{k,p}(R^n)$.

If Δ is the usual Laplacian, we denote by $(I-\Delta)^{\alpha/2}$ the operator given by

$$((I-\Delta)^{\alpha/2} f)^\wedge (\xi) = (1 + 4\pi^2 |\xi|^2)^{\alpha/2} \hat{f}(\xi)$$

and by $(-\Delta)^{\alpha/2}$ the operator

$$((-\Delta)^{\alpha/2} f)^\wedge (\xi) = (2\pi |\xi|)^\alpha \hat{f}(\xi).$$

LEMMA 1.1 [8 , p. 135]. The convolution operator $(I-\Delta)^{-k/2}$ takes $L^p(R^n)$ isomorphically onto $W^{k,p}(R^n)$ if $1 < p < \infty$ and k is a positive integer.

We also recall the following result on the operators $(-\Delta)^{-\alpha/2}$.

LEMMA 1.2 [8, p. 119]. Let $0 < \alpha < n$, $1 < p < q < \infty$ and $1/q = 1/p - \alpha/n$. Then

---

(*) The results of this article were presented by A. Bonami in the conference on Harmonic Analysis at Cortona

$$\|(-\Delta)^{-\alpha/2}f\|_{L^q} < C_{p,q}\|f\|_{L^p}$$

for some constant $C_{p,q} > 0$.

Let $L(q,p)$ stand for the Lorentz space on R^n (see [9] for details). That is, for a measurable function f on R^n, if we define the distribution function $f_*(\lambda)$ by

$$f_*(\lambda) = \left|\left\{x : |f(x)| > \lambda\right\}\right| \quad \text{for all} \quad \lambda > 0,$$

(where $|E|$ is the Lebesgue measure of the set E), and the non increasing rearrangement of f by

$$f^*(t) = \inf\left\{ s : f_*(s) < t\right\}$$

then $f \in L(q,p)$ if it satisfies

$$\|f\|^*_{qp} = \left\{p/q \int_0^\infty (t^{1/q} f^*(t))^p \frac{dt}{t}\right\}^{1/p} < \infty$$

for $1 \le p < \infty$, $1 \le q < \infty$. The space $L(p,\infty)$ is the space of measurable functions f with

$$\|f\|^*_{p,\infty} = \sup_{t>0} t^{1/p} f^*(t) < \infty.$$

The Sobolev embedding theorem [8, p. 124] asserts that for $1 \le p < \infty$, $1/q = 1/p - k/n$ and $q < \infty$, the Sobolev spaces $W^{k,p}(R^n)$ are continuously embedded in $L^q(R^n)$.

By lemmas 1.1 and 1.2 a sharper version is known in the case of $p > 1$, that $W^{k,p}$ is contained in $L(q,p)$, where $\frac{1}{q} = \frac{1}{p} - \frac{k}{n}$. For, by interpolation, $(-\Delta)^{-k/2}$ maps L^p into $L(q,p)$. Since the operator $(-\Delta)^{k/2}(I-\Delta)^{-k/2}$ is given by a bounded measure [8, p. 133] by lemma 1.1 we get the inclusion $W^{k,p} \subset L(q,p)$, $p > 1$.

In the case $p = 1$, $n \geq 1$, the operator $(I-\Delta)^{-k/2}$ maps L^1 into $W^{k,1}$ only when k is even [8, p. 160]. Nevertheless $(I-\Delta)^{-k/2}$ maps $H^1(R^n)$ into $W^{k,1}$ for all k. If we define $\tilde{W}^{k,1}(R^n) = \left\{f \in W^{k,1}(R^n) : \partial^\alpha f \in H^1(R^n),\ |\alpha| \le k\right\}$, then we find

LEMMA 1.3. The convolution operator $(I-\Delta)^{-k/2}$ maps $H^1(R^n)$ isomorphical onto $\widetilde{W}^{k,1}(R^n)$.

Proof. For any $f \in \widetilde{W}^{k,1}$, $(I-\Delta)^{k/2}f$ belongs to H^1, if k is even. When k is odd, we write $(I-\Delta)^{k/2} = (I-\Delta)^{k-1/2}(I-\Delta)^{1/2}$ and notice that $(I-\Delta)^{1/2}$ can be written as $\mu + (-\Delta)^{1/2}\nu$ for some bounded measures μ and ν [8, p. 133]. Since $(-\Delta)^{1/2} = \sum_{j=1}^{n} R_j \frac{\partial}{\partial x_j}$, where R_j are the Riesz operators, we obtain, for any $f \in \widetilde{W}^{k,1}$,

$$(I-\Delta)^{k/2}f = (I-\Delta)^{k-1/2}\mu * f + \nu * \sum_{j=1}^{n} R_j (I-\Delta)^{k-1/2} \frac{\partial f}{\partial x_j}.$$

This belongs to H^1, by the definition of $\widetilde{W}^{k,1}$ and the fact that R_j are multipliers for H^1.

The reverse inclusion can easily be seen.

This enables us to observe, as before, that $\widetilde{W}^{k,1}$ is continuously embedded in $L(q,1)$ where $\frac{1}{q} = 1 - \frac{k}{n}$. For, the Riesz potential $(-\Delta)^{-k/2}$ maps $H^1(R^n)$ into $L(q,1)$.

We cannot avail of similar tools in the case of $W^{k,1}$. However, the embedding result is true as proved in [5].

THEOREM 1.4 [5]. The Sobolev space $W^{k,1}(R^n)$ is continuously embedded in the Lorentz space $L(q,1)$, where $\frac{1}{q} = 1 - \frac{k}{n}$ $(1 \leq k < n)$.

In the proof of theorem 1.4, we make use of a lemma of Bourgain [2]:

Take a real valued C^1 function ξ on $\mathbb{R}$ which is 0 on $]-\infty, 0]$, 1 on $[1, +\infty[$ and such that $0 \leq \xi \leq 1$. Let us define $\xi_\lambda^\varepsilon(t) = \xi(\varepsilon^{-1}(t-\lambda))$. Then, for any real valued f in $W^{1,1}(\mathbb{R}^n)$, $\xi_\lambda^\varepsilon \circ f$ belongs to $W^{1,1}(\mathbb{R}^n)$ as well as $\xi_\lambda^\varepsilon \circ (-f)$, and

$$(1) \qquad \int_0^\infty \{\|\xi_\lambda^\varepsilon \circ f\|_{W^{1,1}} + \|\xi_\lambda^\varepsilon \circ (-f)\|_{W^{1,1}}\} d\lambda \leq C\|f\|_{W^{1,1}}.$$

In order to prove the theorem 1.4 when $k = 1$, it is then enough to have the estimate :

$$\{f_*(t+\varepsilon)\}^{1/q} = \|\chi_{f>t+\varepsilon} + \chi_{-f>t+\varepsilon}\|_q$$

$$\leq \|\xi_\lambda^\varepsilon \circ f\|_q + \|\xi_\lambda^\varepsilon \circ (-f)\|_q$$

$$\leq C_q\left\{\|\xi_\lambda^\varepsilon \circ f\|_{W^{1,1}} + \|\xi_\lambda^\varepsilon \circ (-f)\|_{W^{1,1}}\right\}.$$

The inequality (1) is given by Bourgain [2] to give, in the case of the torus, a Hardy type inequality,

(2) $$\sum_{m\in Z^n} \frac{|\hat{f}(m)|}{(1+|m|)^{n-k}} \leq C\|f\|_{W^{k,1}(\mathbf{T}^n)}$$

for functions $f \in W^{k,1}(\mathbf{T}^n)$. His proof can also be adapted for R^n, $n \geq 2$. This Hardy type inequality for functions of $W^{k,1}(R^n)$, $n > 2$, follows, in fact, as a consequence of Theorem 1.4.

COROLLARY 1.5 [5]. Let $n > 2$. Assume $f \in W^{k,1}(R^n)$, $k \geq 1$. Then

$$|\xi|^{k-n}\hat{f}(\xi) \text{ and } (1+|\xi|)^{k-n}\hat{f}(\xi)$$

are integrable.

In particular, taking $k = n$, $W^{n,1}$ is continuously embedded in $\mathcal{F}L^1(R^n)$.

REMARK 1.6. By using lemmas 1.1 and 1.2 and interpolation, whenever $1 < p < \infty$, we see that the operator $(I-\Delta)^{\alpha/2}$ is bounded from $W^{k,p}$ into $L(q, p)$, where $\frac{1}{q} = \frac{1}{p} - \frac{(k-\alpha)}{n}$ and $0 \leq \alpha \leq k$.

Again, using lemma 1.3 and [8, p. 237, 4.7] we observe that $(I-\Delta)^{\alpha/2}$ is bounded from $\tilde{W}^{k,1}$ into $L(q,1)$, for $\frac{1}{q} = 1 - \frac{(k-\alpha)}{n}$, $0 \leq \alpha \leq k$.

Seeking a similar result for $W^{k,1}$, we observe the following consequence of theorem 1.4.

COROLLARY 1.7. For $k \geq 2$, $0 \leq \alpha \leq k-1$, the operator $(I-\Delta)^{\alpha/2}$ is bounded from $W^{k,1}$ into $L(q,1)$, where $\frac{1}{q} = 1 - (\frac{k-\alpha}{n})$.

Proof. Let α be an integer such that $0 \leq \alpha \leq k-1$. Whenever α is even, evidently $(I-\Delta)^{\alpha/2} f \in W^{k-\alpha,1}$ for $f \in W^{k,1}$. Now using theorem 1.4, we get the result.

If α is an odd integer, write $(I-\Delta)^{\alpha/2} = (I-\Delta)^{(\alpha-1)/2}(I-\Delta)^{1/2}$. Observe that

$$(I-\Delta)^{1/2} = (I-\Delta)^{-1/2} - (-\Delta)^{1/2}(I-\Delta)^{-1/2} \sum_{j=1}^{n} R_j \frac{\partial}{\partial x_j}. \tag{3}$$

For $f \in W^{k,1}$, $(I-\Delta)^{(\alpha-1)/2} \frac{\partial f}{\partial x_j} \in W^{k-\alpha,1} \hookrightarrow L(q,1)$ and R_j maps $L(q,1)$ into itself. Since $(I-\Delta)^{-1/2}$ is an L^1 function and $(-\Delta)^{1/2}(I-\Delta)^{-1/2}$ is a bounded measure, from (3) we get $(I-\Delta)^{\alpha/2} f \in L(q,1)$.

For any real α, $0 \leq \alpha \leq k-1$, we write $(I-\Delta)^{\alpha/2} = (I-\Delta)^{\alpha/2-(\frac{k-1}{2})}(I-\Delta)^{(k-1)/2}$. By the integer case, we see that $(I-\Delta)^{k-1/2} f \in L(\frac{n}{n-1},1)$. By interpolation, we know that $(I-\Delta)^{\alpha/2-(\frac{k-1}{2})} : L(\frac{n}{n-1},1) \to L(q,1)$ where $\frac{1}{q} = 1 - (\frac{k-\alpha}{n})$.

This proves the corollary.

It will be of interest to ask the same question for functions in $W^{1,1}$, for $0 < \alpha < 1$.

FOURIER MULTIPLIERS OF $W^{k,1}$.

Now we turn to the problem of multipliers on $W^{k,1}(R^n)$.

By a multiplier $T : W^{k,p} \to W^{k,p}$ we mean a bounded linear transformation which commutes with translations. Equivalently, we call a function $m(\xi)$ on R^n a Fourier multiplier on $W^{k,p}$ if for each $f \in W^{k,p}$, $m(\xi)\hat{f}(\xi)$ is the Fourier transform of a function in $W^{k,p}$.

If $p > 1$, by means of the isomorphism $(I-\Delta)^{-k/2}$ (which commutes with translations) between L^p and $W^{k,p}$ we see that the multipliers on $W^{k,p}$ are the same as the L^p-multipliers. Again, $(I-\Delta)^{-k/2}$ serves as an isomorphism between $\tilde{W}^{k,1}$ and H^1 (Lemma 1.3) ; the multipliers of $\tilde{W}^{k,1}$ are the multipliers of H^1.

On the real line R, the differential operator $(\frac{d}{dx}+1)^k$ maps $W^{k,1}(R)$ isomorphically onto $L^1(R)$ and hence the multipliers are given by bounded measures on the real line.

Hence, in the case of $n \geq 2$, the natural question to ask would be if there exists a multiplier on $W^{k,1}(R^n)$ which is not given by a bounded measure. In [6] we show that $M(R^n)$, the space of bounded measures on R^n, $n \geq 2$, is strictly contained in the space of multipliers on $W^{k,1}$. We recall [6]:

THEOREM 2.1. Let $B^k = \{$integrable distributions T such that $\partial^\alpha T$ are of order $\leq k$ for $|\alpha| \leq k\}$. Then every distribution T belonging to B^k defines a multiplier, by convolution, on $W^{k,p}(R^n)$, for $1 \leq p < \infty$.

Hence B^k is a class of pseudomeasures which coincide with M(R) on the real line. The following proposition asserts that the space B^k of multipliers for $W^{k,1}(R^n)$, $n \geq 2$, is larger than $M(R^n)$.

PROPOSITION 2.2 [6]. There exists a distribution in B^1 which is not a bounded measure on R^n, if $n \geq 2$.

The proof of the above proposition relies on a theorem of Ornstein [4].

THEOREM 2.3 (Ornstein). Let $B, D_1, D_2, \ldots, D_\ell$ be a set of linearly independent linear differential operators homogeneous of degree m. Then there exists a sequence of C^∞ functions $\{\varphi_k\}$ with support contained in the unit cube, such that $\|B\varphi_k\|_1 \geq k$, and $\|D_i\varphi_k\|_1 \leq 1$ for $i = 1, 2, \ldots \ell$.

We shall write this theorem in the following version, suitable to our purposes.

THEOREM 2.4. Let $B, D_1, D_2 \dots D_\ell$ be as in theorem 2.3. Then there exists $f \in L^1$ such that $D_i f \in L^1$ $(i = 1, 2, \dots \ell)$ and $Bf \notin L^1$.

We indicate the proof of proposition 2.2, to show the efficacy of Ornstein's theorem. In fact, we use also this theorem to obtain the non multiplier results (Proposition 2.10).

Proof of proposition 2.2. Take a suitable choice of linearly independent, homogeneous differential operators $B, D_1, \dots D_\ell$ to apply theorem 2.4. For example, let $B = \frac{\partial^n}{\partial x_1^{n-1} \partial x_2}$ and $D_j = \frac{\partial^n}{\partial x_1^{n-1} \partial x_j}$, $j = 1, 3, 4 \dots n$. Then there exists $f \in L^1(R^n)$ such that $D_j f \in L^1$ but $Bf \notin L^1(R_n)$. Let $S = Bf$. Clearly, $\frac{\partial S}{\partial x_j} = \frac{\partial}{\partial x_2}(D_j f)$ for all $j \neq 2$ and $\frac{\partial S}{\partial x_2} = \frac{\partial}{\partial x_1}(D_1 f)$. Hence $\frac{\partial S}{\partial x_j}$ are integrable distributions of order ≤ 1.

To obtain an element T in B^1, we put $T = \varphi S$ where $\varphi \in \mathcal{D}(R^n)$ which is $\equiv 1$ on a relatively compact open set on which S is not a measure. For details, see [6].

We also have a necessary condition on the multipliers of $W^{k,1}(R^n)$.

THEOREM 2.5 [6]. Every multiplier on $W^{k,1}$ is an integrable distribution of order ≤ 1.

The condition $T \in B^k$ can only be sufficient to ensure a multiplier on $W^{k,1}(R^n)$. For, we have multipliers on $W^{1,1}$ which are not in B^1. We state the result for the torus T^2.

THEOREM 2.6. Let S be a pseudomeasure on T^2 such that $\frac{\partial S}{\partial x}$ and $\frac{\partial S}{\partial y}$

are pseudomeasures ; then S is a multiplier of $W^{1,1}(T^2)$. Moreover, there exists such an S with this property such that at least one of $\frac{\partial S}{\partial x}$ and $\frac{\partial S}{\partial y}$ is not a distribution of order ≤ 1.

Proof. Clearly $L^2 \subsetneq L^1$, $W^{1,2} \subsetneq W^{1,1}$ on T^2 and by Sobolev embedding theorem, $W^{1,1} \subsetneq L^2$. Now if $f \in W^{1,1}$, then $S * f \in L^2$, $\frac{\partial}{\partial x}(S * f) = \frac{\partial S}{\partial x} * f$, $\frac{\partial}{\partial y}(S * t) = \frac{\partial S}{\partial y} * f$ are in L^2. Hence $S * f \in W^{1,1}$. The continuity is easily seen. Let P be a parametrix of the Cauchy-Riemann operator $\frac{\partial}{\partial \bar{z}}$ on T^2. For example,

$$P(x,y) = \sum_{\substack{m \neq 0 \\ m \in Z^2}} -\frac{i}{\pi} \frac{1}{m_1 + i m_2} \exp(2\pi i(m_1 x + m_2 y)),$$

so that $\frac{\partial P}{\partial \bar{z}} = \delta - 1$. As P maps L^2 into $W^{1,2}$, we can take $S = P * \sigma$ for a suitable pseudomeasure σ which is not of order ≤ 1.

Theorem 2.6 enables us to obtain a new class of multipliers on $W^{1,1}(R^2)$ by extension. The details would appear elsewhere.

In the sequel too, we restrict our attention to $W^{1,1}$.

NON MULTIPLIERS FOR THE SPACE $W^{1,1}$.

By means of theorem 2.5, we see that any multiplier on $W^{1,1}(R^n)$ is of the form

$$T = \mu_o + \sum_{i=1}^{n} \frac{\partial \mu_i}{\partial x_i}$$

where $\mu_i \in M(R^n)$, $j = 0, 1, \ldots, n$. Hence the Fourier transforms of measure being continuous at the origin, certain natural operators like the Riesz operators and singular integral operators are not multipliers of $W^{1,1}$. We shall see that even a reasonable behaviour at the origin does not ensure a Fourier multiplier. For example, $\frac{\xi_i \xi_j}{1+|\xi|^2}$ fa to be a Fourier multiplier for $W^{1,1}$ (corollary 2.11).

To study this phenomenon, we introduce the related space of homogeneous type $\mathring{W}^{1,1}$ as follows.

If $\mathcal{D}$ is the usual space of C^∞ functions with compact support in R^n, we shall denote by $\mathring{W}^{1,1}(R^n)$ the completion of $\mathcal{D}$ under

$$\|\varphi\|_{\mathring{W}^{1,1}} = \sum_{i=1}^{n} \left\|\frac{\partial \varphi}{\partial x_i}\right\|_{L^1}, \qquad \varphi \in \mathcal{D}.$$

Using the Sobolev embedding theorem we see that $\mathring{W}^{1,1}$ is a space of distributions and in fact, $\mathring{W}^{1,1} \subsetneq L^{n/n-1}$ [8, p. 127]. $\mathring{W}^{1,1}$ can also be defined as distributions T with $\frac{\partial T}{\partial x_i}$ in L^1, $i = 1 \ldots n$, modulo constant functions.

Clearly $W^{1,1}$ is contained in $\mathring{W}^{1,1}$. We would like to compare the multipliers of $W^{1,1}$ and $\mathring{W}^{1,1}$.

LEMMA 2.7. Suppose that $\dfrac{P_m(\xi)}{(1+|\xi|^2)^{m/2}}$ is a Fourier multiplier of $W^{1,1}(R^n)$ where $P_m(\xi)$ is a homogeneous polynomial of degree m. Then $\dfrac{P_m(\xi)}{|\xi|^m}$ is a Fourier multiplier of $\mathring{W}^{1,1}(R^n)$.

Proof. For a function φ define φ_ε by $\varphi_\varepsilon(x) = \varepsilon^{1-n}\varphi(x/\varepsilon)$, $\varepsilon > 0$. We have $\|\varphi_\varepsilon\|_{L^1} = \varepsilon\|\varphi\|_{L^1}$ and $\left\|\frac{\partial \varphi_\varepsilon}{\partial x_i}\right\|_{L^1} = \left\|\frac{\partial \varphi}{\partial x_i}\right\|_{L^1}$ so that for $\varphi \in W^{1,1}$, we have $\|\varphi_\varepsilon\|_{W^{1,1}} = \varepsilon\|\varphi\|_{L^1} + \Sigma\left\|\frac{\partial \varphi}{\partial x_i}\right\|_{L^1}$.

Let T_ε be the convolution operator whose symbol is $\omega_\varepsilon(\xi) = \omega(\xi/\varepsilon)$ where $\omega(\xi) = \dfrac{P_m(\xi)}{(1+|\xi|^2)^{m/2}}$. We verify that

$$T_1(\varphi_\varepsilon) = (T_\varepsilon \varphi)_\varepsilon. \tag{4}$$

Now since T_1 is a multiplier of $W^{1,1}$ by hypothesis, we have

$$\|T_1\varphi_\varepsilon\|_{W^{1,1}} \le C\|\varphi_\varepsilon\|_{W^{1,1}} \qquad \text{for} \quad \varphi \in \mathcal{D}$$

by (4) $\|(T_\varepsilon \varphi)_\varepsilon\|_{W^{1,1}} \leq C\|\varphi_\varepsilon\|_{W^{1,1}} = C(\varepsilon\|\varphi\|_{L^1} + \Sigma\|\frac{\partial\varphi}{\partial x_i}\|_{L^1})$

and hence,

$$\Sigma\|\frac{\partial T_\varepsilon\varphi}{\partial x_i}\|_{L^1} = \Sigma\|\frac{\partial(T_\varepsilon\varphi)_\varepsilon}{\partial x_i}\|_{L^1} \leq C(\varepsilon\|\varphi\|_{L^1} + \Sigma\|\frac{\partial\varphi}{\partial x_i}\|_{L^1})$$

by letting ε tend to 0, we get

$$\|T\varphi\|_{\mathring{W}^{1,1}} \leq C\,\Sigma\|\frac{\partial\varphi}{\partial x_i}\|_{L^1}$$

where T is the convolution operator with symbol $\frac{P_m(\xi)}{|\xi|^m}$ on noting that $\omega_\varepsilon(\xi)$ tend to $\frac{P_m(\xi)}{|\xi|^m}$ as ε tends to 0.

REMARK 2.8. The above argument shows in essence the following. Let ω be a function such that $\Omega(\xi) = \lim_{r\to\infty} \omega(r\xi)$ exists for $\xi \neq 0$. If ω is a Fourier multiplier of $W^{1,1}$ then Ω is a Fourier multiplier of $\mathring{W}^{1,1}$. By Theorem 2.9 to follow we see that Ω must be constant.

THEOREM 2.9 [1]. Let w be a homogeneous function of degree 0 on R^n, C^∞ outside the origin. Then w is a Fourier multiplier of $\mathring{W}^{1,1}$ if and only if w is a constant.

We shall first prove a special case of Theorem 2.9.

PROPOSITION 2.10. Suppose that $\frac{P_m(\xi)}{|\xi|^m}$ is a Fourier multiplier of $\mathring{W}^{1,1}(R^n)$ where $P_m(\xi)$ is a homogeneous polynomial of order m. Then $P_m = C|\xi|^m$, where C is a constant.

Proof. We can restrict ourselves to the case when m is even : for if $\frac{P_m(\xi)}{|\xi|^m}$ was a Fourier multiplier of $\mathring{W}^{1,1}$, so would be $\frac{P_m^2}{|\xi|^{2m}}$. Let then $m = 2\ell$. We shal show on the basis of our hypothesis that the polynomials

$$\xi_1 P_{2\ell},\ \xi_1 |\xi|^{2\ell},\ \ldots.\ \xi_n |\xi|^{2\ell}$$

are linearly dependent. Then from the relation $C_1 \xi_1 P_{2\ell} + \sum_{j=1}^{n} C_j |\xi|^{2\ell} \xi_j = 0$,

it is easy to see that $P_{2\ell}(\xi) = C|\xi|^{2\ell}$.

To prove the linear dependence of these polynomials, suppose they were not. Then the differential operators $\frac{\partial}{\partial x_1} \circ P_{2\ell}(D)$, $\frac{\partial}{\partial x_1} \circ \Delta^{\ell}, \ldots.$ $\frac{\partial}{\partial x_n} \circ \Delta^{\ell}$ would be linearly independent. So we could find by theorem 2.5, a function $f \in L^1$ such that $\frac{\partial}{\partial x_j} \Delta^{\ell} g \in L^1$, $j = 1, 2, \ldots n$ and $\frac{\partial}{\partial x_1} P_{2\ell}(D) g$ does not belong to L^1. Put $f = \Delta^{\ell} g$. Then $g \in \overset{\circ}{W}{}^{1,1}$. (In fact $f \in W^{1,1}$. For, writing $g = P * \Delta^{\ell} g + \varphi * g$ for a parametrix P of Δ^{ℓ} in $W^{2\ell-1,1}$ [7] and $\varphi \in \mathcal{D}$, we get $\frac{\partial g}{\partial x_i} = P * \frac{\partial}{\partial x_i}(\Delta^{\ell} g) + \frac{\partial}{\partial x_i} * g$. This shows that $g \in W^{2\ell,1}$ so that $f \in L^1$).

If S is the multiplier corresponding to $\frac{P_{2\ell}(\xi)}{|\xi|^{2\ell}}$, we have $Sf = P_{2\ell}(D) g$.

Since $\frac{\partial}{\partial x_1} P_{2\ell}(D) g$ is not in L^1, this contradicts the assumption that S is a multiplier.

COROLLARY 2.11. If $\frac{P_m(\xi)}{(1+|\xi|^2)^{m/2}}$ is a Fourier multiplier of $W^{1,1}(R^n)$. Then $P_m(\xi) = C|\xi|^m$ for some constant C. In particular, $\frac{\xi_i \xi_j}{1+|\xi|^2}$ cannot be a Fourier multiplier on $W^{1,1}(R^n)$.

The proof is a consequence of lemme 2.7 and proposition 2.10.

<u>Proof of theorem 2.9</u>. For simplicity we shall prove the theorem for the case $n = 2$.

Let $\sum_{k=-\infty}^{\infty} a_k e^{ik\theta}$ be the Fourier expansion of ω restricted to the unit circle. If ω is the Fourier multiplier of $\overset{\circ}{W}{}^{1,1}$, then $\omega(e^{i(\theta-\tau)})$ is again a Fourier multiplier of the same norm and so $\omega * e^{ik\theta} = \frac{1}{2\pi}\int_0^{2\pi} \omega(e^{i(\theta-\tau)}) e^{ik\tau} d\tau$ is a Fourier multiplier on $\mathring{W}^{1,1}(R^2)$. Since $\omega * e^{ik\theta} = a_k e^{ik\theta}$ and

$$e^{ik\theta} = \begin{cases} \dfrac{z^k}{|z|^k} & \text{if} \quad k > 0 \\ \dfrac{\bar{z}^{-k}}{|z|^k} & \text{if} \quad k < 0 , \end{cases}$$

$\omega * e^{ik\theta}$ is of the form $\dfrac{P_m(\xi)}{|\xi|^m}$. We now apply proposition 2.10 to conclude that $a_k = 0$ for all $k \neq 0$. Hence ω is a constant.

For $n \geq 3$ one uses the expansion in spherical harmonics on the unit sphere.

REMARK 2.12. We observe, for $\mathring{W}^{1,1}(R^2)$, a relation to a space defined by second order Riesz operators. Let $K = \left\{f \in L^1(R^2) : \text{p. v. } \frac{1}{z^2} * f \in L^1(R^2)\right\}$ with the obvious norm. This space has been considered by J. Garcia-Cuerva in connection with a question by C. Fefferman [3]. In particular, he showed that K does not coincide with H^1.

This space K fits into our setting as one verifies that the isomorphism $\frac{\partial}{\partial \bar{z}}$ takes $\mathring{W}^{1,1}(R^2)$ onto K. Hence the multipliers of $\mathring{W}^{1,1}(R^2)$ and of K are the same. By proposition 2.10 the operator $R_1 R_2$, for example, is not a multiplier on $\mathring{W}^{1,1}$ and hence not on K. But one knows that $R_1 R_2$ is a multiplier on H^1. Hence $H^1 \neq K$.

On the other hand, every multiplier of $\mathring{W}^{1,1}$ is a multiplier on H^1. This follows from the inclusion

$$(-\Delta)^{-1/2} H^1 \hookrightarrow W^{1,1}$$

and the fact that a multiplier from H^1 into L^1 in fact takes H^1 into H^1.

References

[1] BONAMI, A. and POORNIMA, S. Some nonmultipliers for Sobolev spaces. Preprint, Orsay (1982).

[2] BOURGAIN, J. A Hardy inequality in Sobolev spaces. Preprint (1981).

[3] GANDULFO, A., GARCIA-CUERVA, J. and TAIBLESON, M. Conjugate system characterization of H^1 : counterexamples for the Euclidean plane and local fields. Bull. Amer. Math. Soc. 82 (1976), 83-85.

[4] ORNSTEIN, D. A non inequality for differential operators in the L^1- norm. Arkiv. Rat. Mech. Anal. 11 (1962), 40–49.

[5] S. POORNIMA An embedding theorem for the Sobolev space $W^{k,1}$. To appear in Bulletin des Sciences Math.

[6] S. POORNIMA Multipliers of Sobolev spaces. J. Funct. Anal. 45 (1982), 1-28.

[7] SCHWARTZ, L. Théorie des distributions. Paris, Hermann (1966).

[8] STEIN, E. M. Singular integrals and differentiability properties of functions. Princeton (1970).

[9] STEIN, E. M. and WEISS, G. Introduction to Fourier analysis on euclidean spaces. Princeton (1971).

Interval averages of H^1-functions
and BMO norm of inner functions

N. DANIKAS and V. NESTORIDIS

Abstract

The main result of this paper states that every value
of an H^1-function taken at an interior point is equal
to the average of the function on a boundary interval.
This fact is used in order to find the exact value of
the BMO norm of inner functions.

1. Introduction

We denote by D the open unit disc of the complex
plane $\mathbb{C}$ and by T the unit circle. For any subinterval (arc) I
of T with length $|I|$ satisfying $0 < |I| \leqslant 2\pi$ and for any
function φ integrable on T with respect to the Lebesgue
measure $d\theta = |d\mathfrak{z}|$, we denote by φ_I the average of φ on I :

$$\varphi_I = \frac{1}{|I|} \int_I \varphi(\mathfrak{z}) |d\mathfrak{z}| \quad .$$

For $1 \leqslant p < \infty$ we denote by $_p\|\varphi\|$ the usual ([3]) non-con-
formaly invariant p-BMO (semi-)norm of the function φ on T :

$$_p\|\varphi\| = \sup_I \left\{ \frac{1}{|I|} \int_I |\varphi(\mathfrak{z}) - \varphi_I|^p |d\mathfrak{z}| \right\}^{\frac{1}{p}} \quad .$$

We recall that a function φ is called inner, if φ
is bounded holomorphic in D and has unimodular boundary values;

that is $|\varphi(e^{i\theta})| = 1$ a.e. By the BMO norms of an inner function φ we mean the BMO norms of the boundary function $\varphi(e^{i\theta})$ on T.

We show that, for $p \in [1, 2]$ and φ a non-constant inner function, the exact value of ${}_p\|\varphi\|$ is 1. The main step towards this end is a property of H^1-functions which, despite the simplicity of its statement has, as far as we know, passed unnoticed. This result appears to be of independent interest and its precise statement is the following:

<u>Theorem 1</u>. Let f be an H^1-function and let α be a value of f taken at a point of the open unit disc D. Then there is an interval $I \subset T$ of length $|I|$, $0 < |I| \leqslant 2\pi$, such that

$$\alpha = \frac{1}{|I|} \int_I f(\zeta)\,|d\zeta| \quad .$$

The proof of the theorem, which is given in Section 2, is based on a careful examination of the graphs of the functions

$$\Phi_\varepsilon(e^{ix}) = \frac{1}{2\varepsilon} \int_{x-\varepsilon}^{x+\varepsilon} f(e^{i\theta})\,d\theta \ , \ 0 < \varepsilon \leqslant \pi$$

and showing by contradiction that at least one of these graphs passes through the point α.

Using Theorem 1 and the fact that a non-constant inner function takes arbitrarily small values in D, we obtain the following

<u>Theorem 2</u>. For every $p \in [1, 2]$ and every non-constant inner function φ, the p-BMO norm of φ is exactly equal to 1; that is ${}_p\|\varphi\| = 1$.

Sections 3 and 4 contain comments and remarks pertaining to Theorems 1 and 2. In this paper we mainly restrict ourselves to the one-dimensional periodic case. Our results for this case appear to be fairly complete.

Theorem 1 can be easily extended to the case of functions in the half-plane. Extensions for functions of several complex variables are possible, as E. Stein suggested. The case of functions in the unit ball of $\mathbb{C}^n$ is not immediate and presents some particularities. The extension for functions in the polydisc is rather easy. These extensions will be examined in detail, we hope, in a subsequent paper; some preliminary results, however, will be briefly sketched in Section 5.

In Section 6 we include some elementary results, which follow from arguments used in the proof of Theorem 1. More precisely, the results are that every value of a holomorphic function is equal to the average of the function on a segment, or on an arc, or on a triangle etc.

We have discussed the subject matter of this paper with S. Pichorides on several occasions. This and the many helpful talks we had with him on related topics in the past year gives us the opportunity to express our thanks to him. We would also like to thank E. Galanis, A. Mallios and Ch. Pommerenke for their encouragement during the preparation of this paper.

2. Proofs

We now give the proofs of Theorems 1 and 2 that have been stated in the introduction.

Proof of Theorem 1. Substracting a constant from f we may assume $\alpha = 0$. Therefore we suppose that f vanishes at some point z_0 of D and assuming that $f_I \neq 0$ for all I's we will arrive at a contradiction.

For $\varepsilon \in (0,\pi]$ and $e^{ix} \in T$ we denote by $I_{\varepsilon,x}$ the arc of T with mid-point e^{ix} and length 2ε. Next we consider the map $\Phi : (0,\pi] \times T \to \mathbb{C} \setminus \{0\}$ defined as follows:

$$\Phi(\varepsilon, e^{ix}) = f_{I_{\varepsilon,x}} = \frac{1}{2\varepsilon} \int_{x-\varepsilon}^{x+\varepsilon} f(e^{i\theta})\, d\theta .$$

For a fixed $\varepsilon \in (0,\pi]$, we denote by Φ_ε the slice map from T into $\mathbb{C} \setminus \{0\}$ defined by $\Phi_\varepsilon(e^{ix}) = \Phi(\varepsilon, e^{ix})$. Note that $\Phi_\varepsilon = K_\varepsilon * f$, where $K_\varepsilon = \frac{1}{2\varepsilon} \chi[-\varepsilon,\varepsilon]$ is an approximate identity as $\varepsilon \to 0$.

The map Φ takes values in $\mathbb{C} \setminus \{0\}$ because of the assumption $f_I \neq 0$ for all I's. Since f is integrable on T, Φ is a continuous function of (ε, e^{ix}). Therefore the closed curves $\Phi_\varepsilon : T \to \mathbb{C} \setminus \{0\}$, for $\varepsilon \in (0,\pi]$, are homotopic in $\mathbb{C} \setminus \{0\}$. Hence they have the same winding number with respect to 0. This winding number is zero, because for $\varepsilon = \pi$ the curve Φ_π reduces to the point $\Phi_\pi(e^{ix}) = f_{I_{\pi,x}} = f(0) \neq 0$ for all $e^{ix} \in T$.

Next we observe that each Φ_ε has an extension holomorphic in D and continuous on $\overline{D}$; indeed $\widehat{\Phi}_\varepsilon(n) = \widehat{f * K_\varepsilon}(n) = \hat{f}(n)\widehat{K}_\varepsilon(n) = 0$ for all $n<0$, as f is holomorphic. By the argument principle the number of zeros of Φ_ε in D is equal to the winding number with respect to 0 of the curve $\Phi_\varepsilon : T \to \mathbb{C} \setminus \{0\}$. This winding number has been found to be zero, and therefore $\Phi_\varepsilon(z) \neq 0$ for all $z \in D$.

Let now $\varepsilon \to 0$; since K_ε is an approximate identity and f is integrable on T, we conclude $\Phi_\varepsilon \to f$ in L^1 and hence in H^1. The H^1 convergence implies the uniform convergence on compacta of D.

Since $\Phi_\varepsilon(z) \neq 0$ for all $z \in D$ and $\Phi_\varepsilon \to f$ uniformly on compacta as $\varepsilon \to 0$, Hurwitz's theorem implies that either

$f \equiv 0$ or $f(z) \neq 0$ in D. But $f \equiv 0$ contradicts our assumption $f_I \neq 0$, while $f(z) \neq 0$ in D contradicts our hypothesis that f vanishes at some point of D. The proof is now complete.

We now pass to the proof of Theorem 2 :

Proof of Theorem 2 . For φ unimodular, it is easy to verify that

$$_2\|\varphi\| = \left\{\sup_I (1 - |\varphi_I|^2)\right\}^{\frac{1}{2}} = \left\{1 - \inf_I |\varphi_I|^2\right\}^{\frac{1}{2}} \leqslant 1 .$$

As $_1\|\varphi\| \leqslant {}_p\|\varphi\| \leqslant {}_2\|\varphi\|$ for $1 \leqslant p \leqslant 2$, it suffices to show that $_1\|\varphi\| \geqslant 1$.

On applying the triangular inequality and taking into account that φ is unimodular we obtain

$$_1\|\varphi\| = \sup_I \frac{1}{|I|} \int_I \left|\varphi(e^{i\theta}) - \varphi_I\right| d\theta \geqslant \sup_I \left(1 - |\varphi_I|\right) = 1 - \inf_I |\varphi_I| .$$

Theorem 1 and the trivial fact that the inner functions are in H^1 yield

$$0 \leqslant \inf_I |\varphi_I| \leqslant \inf_{z \in D} |\varphi(z)| .$$

But every non constant inner function takes àrbitrarily small values (see §4, b)) . Therefore $\inf_{z \in D} |\varphi(z)| = 0$. It follows $\inf_I |\varphi_I| = 0$ and hence $1 \leqslant {}_1\|\varphi\|$, which completes the proof .

3. Comments and supplementary results

a) Consider the BMO distances

$$d_p(\varphi) = \sup_I \frac{1}{|I|} \int_I |\varphi(e^{i\theta}) - \varphi_I|^p d\theta \ , \quad 0<p<1 .$$

If φ is a non constant inner function, then $d_p(\varphi)=1$, because

$$1 = {}_1\|\varphi\|^p \geq d_p(\varphi) \geq \sup_I \left(1-|\varphi_I|^p\right) = 1 - \inf_I |\varphi_I|^p = 1 .$$

Further it is not true that ${}_p\|\varphi\| = 1$ for all $p>2$ and φ non constant inner function; as A. Baernstein II suggested, for each such φ with $\varphi(0)\neq 0$ we have $\lim_{p\to\infty} {}_p\|\varphi\| \geq 1+|\varphi(0)| > 1$. Indeed

$${}_p\|\varphi\| \geq \left\{ \frac{1}{2\pi}\int_0^{2\pi} |\varphi(e^{i\theta})-\varphi(0)|^p d\theta \right\}^{\frac{1}{p}} \xrightarrow[p\to\infty]{} \|\varphi-\varphi(0)\|_\infty = 1+|\varphi(0)| .$$

For the conformaly invariant BMO norms

$$N_p(\varphi) = \sup_{z\in D} \left\{ \frac{1}{2\pi}\int_T |\varphi(\zeta)-\varphi(z)|^p \frac{1-|z|^2}{|\zeta-z|^2} |d\zeta| \right\}^{\frac{1}{p}} \quad (p\geq 1)$$

we have exactly the same results, but with simpler proofs, as we do not need Theorem 1 (see [2] , p.10).

b) Theorem 1 does not generally hold for Poisson integrals of L^1-functions. Here is a counterexample :

Consider the L^1-function f with $f(e^{i\theta})=0$ for $\theta\in(0,\frac{\pi}{2})\cup(\pi,\frac{3\pi}{2})$, $f(e^{i\theta})=i$ for $\frac{\pi}{2}<\theta<\pi$ and $f(e^{i\theta})=1$ for $\frac{3\pi}{2}<\theta<2\pi$. Then all f_I's are on the segments $[0,i]$ and $[0,1]$ or in the closed triangle with vertices 0, $\frac{2i}{3}$, $\frac{2}{3}$. Therefore inside the open triangle with vertices 0, i, 1 there are no f_I's that are very close to i . On the other hand all values of the Poisson integral of f lie in the open triangle $\langle 0, i, 1\rangle$ because the weights of 0 , i and 1 are all strictly positive. We may also have the mass of i as close to one as we wish. Therefore inside the open triangle $\langle 0, i, 1\rangle$ there are Poisson integral-values of f very close to i . Such values are not of the form f_I .

The above counterexample can easily be modified to give an f continuous on T . Finally we note that f in the counterexample can not be choosen to take only real values; if f is real, the intermediate value theorem easily implies the result of Theorem 1 . In the holomorphic case the intermediate value theorem has been replaced by the argument principle .

c) It is possible for an f_I not to be a value of f , but merely a value in the closure of the convex hull of the set f(D) . An obvious example is a conformal mapping of D onto a non-convex polygon.

Even in the case of conformal mappings, it is possible to have two different I's , say I_1 and I_2 , such that $f_{I_1} = f_{I_2} \in f(D)$. For example this happens if f is the symmetric conformal mapping of D onto the symmetric with respect to the real axis polygon in the figure below :

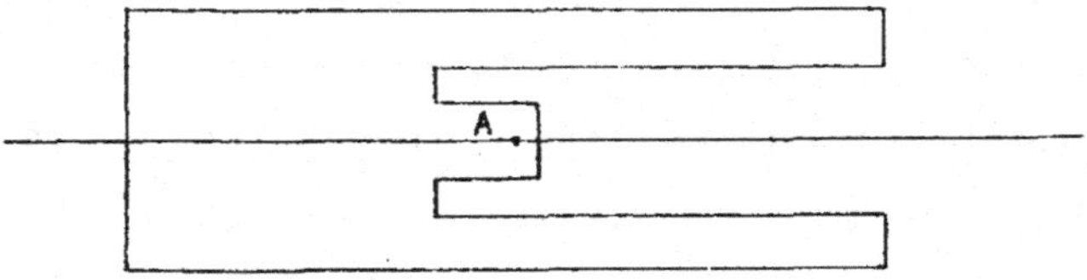

Indeed, first observe that $f_I \in \mathbb{R}$ for all intervals I symmetric with respect to the real axis. Therefore, considering only such intervals, the intermediate value theorem applies. Hence, for a suitable real A in the polygon, there exist at least two distinct symmetric intervals I_1, I_2 such that $f_{I_1} = f_{I_2} = A$.

d) Comparing the number of I's such that $f_I = \alpha$ to the number of the zeros of $f-\alpha$ one can see by examples that the former may be much larger. On the other hand we do not have an example with the number of I's strictly less than the number of zeros of $f-\alpha$. Propositions 3 and 4 below furnish some information towards this direction. Proposition 3 covers the case of an infinite number of zeros. Proposition 4 is a slightly stronger version of a special case of Theorem 1, which can also be used to transfer Theorem 1 from the disc to the half-plane.

Proposition 3. Let $f \in H^1$ and suppose that the value α is taken by f an infinite number of times in D. Then there is a sequence I of subintervals (arcs) of T with $0 < |I_n| \longrightarrow 0$, such that $\alpha = f_{I_n}$ for all n.

Proposition 4. Let $f \in H^1$ and suppose that the function $f-f(0)$ has more than one zero in D, multiplicity taken into account. Then there is a proper subarc I of T, $0 < |I| < 2\pi$, such that $f(0) = f_I$.

We now give the proofs of Propositions 3 and 4. First we prove Proposition 3:

We may again assume $\alpha = 0$. If the statement of Proposition 3 is false, then with the same notation as in the proof of Theorem 1, there is an $\varepsilon_0 > 0$ such that $\Phi_\varepsilon(e^{ix}) \neq 0$ for all $e^{ix} \in T$ and $\varepsilon < \varepsilon_0$. Therefore, for $0 < \varepsilon < \varepsilon_0$ the curves Φ_ε are homotopic in $\mathbb{C} \setminus \{0\}$ and they have the same winding number k with respect to 0. By the argument principle each Φ_ε $(0 < \varepsilon < \varepsilon_0)$ vanishes in D exactly k times.

If $f \equiv 0$ the proposition is obvious, so we may assume $f \not\equiv 0$. Since f vanishes an infinite number of times, we can find a disc D_r with center 0 and radius $r \in (0, 1)$, such that f has $m > k$ zeros in D_r and no zero on $|z| = r$.

The uniform convergence on compacta of φ_ε to f, as $\varepsilon \to 0$, implies that for ε close to zero, f and φ_ε have the same number of zeros in D_r. As f has m zeros in D_r, so does φ_ε. But this is impossible, since φ_ε vanishes in $D \supset D_r$ $k < m$ times. The proof is now complete.

To prove Proposition 4, we may assume $f(0) = 0$ and therefore, $f = zg$. As S. Pichorides suggested, the primitive F of g is one to one on T if and only if the statement of Proposition 4 is false. In such a case, by a standard argument, F is also one to one in D. Therefore $g = F'$ vanishes nowhere and f vanishes only once, contradicting the hypothesis. Hence Proposition 4 holds.

e) If in the proof of Theorem 1 we replace K_ε by other "nice" approximate identities, the proof goes through; we conclude that every value of f is a value of some $f * K_\varepsilon$ at some point of T.

It is not difficult to find sufficient conditions for K_ε in order that the proof of Theorem 1 still remains valid. To simplify the situation we restrict ourselves to the disc algebra A(D) and we have the following result:

"Consider any path $(0, 1] \ni t \to \mu_t \in M(T)$ in the space of measures on T, such that

(i) μ_t is an approximate identity as $t \to 0$,

(ii) $t \to \mu_t$ is continuous on $(0, 1]$ in norm and

(iii) μ_1 is the measure $\frac{\chi_T}{2\pi} d\theta$.

Then for every $f \in A(D)$ and every value α of f taken in D (that is $\alpha = (P_r * f)(\theta)$ for some r and θ), there are t and θ', so that $\alpha = (\mu_t * f)(\theta')$. "

This may be interpreted as a sort of minimality of the Poisson kernel P_r among the "paths" satisfying properties (i), (ii) and (iii).

The proof of Proposition 3 also works if we replace K_ε by μ_t. In that proof we have not used the fact that $K_n = \frac{\chi_I}{2\pi}$; therefore, requirement (iii) may be dropped if we suppose that α is taken by f infinitely many times. More precisely we obtain the following slightly stronger version of the previous result :

" Suppose the path $(0, 1] \ni t \to \mu_t \in M(T)$ satisfies conditions (i) and (ii) above and that $f \in A(D)$. If $\alpha = (P_r * f)(\theta)$ holds for infinitely many (r, θ), then there is a sequence (t_n, θ_n) with $t_n \to 0$, such that $\alpha = (\mu_{t_n} * f)(\theta_n)$. "

4. Remarks

a) In the space of BMO analytic functions the norms ${}_p\| \ \|$ and N_p $(1 \leqslant p)$ are equivalent ([1]). To evaluate the best constants relating these norms, it is useful to guess which functions "pick" the best constants. Theorem 2 shows that, for $1 \leqslant p \leqslant 2$, such functions are certainly not among the inner ones.

The calculation of the exact value of the conformaly invariant norm N_p of an inner function is simple for $1 \leqslant p \leqslant 2$ (see [2] , p. 10). For the inequality $1 \leqslant N_p(\varphi)$, it suffices to apply the triangular inequality and to take into account the fact that non constant inner functions assume arbitrarily small values. It was not easy to immitate the above argument in the case of the non-conformaly invariant norms $_p\| \ \|$, and this led us to Theorem 1.

b) In the proof of Theorem 2 we use the known fact that a non constant inner function takes arbitrarily small values in D : In the opposite case $\frac{1}{\varphi}$ would be an H^∞-function, bounded by $\sup_{|z|=1} \frac{1}{|\varphi(z)|} = 1$, which in turn implies that φ is constant (see also [4], p. 76).

c) We now make some remarks on the proof of Theorem 1:

1) The holomorphic extension of $\Phi_\varepsilon(e^{ix})$ is given by

$$\Phi_\varepsilon(z) = \frac{1}{2\varepsilon}\int_{-\varepsilon}^{\varepsilon} f(ze^{it})\,dt$$

. From this we can directly verify that Φ_ε is continuous on $\bar{D}$ and, using Morera's and Fubini's theorems, holomorphic in D. Further we can avoid the H^1-convergence argument: Restricting ourselves to a compact subset of D, we can obtain directly the uniform convergence (on compacta) $\Phi_\varepsilon \to f$, as $\varepsilon \to 0$. To this end it suffices to apply the (elementary) approximate identity argument to continuous rather than L^1-functions.

2) The proof of Theorem 1 becomes simpler in the following special case:

Suppose f is in the disc algebra and that f does not vanishe anywhere on T. Then, because f is continuous on T, $\Phi_\varepsilon \to f$ uniformly on T, as $\varepsilon \to 0$. Therefore f and Φ_ε have the same winding number with respect to 0, which is zero. The argument principle applied to f implies that $f(z) \neq 0$ for all $z \in D$, contradicting the hypothesis.

In the above special case we do not need the observation that Φ_ε is holomorphic in D.

5. Some extensions

a) Extension to the half-plane. Suppose f is in H^1 of the half-plane. Then every value α of f taken in the open half-plane is equal to the average f_I of f on a boundary interval I of length $|I|$, $0<|I|<\infty$.

Since $\int_{-\infty}^{+\infty} f(x)\,dx = 0$ for every H^1-function, we distinguish two cases in the proof : $\alpha \neq 0$ and $\alpha = 0$.

In the case $\alpha \neq 0$, we define $\Phi_\varepsilon(x) = \frac{1}{2\varepsilon}\int_{x-\varepsilon}^{x+\varepsilon} f(t)\,dt$ for $0<\varepsilon<\infty$. and the proof is similar as in the disc. In the case $\alpha = 0$ we may turn buck to the disc using a conformal transformation, and then Proposition 4 completes the proof.

Because inner functions are not in H^1 of the half--plane, we can not use the above statement to evaluate the non-conformaly invariant BMO norms of inner functions. However the proof of Proposition 3 extends to the case of H^∞-functions on the half-plane, that are continuous at ∞ ; this yields the following partial result : If φ is an infinite Blaschke product in the half-plane which is continuous at ∞, then ${}_p\|\varphi\| = 1$ for all $p \in [1, 2]$.

b) Extension to the polydisc. A routine induction argument extends Theorems 1 and 2 in a natural way in the case of holomorphic functions in the polydisc $D_n = D \times \ldots \times D$. Every value α of an H^1-function f is equal to the average f_I of f on a generalized boundary interval $I = I_1 \times I_2 \times \ldots \times I_n$ such that $|I_1| = |I_2| = \ldots = |I_n|$. More generally we may require that $|I_1|, |I_2|, \ldots, |I_n|$ depend on one parameter. The precise statement is the following :

Theorem 1. Suppose $\varphi_j : (0, n] \rightarrow (0, n]$, $j = 1, \ldots, n$

are any given continuous functions such that $\varphi_j(n)=n$ and $\lim_{\varepsilon\to 0}\varphi_j(\varepsilon)=0$ for all $j=1,\ldots,n$. Then, for any value α taken in D_n by an H^1-function f, there existe $\varepsilon>0$ and $I=I_1\times\ldots\times I_n$ with $|I_j|=2\varphi_j(\varepsilon)$ for all $j=1,\ldots,n$, such that $\alpha=f_I$.

It follows easily, that in the case $n>1$ and $\alpha\neq f(0,\ldots,0)$ the set of I's such that $\alpha=f_I$ has the power of $\underline{c}$.

For the proof of Theorem $1'$ we consider the functions $\Phi_\varepsilon(e^{ix_1},\ldots,e^{ix_n})=f_{I_1\times\ldots\times I_n}$, where I_j has length $2\varphi_j(\varepsilon)$ and midpoint e^{ix_j}, $j=1,\ldots,n$. As each Φ_ε has an extension in the polydisc algebra, a repeated use of the argument principle completes the proof.

Obviously Theorem $1'$ implies that $_p\|\varphi\|=1$ $(1\leqslant p\leqslant 2)$ for any non-constant inner function φ.

c) <u>Extension to the ball</u> . Let $B=B_n$ be the open unit ball of $\mathbb{C}^n$ ($n>1$), and S the boundary of B with the Haar-measure $d\sigma$ on it. For $\mathfrak{z}\in S$ and $\varepsilon>0$ we denote by $S(\mathfrak{z},\varepsilon)$ the set of x in S such that the Euclidean distance from x to $\mathfrak{z}$ is less than ε. For $f\in L^1(d\sigma)$ we denote by $f_{S(\mathfrak{z},\varepsilon)}$ the $d\sigma$-average of f on $S(\mathfrak{z},\varepsilon)$. The BMO norms $_p\|f\|$ are defined in the usual way.

The analogous statement of Theorem 1 is the following

Theorem $1''$. Suppose f is an H^1-function in B , $\alpha\in f(B)$ and $\varepsilon_0>0$. Then there are $\mathfrak{z}\in S$ and $\varepsilon\in(0,\varepsilon_0)$, such that $\alpha=f_{S(\mathfrak{z},\varepsilon)}$.

It follows immidiately that $_p\|\varphi\|=1$ for all $p\in[1,2]$ and φ inner non-constant . We mention that non-constant inner functions in B exist, as A.B. Alexandrof and others proved recently .

The above statements remain true, with similar proofs, if we replace $S(\zeta,\varepsilon)$ by the non-isotropic neighbourhoods ([5], p. 65).

To prove Theorem 1″ we consider the functions $\Phi_\varepsilon(\zeta) = f_{S(\zeta,\varepsilon)}$ $(0<\varepsilon\leqslant 2)$ on S. In contrast to all previous cases we do not need the fact $\Phi_2 = \Phi_\varepsilon$=constant, since one can easily show the following :

Let g be continuous on $\bar{B}$ and holomorphic in at least one of two variables. If g does not vanish on S, then g does not vanish in B.

For each $k=1,\dots,n$ there is an extension ${}_k\Phi_\varepsilon$ of Φ_ε, which is continuous on B and holomorphic in the kth variable z_k. The proof of the existence of such a ${}_k\Phi_\varepsilon$ hinges on the following facts :

i) For $\zeta = \left(z_1,\dots,z_{k-1},\sqrt{1-|z_1|^2-\dots-|z_{k-1}|^2-|z_{k+1}|^2-\dots-|z_n|^2}\; e^{it}, z_{k+1},\dots,z_n\right)$ we have the factorisation $d\sigma(\zeta) = d\mu(z_1,\dots,z_{k-1},z_{k+1},\dots,z_n)\frac{dt}{2\pi}$,

where $d\mu$ is a probability measure on the ball of $\mathbb{C}^n$, and

ii) For $(z_1,\dots,z_n)\in S$ and $(\zeta_1,\dots,\zeta_n)\in S$, the set of e^{it} such that the Euclidean distance of $(z_1,\dots,z_{k-1},|z_k|e^{it},z_{k+1},\dots,z_n)$ from $(\zeta_1,\dots,\zeta_{k-1},|\zeta_k|e^{i\theta},\zeta_{k+1},\dots,\zeta_n)$ is less than ε, is either empty or an arc with mid-point $e^{i\theta}$ and length independent of θ. We also notice that ${}_k\Phi_\varepsilon \longrightarrow f$ uniformly on compact subsets of B, as $\varepsilon\longrightarrow 0$.

With the above discussion in mind the proof of Theorem 1″ runs as follows : We assume $\alpha = f(w_1,\dots,w_n) = 0$ and suppose $f_{S(\zeta,\varepsilon)} \neq 0$ for all $\zeta\in S$ and $\varepsilon\in(0,\varepsilon_0)$. Then Φ_ε does not vanish on S for $0<\varepsilon<\varepsilon_0$, which implies that ${}_k\Phi_\varepsilon$ does not vanish in B. Since ${}_k\Phi_\varepsilon \to f$ as $\varepsilon\to 0$ uniformly on compacta, Hurwitz's theorem implies $f(w_1,\dots,w_{k-1},z,w_{k+1},\dots,w_n) = 0$ for all z. This holds for all $k = 1,\dots,n$ and so $f \equiv 0$, in contradiction to the fact that $f_{S(\zeta,\varepsilon)} \neq 0$.

6. Holomorphic convolutions

An essential point in the proof of Theorem 1 is that for $f \in H^1$ the convolution $\Phi_\varepsilon(e^{i\theta}) = f * K_\varepsilon = \frac{1}{2\varepsilon}\int_{\theta-\varepsilon}^{\theta+\varepsilon} f(e^{it})\,dt$ is of holomorphic type. The holomorphic extension is given by $\Phi_\varepsilon(z) = \frac{1}{2\varepsilon}\int_{-\varepsilon}^{\varepsilon} f(ze^{it})dt$. The above situation may be described as follows : We start with the holomorphic function f and the measure $d\mu_\varepsilon = \chi_{(-\varepsilon,\varepsilon)}\frac{d\theta}{2\varepsilon}$; then $\Phi_\varepsilon(z)$ is defined by the "multiplicative" convolution of f and μ_ε ; that is, using a rotation and a homothety with center 0 we transport μ_ε from the point $1 = e^{i0}$ to the point $z = |z|e^{i\theta}$; then we integrate f with respect to this new measure and the result is $\Phi_\varepsilon(z)$, the value of the convolution of f and μ_ε at the point z . We note that Φ_ε is holomorphic .

An analogous situation holds in the case of the polydisc. In the case of the half-plane, Φ_ε is also holomorphic and is defined by an "additive" convolution , that is, we use the group of translations .

In the case of the ball, the function Φ_ε defined on the boundary S has an extension ${}_k\Phi_\varepsilon$ which is continuous with respect to all variables and holomorphic with respect to the kth variable . In order to evaluate ${}_k\Phi_\varepsilon(z_1,\dots,z_n)$ we convolve the holomorphic function f with the measure $d\lambda_\varepsilon = \chi_{S((1,0,\dots,0),\varepsilon)}\frac{d\sigma(\zeta)}{\sigma[S((1,0,\dots,0),\varepsilon)]}$. More precisely, we first use a "rotation" of $\mathbb{C}^n$ to transfer $d\lambda_\varepsilon$ from $(1,0,\dots,0)$ to the point $M = \left(z_1,\dots,z_{k-1},\frac{z_k}{|z_k|}\sqrt{1-|z_1|^2-\dots-|z_{k-1}|^2-|z_{k+1}|^2-\dots-|z_n|^2}, z_{k+1},\dots,z and we obtain a new measure $d\nu_{\varepsilon, z_1, \dots, z_n}$. This last measure does not depend on the rotation we selected, provided it maps $(1,0,\dots,0)$ to M. Then we transform $d\nu_{\varepsilon, z_1, \dots, z_n}$ according to the transformation

$$S \ni (\zeta_1, \dots, \zeta_n) \longrightarrow \left(\zeta_1, \dots, \zeta_{k-1}, \zeta_k \frac{|z_k|}{\sqrt{1-|z_1|^2-\dots-|z_{k-1}|^2-|z_{k+1}|^2-\dots-|z_n|^2}}, \zeta_{k+1}, \dots, \zeta_n\right).$$

We thus obtain a probability measure $d\rho_{\varepsilon, k, z_1, \dots, z_n}$ supported in a neighbourhood of $(z_1, \dots, z_n)$. The value of ${}_k\Phi_\varepsilon(z_1, \dots, z_n)$ is given by $\int f \, d\rho_{\varepsilon, k, z_1, \dots, z_n}$.

At this point, it is natural to ask how to define the convolution of a function on an open subset of $\mathbb{C}^n$ with a measure of $\mathbb{C}^n$ with respect to a group of transformations, and under which conditions analyticity is preserved. The answer to these questions appears to give some new information. For example, if we restrict ourselves to the complex plane and to the usual groups of translations and rotation-homotheties with center 0, the additive and multiplicative convolutions respectively are well defined , and analyticity is preserved. More precisely, if f is holomorphic in the open set $G \subset \mathbb{C}$ and if $d\mu$ is a measure with support bounded by $b \in (0, +\infty)$, then the additive convolution $(f * d\mu)(z) = \int f(z-\zeta)\, d\mu(\zeta)$ is holomorphic in the open set $G_b = \{z \in G : \operatorname{dist}(z, \partial G) > b\}$. Respectively if $d\mu$ is a measure on $\mathbb{C}$ supported in the disc with center 1 and radius b, $0 < b < 1$, then the multiplicative convolution $(f * d\mu)(z) = \int f(z\zeta^{-1})\, d\mu(\zeta)$ is holomorphic in the set ${}_bG = \left\{z \in G : \operatorname{dist}(z, \partial G) > |z| \frac{b}{1-b}\right\}$, provided f is holomorphic in G. Further if we let $d\mu$ move continuously to the Dirac measure at 0 (in the case of the additive convolution), or to the Dirac mass at 1 (in the case of the multiplicative convolution), we obtain a family of holomorphic functions which converge to f uniformly on compacta. Therefore for every $z_0 \in G$ the value $f(z_0)$ is also taken by the convolutions $f * d\mu$ at points close to z_0. More precisely we state the following :

A : Suppose that the measures $d\mu_\varepsilon$ ($\varepsilon > 0$) are supported in the discs of the complex plane with center 0 and radius b_ε, and have the properties

i) $b_\varepsilon \to 0$ as $\varepsilon \to 0$

ii) $\sup_\varepsilon \|\mu_\varepsilon\| < \infty$ and

iii) $\mu_\varepsilon(\mathbb{C}) \to 1$ as $\varepsilon \to 0$.

Suppose also that $G \subset \mathbb{C}$ is a region and $z_0 \in G$ and $r > 0$ are given. If f is a non-constant holomorphic function in G, then there exists $\varepsilon_0 > 0$ such that for every $\varepsilon \in (0, \varepsilon_0)$ there is a $z_\varepsilon \in G$ with $|z_0 - z_\varepsilon| < r$, so that $f(z_0) = \int f(z_\varepsilon - \zeta)\, d\mu_\varepsilon(\zeta)$.

B : If the measures $d\mu_\varepsilon$ ($\varepsilon > 0$) are supported in the discs $|z-1| < b_\varepsilon$, then under the same hypotheses as in A we have $f(z_0) = \int f(z_\varepsilon \zeta^{-1})\, d\mu_\varepsilon(\zeta)$.

Applying A and B for some particular choises of μ_ε, we obtain the following results, which, despite their simplicity, appear to be new at least as far as we know.

C : Suppose f is holomorphic in an open set G, and suppose a direction $\vec{n}$ is given. Then for every

$z_0 \in G$, there is a segment I parallel to $\vec{n}$, as close to z_0 and as small as we please, so that $f(z_0) = f_I = \frac{1}{|I|} \int_I f(\zeta)|d\zeta|$.

D : Suppose f is holomorphic in an open set G, and suppose a triangle Δ is given. Then for every $z_0 \in G$, there is a triangle I with sides parallel to those of Δ, so that $f(z_0)$ is equal to the average of f on I with respect to the area measure. The triangle I can be choosen arbitrarily small and close to z_0.

E : Suppose f is holomorphic in an open set G and $z_0 \in G$. Then $f(z_0)$ is equal to the average of f on circular arcs with prescribed radius. These arcs may be choosen arbitrarily small and close to z_0.

F : Suppose f is holomorphic in an open set G and $z_0 \in G$. Then $f(z_0)$ is equal to the average of f on circular arcs with any prescribed center. Again these arcs may be choosen arbitrarily small and close to z_0.

BIBLIOGRAPHY

[1] A. BAERNSTEIN II, Analytic functions of bounded mean oscillation, Durham Conference 1979.

[2] N. DANIKAS, Untersuchungen über analytische Funktionen von beschränkter mittlerer Oszillation, Dissertation, Technische Universität Berlin 1981.

[3] C. FEFFERMAN and E. STEIN, H^p spaces of several variables, Acta Math. 129 (1972), 137-193.

[4] J. GARNETT, Bounded analytic functions, Academic Press 1981.

[5] W. RUDIN, Function theory in the unit ball of $\mathbb{C}^n$, Springer Verlag

V. Nestoridis
Kalavryton 37, Dasos Haidariou
ATHENS
GREECE

N. Danikas
Epidaurou 1, Chalandri
ATHENS
GREECE

Athens University and
University of Crete

An H^1 function with non-summable Fourier expansion

E.M. Stein

Dedicated to the memory of Salomon Bochner

1. Introduction

The purpose of this note is to show the existence of an H^1 function on the n-torus, whose Fourier series is almost-everywhere non-summable with respect to the means of the critical index $\frac{n-1}{2}$.

The background for the problem treated here is as follows. In the case $n = 1$ it is known (going back to a construction of Kolmogorov) that there exists an L^1 function whose Fourier series diverges almost everywhere, and this can be modified to yield an H^1 function with similar properties (see [9], Chapter 8). In the other direction the theorem of Carleson-Hunt-Sjölin guarantees the convergence almost everywhere wherever $f \in L \log L(\log \log L)$; see [4].

It was Bochner ([1]) who first pointed out that when $n > 1$, summability at the "critical index" $\frac{n-1}{2}$ was the correct analogue of convergence, for phenomena near L^1. In this sense versions of several of the above results are known in the case of general n. Thus there exists an L^1 function on the n-torus T^n whose Fourier series is not summable almost everywhere at the critical index ([6]): on the other hand ([5]), whenever $f \in L(\log L)^2$, its Fourier series is summable almost everywhere. A problem that remained therefore was what happens for functions in $H^1(T^n)$, $n > 1$. Of related interest is the fact that the sharp results for summability of $H^p(\mathbb{R}^n)$ with $p < 1$, have recently been determined (see [7]), and summability almost everywhere at the critical index was shown to hold for certain "block spaces" (see [3]).

In the case $n = 1$ there is in principle an easy way of going from existence of an L^1 function to an H^1 function whose Fourier series diverges almost everywhere. For suppose P is a trigonometric polynomial (of degree N) with (say) $\|P\|_{L^1} \leq 1$, and the partial sums of P relatively large. Then $e^{2\pi iNx}P(x)$ has H^1 norm bounded, and its partial sums are still large. This kind of special argument breaks down when $n > 1$. Nevertheless combining an idea that goes back to Bochner [1] with some techniques used in [6], one can show the following.

We write $f \sim \sum a_m e^{2\pi i m\cdot x}$ for the Fourier series of a function f on the n-torus, and $\tilde{S}_R$ is defined by $\tilde{S}_R(f)(x) = \sum_{|m|<R} a_m\left(1-\frac{|m|^2}{R^2}\right)^{\frac{n-1}{2}} e^{2\pi i m\cdot x}$.

<u>Theorem</u>. <u>There exists an</u> $f \in H^1(T^n)$ <u>so that</u>

$$\limsup_{R\to\infty} |\tilde{S}_R(f)(x)| = \infty \text{ a.e.}$$

The proof of the theorem can be reduced to a corresponding fact in the non-periodic case. We define the $\mathbb{R}^n$ analogue of $\tilde{S}_R$ by

$$S_R(f)(x) = \int_{\mathbb{R}^n} \hat{f}(\xi)(1-|\xi|^2/R^2)_+^{\frac{n-1}{2}} e^{2\pi i x\cdot\xi}\, d\xi$$ whenever $f \in L^1(\mathbb{R}^n)$, with $\hat{f}$ the Fourier transform of f.

The main point of the construction is the following lemma.

Lemma

For each $N > 0$ there exists an $f = f_N \in H^1(\mathbb{R}^n)$ and a set $E = E^N$, E^N contained in the unit cube $Q = \{x \mid 0 \leq x_j < 1\}$, so that

(i) $\|f\|_{H^1(\mathbb{R}^n)} \leq 1$

(ii) $\sup_R |S_R(f)(x)| \geq c \log N$, whenever $x \in E^N$

(iii) $m(E^N) \geq 1/2$.

For simplicity of notation we shall give the proof in the case $n = 2$; the modifications required for the general case $n > 2$ will be indicated below. We use the fact that $S_R(f) = f*K_R$, where

$$K_R(x) = c\,\frac{\cos(2\pi R|x|)}{|x|^2} + O\left(\frac{1}{R|x|^3}\right), \tag{2.1}$$

whenever $R|x| \geq 1$. (See [8], pp. 158 and 171).

Let $a_1, \ldots, a_{N^2}$ be an enumeration of the N^2 points in the unit cube Q with coordinates $(\ell_1/N, \ell_2/N)$, where ℓ_i or integers with $0 \leq \ell_i < N$. Then for a full set E_0, (with $m(^c E_0)=0$) whenever $x \in E_0$, the N^2 numbers $\{|x-a_\ell|\}_{\ell=1}^{N^2}$ are linearly independent over the rationals; (see [1], and [8], p. 271). In particular for $x \in E_0$, $|x-a_\ell| \neq 0$. Using the linear independence and Kronecker's theorem it is not difficult to see that for each $x \in E_0$ there exists a large enough M_x so that the following holds: Whenever y_ℓ, $\ell = 1,\ldots,N^2$ are N^2 given real numbers, then there exists an R with $\sup_\ell \frac{1}{|x-a_\ell|} \leq R \leq M_x$, so that

$$(2.2)\qquad \begin{cases} R\,|x-a_\ell|-y_\ell = \delta_\ell \mod 1 \\ \text{with } |\delta_\ell| \le 1/10 \end{cases} .$$

Hence we can find an $E_1 \subset Q \cap E_0$, with $m(E_1) \ge 3/4$ and an M, so that whenever $x \in E_1$, $M_x \le M$, and as a result (2.3) holds for some R, with $\sup_\ell \frac{1}{|x-a_\ell|} \le R \le M$.

As a consequence, whenever $x \in E_1$,

$$(2.3)\qquad \sup_{M\le R\le 2M} \frac{1}{N^2} \sum_{\ell=1}^{N^2} \left|\frac{\cos(2\pi R|x-a_\ell|)}{|x-a_\ell|^2}\right| \ge c/N^2 \sum_{\ell=1}^{N^2} \frac{1}{|x-a_\ell|^2} \ge c \log N ,$$

and this follows from (2.2) by writing $R|x-a_\ell| = (R-M))|x-a_\ell| + M|x-a_\ell|$, noting that $\sup_\ell \frac{1}{|x-a_\ell|} \le M$, for $x \in E_1$, and taking $y_\ell = -M|x-a_\ell|$.

We next choose a triple of functions, $\hat\phi$, $\hat\psi$, and $\hat\alpha$, each in $C_0^\infty(\mathbb{R}^2)$ with the following additional properties:

(1) $\hat\phi(\xi) = 0$, for $|\xi| \le \frac14$ or $|\xi| \ge 4$; $\hat\phi(\xi) = 1$ when $\frac12 \le |\xi| \le 2$.

(2) $\hat\psi(\xi) = 1-\hat\phi(\xi)$, for $|\xi| \le 2$, and $\hat\psi(\xi) = 0$ for $|\xi| \ge \frac12$.

(3) $\hat\alpha(\xi) = (1-|\xi|^2)^{\frac12}$, for $|\xi| \le \frac12$.

We let ϕ, ψ, and α denote the inverse Fourier transforms of $\hat\phi$, $\hat\psi$, and $\hat\alpha$ respectively, and set $\hat\phi_M(\xi) = \hat\phi(\xi/M)$; thus $\phi_M(x) = \phi(Mx)M^2$, with similar notation for $\hat\psi_M$, $\hat\alpha_M$ etc.

Observe that $\phi \in H^1(\mathbb{R}^2)$, because $\hat\phi$ vanishes near the origin and is in C_0^∞; hence ϕ_M also belongs to $H^1(\mathbb{R}^2)$, and $\|\phi_M\|_{H^1} = \|\phi\|_{H^1} = c$.

Also $\psi_M \in L^1(\mathbb{R}^2)$, and $\|\psi_M\|_{L^1} = \|\psi\|_{L^1} = c'$. Next if $M \leq R \leq 2M$ then

$$(1-|\xi|^2/R^2)_+^{\frac{1}{2}} = (1-|\xi|^2/R^2)_+^{\frac{1}{2}}\hat{\phi}(\xi/M) + (1-|\xi|^2/R^2)_+^{\frac{1}{2}}\hat{\psi}(\xi/M) ,$$

since $\hat{\phi}(\xi) + \hat{\psi}(\xi) = 1$ whenever $|\xi| \leq 2$.

Also by (2) and (3)

$$(1-|\xi|^2/R^2)_+^{\frac{1}{2}} \hat{\psi}(\xi/M) = \hat{\alpha}(\xi/R)\hat{\psi}(\xi/M) .$$

So with $M \leq R \leq 2M$,

(2.4) $$K_R = K_R * \phi_M + \alpha_R * \psi_M .$$

Now write $f = f_N = \frac{1}{N^2}\sum_{\ell=1}^{N^2} \phi_M(x-a_\ell)$, and

$$g = g_N = \frac{1}{N^2}\sum_{\ell=1}^{N^2} \psi_M(x-a_\ell) .$$

Then $f \in H^1(\mathbb{R}^2)$, with $\|f\|_{H^1} \leq c$, while $g \in L^1(\mathbb{R}^2)$ with $\|g\|_{L^1} \leq c'$

However (2.4) and (2.1) imply that (whenever $M \leq R \leq 2M$, $x \in E_1$) .

(2.5) $$S_R(f) = c/N^2 \sum_{\ell=1}^{N^2} \frac{\cos(2\pi R|x-a_\ell|)}{|x-a_\ell|^2} - \alpha_R * g + O(E_R) .$$

Here $E_R = \frac{R^2}{(1+|x|R)^3} * d\mu$, where the measure $d\mu$ is given by

$d\mu = \frac{1}{N^2}\sum_{\ell=1}^{N^2} \delta(x-a_\ell)$, with δ = Dirac delta mass.

Observe that in our construction if N is given, then M is determined by it. With these now fixed we can take the supremum over R, with $M \leq R \leq 2M$, in (2.5). In order to apply (2.3) we need only estimate $\sup_R |\alpha_R * g(x)|$ and $\sup_R |E_R(x)|$. However both these extensions are dominated by a fixed multiple of the standard maximal function of an integrable function, and a finite measure, each having bounded norm. Thus their sum is less than c'', except in a set E_2 of measure $\leq \frac{1}{4}$. Therefore the lemma is proved with $E^N = E_1 \cap {}^cE_2$, after multiplying f by a suitable constant to sinure $\|f\|_{H^1} \leq 1$.

For the case of general n we use the fact that

$$K_R(x) = c_n \frac{\cos 2\pi R|x|}{|x|^n} + O\left(\frac{1}{R|x|^{n+1}}\right), \text{ when } |x|R \geq 1$$

if n is even, with a similar formula where the first term is replaced by $c_n \frac{\sin 2\pi R|x|}{|x|^n}$ when n is odd; then the argument goes through as in the case $n = 2$, with only the most obvious changes.

3. Proof of the theorem

We first recall the link between $H^1(\mathbb{R}^n)$ and $H^1(T^n)$. If f is given in $\mathbb{R}^n$ then its H^1 norm is defined by

$$\|f\|_{H^1(\mathbb{R}^n)} = \|f\|_{L^1(\mathbb{R}^n)} + \sum_{j=1}^{n} \|R_j(f)\|_{L^1(\mathbb{R}^n)} \quad \text{where } \{R_j\} \text{ are the Reisz}$$

transforms. Similarly for f given in the torus T^n, then

$$\|f\|_{H^1(T^n)} = \|f\|_{L^1(T^n)} + \sum_{j=1}^{n} \|\tilde{R}_j f\|_{L^1(T^n)} \quad \text{where } \{\tilde{R}_j\} \text{ are the periodized Riesz}$$

transforms.

(For a discussion of periodization related to multiplier operators and arguments of the type that follow, see [8], Chapter VII, §3 and [2].

Now with a function f in $\mathbb{R}^n$ consider its periodization $\tilde{f}$ given by

$$\tilde{f}(x) = \sum_{m \in \mathbb{Z}^n} f(x+m) ,$$

and also its dilated periodization $\tilde{f}_\varepsilon(x) = \varepsilon^{-n} \sum_{m \in \mathbb{Z}^n} f\left(\frac{x+m}{\varepsilon}\right)$, $\varepsilon > 0$. We observe that $\int_{Q_n} |\tilde{f}_\varepsilon(x)|\,dx \le \sum_m \varepsilon^{-n} \int_{Q_m} |f(\frac{x+m}{\varepsilon})|\,dx = \|f\|_{L^1(\mathbb{R}^n)}$ where Q_n is the fundamental cube. Similarly $\|\tilde{R}_j \tilde{f}_\varepsilon\|_{L^1(T^n)} \le \|R_j(f)\|_{L^1(\mathbb{R}^n)}$. Thus

$$\|\tilde{f}_\varepsilon\|_{H^1(T^n)} \le \|f\|_{H^1(\mathbb{R}^n)} .$$

Next $(S_R(f_\varepsilon))^\sim = \tilde{S}_R(\tilde{f}_\varepsilon)$, where $f_\varepsilon(x) = f(x/\varepsilon)\varepsilon^{-n}$, in the sense that both periodic functions have $m^{\underline{th}}$ Fourier coefficient equal to

$\hat{f}(\varepsilon m)(1-|m|^2/R^2)^{\frac{n-1}{2}} e^{2\pi i m\cdot x}$.

Next observe (here it is convenient to take f to be C^∞ with compact support) that

$$\lim_{\varepsilon\to 0} \varepsilon^n \tilde{S}_{R/\varepsilon}(\tilde{f}_\varepsilon)(\varepsilon x) = S_R(f)(x) , \text{ each } x \tag{3.1}$$

and so

$$\sup_R |S_R(f)(x)| \le \overline{\lim_{\varepsilon\to 0}} \sup_R |\varepsilon^n \tilde{S}_{R/\varepsilon}(\tilde{f}_\varepsilon)(\varepsilon x)| . \tag{3.2}$$

Now assume that the conclusion of theorem 1 does not hold. Then it follows by general principles (see Corollary 3 in [6]), that there exists a constant A so that

$$(3.3) \qquad \{m\{x \in Q_n \ , \ \sup_R |\tilde{S}_R(g)(x)| > \lambda\} \le (A/\lambda)\|g\|_{H^1(T^n)} \ .$$

Thus

$$m\{x \in Q_n/\varepsilon \ , \ \sup_R \varepsilon^n|\tilde{S}_{R/\varepsilon}(\tilde{f})(\varepsilon x)| > \lambda\} \le (A/\lambda)\|\tilde{f}_\varepsilon\|_{H^1(T^n)} \le (A/\lambda)\|f\|_{H^1(\mathbb{R}^n)} \ .$$

Since Q_n/ε is a family of sets increasing to $\mathbb{R}^n$ as $\varepsilon \longrightarrow 0$, and because of (3.2) we get as a result

$$m\{x \in \mathbb{R}^n \ , \ \sup_R |S_R(f)(x)| > \lambda\} \le (A/\lambda)\|f\|_{H^1(\mathbb{R}^n)} \ , \ \lambda > 0 \ .$$

This contradicts the lemma and proves the theorem.

From the lemma, and by arguments such as in [6], it is also not difficult to produce an $f \in H^1(\mathbb{R}^n)$ so that $\limsup_{R \to \infty} |S_R(f)(x)| = \infty$ a.e.

References

[1] S. Bochner, "Summation of multiple Fourier series by spherical means", Trans. Amer. Math. Soc. 40 (1936), 175-207.

[2] C.E. Kenig and P.A. Tomas, "Maximal operators defined by Fourier multipliers", Studia Math. 68 (1980), 79-83.

[3] S.Z. Lu, M. Taibleson, and G. Weiss, "On the almost-everywhere convergence of Bochner-Riesz means of multiple Fourier series", Harmonic Analysis Proceedings, Minneapolis, 1981, Lecture Notes in Math. #908, (1982), 311-318.

[4] P. Sjölin, "Convergence almost everywhere of certain singular integrals and multiple Fourier series", Arkiv for Mat. 9 (1971), 65-90.

[5] E.M. Stein, "Localization and summability of multiple Fourier series", Acta Math. 100 (1958), 93-147.

[6] __________, "On limits of sequences of operators", Ann. of Math., 74 (1961), 140-170.

[7] E.M. Stein, M. Taibleson, and G. Weiss, "Weak type estimates for maximal operators on certain H^p classes", Rendiconti del Cir. Mat. di Palermo, supplemento, 1981, 81-97.

[8] E.M. Stein and G. Weiss, "Introduction to Fourier analysis on Eulidean spaces", Princeton University Press, 1971.

[9] A. Zygmund, "Trigonometric series", 2nd edition, Cambridge University Press, 1959.

Integral characterization of a space generated by blocks

by

Fernando Soria

0. Introduction:

The spaces of blocks were introduced by M. Taibleson and G. Weiss in [6] in connection with the study of a.e. convergence of Fourier series.

Let T denote the one-dimensional torus which we identify with the interval $[0,2\pi)$ and $|A|$ the Lebesgue measure of the subset A, that is

$$|A| = \frac{1}{2\pi}\int_0^{2\pi} \chi_A(t)dt .$$

For $1 < q \leq \infty$ we say that a function $b(x)$ is a q-block if there exists an interval I such that

i) $b(x) = 0$, outside of I

ii) $\|b\|_q \leq |I|^{\frac{1}{q}-1}$

We define the space B_q as the class of all the functions which can be written as

$$f(x) = \Sigma\, m_k b_k(x) \tag{1}$$

where the b_k are q-blocks and $m = \{m_k\}$ is a sequence of real numbers satisfying

$$M(m) = \Sigma\, |m_k|\,(1+\log^+ 1/|m_k|) < \infty$$

or, equivalently,

$$N(m) = \sum_k |m_k| \{1 + \log(\sum_\ell |m_\ell| / |m_k|)\} < \infty$$

For $f \in B_q$ we define

$$M_q(f) = \inf M(m) , \qquad N_q(f) = \inf N(m) ,$$

the infimum taken over all the possible representations (1). The functional M_q is subadditive (i.e. $M_q(f+g) \leqq M_q(f) + M_q(g)$) and induces on B_q an F-space structure. On the other hand N_q is positive homogeneous and $N_q(f+g) \leqq (1+\log 2)[N_q(f)+N_q(g)]$; it induces on B_q a quasi-Banach space structure. Furthermore, the corresponding topologies are the same.

Observe that the definition of these block spaces extends easily to the real line, to R^n (substituting the intervals by cubes) and to more general topological measure spaces.

In [6] several basic properties of these spaces are shown, including the proper inclusions $B_{q_2} \subset B_{q_1}$ if $1 < q_1 < q_2 \leqq \infty$, as well as the following.

<u>Theorem A</u>: <u>The Fourier series of any</u> $f \in B_q$ <u>converges a.e., where</u> B_q <u>is the block space associated with the torus</u> T, $1 < q \leqq \infty$

The smallest of this classes, B_∞, seems to be the most interesting. It is the only one whose dual cannot be identified with L^∞ and for which the translation operator does not act continuously. (See [3] for more details.)

In [1], R. Fefferman introduced the notion of entropy as a tool for the study of some problems in classical analysis. For A a measurable subset of T we let

$$E(A) = \inf \Sigma\, \varphi(|I_k|) ,$$

the infimum being taken over all sequences of intervals $\{I_k\}$ covering A, where $\varphi(x)$, $0 \leq x \leq 1$, is the increasing and concave function $\varphi(x) = (x/e)\log(e/x)$. Define the <u>entropy</u> of a function f as

$$J(f) = \int_0^\infty E(\{|f| > t\})dt .$$

Denote by J the class of all functions with finite entropy. In [6] it is shown that J is a proper subclass of B_∞. We show here that B_∞ has an entropic-like integral characterization. This is the main theorem of the next section. This will allow us to obtain more information about this particular space (see remarks 2, 3 and 4 and section 3).

The author wishes to express his deep gratitude to Professors M. Taibleson and G. Weiss for their continuous support and invaluable help.

1. Main Results:

Given a function f and $t > 0$ we define

(2) $$\Lambda(t) = \Lambda_f(t) = \inf \Sigma\, t|I_k|(1 + \log^+ \frac{1}{t|I_k|})$$

the infimum taken over all the sequences of intervals $\{I_K\}$ covering the set $\{x \in T : |f(x)| > t\}$.

Theorem 1: The following two statements are equivalent:

(a) f belongs to B_∞ ;

(b) $\|f\|_{B_\infty} = \int_0^\infty \Lambda(t)\,\frac{dt}{t}$ is finite .

Moreover, if this is the case , $M_\infty(f) \sim \|f\|_{B_\infty}$. (*)

We will prove this theorem in section 2. Let us first make some remarks.

Remark 1: $\Lambda(t)/t$ is a non-increasing function for $0 < t < \infty$, and therefore can be viewed as the analog of the distribution function $\lambda(t) = \lambda_f(t) = |\{x : |f(x)| > t\}|$

Remark 2: This integral characterization provides the following simple test to decide when a monotone function is in B_∞ : If f is monotone then $f \in B_\infty$ if and only if

$$\int_0^\infty \lambda(t)(1 + \log^+ \frac{1}{t\lambda(t)}) < \infty .$$

The reason is that for such an f the set $\{|f| > t\}$ is an interval and therefore it is the most efficient covering of itself minimizing (2).

Remark 3: Using the definition of the functional M_∞ it is not hard to see that $M_\infty(t\chi_{\{|f|>t\}}) \leqq \Lambda_f(t)$. On the other hand it is an easy consequence of theorem 1

(*) "$A \sim B$" means as usual $cA \leqq B \leqq CA$ for some constants $0 < c$, $C < \infty$.

that for fixed $t_0 > 0$ we have

$$M_\infty(t_0 \chi_{\{|f| > t_0\}}) \geqq c\Lambda_f(t_0) .$$

We, thus, obtain the corollary

(3) $$M_\infty(f) \sim \int_0^\infty M_\infty(t\chi_{\{|f| > t\}}) \frac{dt}{t} \text{ (*)} .$$

Remark 4: Let A be a measurable subset of T and $f = 2\chi_A$. Then $A = \{|f| > 1\}$ and so $\Lambda(1) \sim M_\infty(\chi_A)$. But also it is clear from the definition that $\Lambda(1) \sim E(A)$, where $E(A)$ is the entropy of A. Therefore we have shown that the entropy and the M_∞ "norms" of the characteristic function of a set are equivalent. We thus obtain the following result

(4) $$J(f) \sim \int_0^\infty M_\infty(\chi_{\{|f| > t\}})\, dt$$

and one can prove, following the argument in [1], pg. 189, that the right hand side of (4) is, in fact, a norm on J.

It is now evident from (3) and (4) why a function with finite entropy must necessarily lie in B_∞.

(*) This was pointed out by M. Taibleson.

2. Proof of theorem 1:

$b \Rightarrow a$. Let k be an integer and put $E_k = \{x : |f(x)| > 2^k\}$. Then,

$$|f| \leq \sum_{k=-\infty}^{\infty} 2^k \chi_{E_k} \quad ;$$

therefore,

$$M_\infty(f) \leq \sum_{k=-\infty}^{\infty} M_\infty(2^k \chi_{E_k}) \leq \sum_{k=-\infty}^{\infty} \Lambda(2^k) \quad .$$

Now remark 1 implies $\|f\|_{B_\infty} \sim \sum_{k=-\infty}^{\infty} \Lambda(2^k)$ and so we have $M_\infty(f) \leq C\|f\|_{B_\infty}$ for some constant C independent of f.

$a \Rightarrow b$. We will need the following.

Lemma 2: Any function $f \in B_\infty$ has a representation $f(x) = \Sigma\, c_k b_k(x)$ with blocks supported on some dyadic interval of T (that is, an interval having the form $[2\pi n 2^{-\ell}, 2\pi(n+1)2^{-\ell})$ for some $\ell \in \{0,1,2,\ldots\}$ and $n \in 0,1,\ldots,2^{-\ell}-1\}$). If we define $M_\infty^d(f) = \inf \Sigma\, |c_k|\,(1 + \log^+ 1/|c_k|)$, with the infimum taken over such dyadic representations, then

$$M_\infty(f) \leq M_\infty^d(f) \leq 4\, M_\infty(f) \quad .$$

The proof is easy and will be omitted.

Let f be a function in B_∞. Then $f = \Sigma\, c_k b_k(x)$ where $|b_k(x)| \leq |I_k|^{-1} \chi_{I_k}(x)$ for some dyadic interval I_k and $c = \{c_k\}$ is a sequence of positive real numbers with $M(c) < \infty$. We show

$$\|f\|_{B_\infty} \leqq \Sigma\, c_k (2 + \log^+ \frac{1}{c_k}) = K$$

and, hence, $\|f\|_{B_\infty} \leqq 2\, M^d_\infty(f)$.

Let $h = \Sigma\, c_k |I_k|^{-1} \chi_{I_k}$. Since $|f| \leqq h$, we have $\|f\|_{B_\infty} \leqq \|h\|_{B_\infty}$; thus it is enough to prove $\|h\|_{B_\infty} \leqq K$.

<u>Lemma 3</u>: <u>If</u> $g = c|I|^{-1}\chi_I + \varphi$ <u>with</u> $c > 0$, I <u>an interval</u>, $\varphi \geqq 0$ <u>and</u> $\operatorname{supp}\varphi \subset I$ <u>then</u> $\|g\|_{B_\infty} \leqq c(2 + \log^+ 1/c) + \|\varphi\|_{B_\infty}$.

<u>Proof</u>: For $t > 0$ we have

$$\{x : g(x) > t\} = \begin{cases} I, & \text{if } 0 < t < c/|I| \\ \{x : \varphi(x) > t - c/|I|\}, & \text{if } t \geqq c/|I| \end{cases}.$$

This and remark 1 imply

$$\Lambda_g(t)/t = \begin{cases} |I|(1 + \log^+ 1/t|I|), & 0 < t < c/|I| \\ \Lambda_\varphi(t - c/|I|) \,/\, (t - c/|I|), & t \geqq c/|I| \end{cases}.$$

Therefore,

$$\|g\|_{B_\infty} \leq \int_0^{c/|I|} |I|(1+\log^+ \frac{1}{t|I|})dt + \int_{c/|I|}^{\infty} \Lambda_\varphi(t - c/|I|) \frac{dt}{t - c/|I|}$$

$$\leq c(2+\log^+ \frac{1}{c}) + \int_0^\infty \Lambda_\varphi(t) \frac{dt}{t}$$

q.e.d.

Lemma 4: If $(\operatorname{supp}\varphi) \cap (\operatorname{supp}\psi) = \emptyset$ then $\|\varphi+\psi\|_{B_\infty} \leq \|\varphi\|_{B_\infty} + \|\psi\|_{B_\infty}$.

The proof is obvious since $\{x : |\varphi(x)+\psi(x)| > t\} = \{x : |\varphi(x)| > t\} \cup \{x : |\psi(x)| > t\}$ and therefore $\Lambda_{\varphi+\psi}(t) \leq \Lambda_\varphi(t) + \Lambda_\psi(t)$.

Let $h_N = \sum_{k=1}^{N} c_k |I_k|^{-1} \chi_{I_k}$. From lemmas 3 and 4 and finite induction we get

$$\|h_N\|_{B_\infty} \leq \sum_{k=1}^{N} c_k(2+\log^+ 1/c_k) .$$

Observe that $h_N(x)$ increases to $h(x)$ as $N \to \infty$. So to finish the proof of theorem 1 we need only the following:

Lemma 5: Let $0 \leq g_1 \leq \dots \leq g_n \leq \dots$ be an increasing sequence of functions and let $g(x)$ be its limit. If $\|g_n\|_{B_\infty} \leq K$ for $n = 1,2,\dots$ then $\|g\|_{B_\infty} \leq K$.

Proof: It is enough to prove that $\Lambda_{g_n}(t) \to \Lambda_g(t)$ as $n \to \infty$ and then use the Lebesgue monotone convergence theorem.

Let us first introduce some terminology. Let (Ω, d) be a metric space and ℓ a positive increasing continuous function defined on $[0,\infty)$. For $\delta > 0$ and

E a Borel subset of Ω define

$$\mu_\delta^{\ell}(E) = \inf\{\Sigma \ell(d(Gi)) : Gi \text{ open}, \ \cup Gi \supset E \text{ and}$$

$$d(Gi) = \text{diameter of } Gi < \delta\}$$

Then $\mu^{\ell}(E) = \lim_{\delta \to 0} \mu_\delta^{\ell}(E)$ defines a Borel measure called the ℓ-Hausdorff measure.

So, if $\Omega = T$, d is the Euclidean distance, $\psi^t(x) = tx(1 + \log^+ 1/tx)$ and f is any function on T we have

$$\Lambda_f(t) = \mu_1^{\psi^t}(E_{t,f}), \quad E_{t,f} = \{x : |f(x)| > t\}.$$

Now observe that for $t > 0$ $E_{t,g_1} \subset E_{t,g_2} \subset \ldots$ and $\cup E_{t,g_n} = E_{t,g}$. Thus $\Lambda_{g_n}(t) \to \Lambda_g(t)$ is true because of the non-trivial fact that for any increasing sequence of sets $\{E_n\}$ and any $\delta > 0$ we have

$$\mu_\delta^{\ell}(\cup E_n) = \lim_n \mu_\delta^{\ell}(E_n)$$

See Rogers [4], theorem 47 and the references mentioned there for a proof.

3. <u>Final remarks</u>:

Theorem 1 can be extended to the space B_∞ associated to the real line (and to R^n with obvious modifications). The only non-trivial step would be lemma 5 which requires some control on the length of the intervals. This can be done in the following way:

Suppose that $f \in B_\infty(\mathbb{R})$. Set $I_0 = [-2,2)$ and $I_k = [2^k, 2^{k+1})$ if $k = 1,2,\ldots$, $[-2^{|k|+1}, -2^{|k|})$ if $k = -1,-2,\ldots$. Let f_k be the restriction of f to I_k. Then it can be shown that $M_q(f) \leqq \sum_{-\infty}^{\infty} M_q(f_k) \leqq C_q M_q(f)$. See [5].

Lemma 5 applies to each f_k and, therefore, we have $\|f_k\|_{B_\infty} \sim M_\infty(f_k)$. Finally, $M_\infty(f) \leqq \|f\|_{B_\infty} \leqq \sum_{-\infty}^{\infty} M_\infty(f_k) \sim M_\infty(f)$.

We can use theorem 1 to find the smallest rearrangement invariant class, $\widetilde{B}_\infty$, containing B_∞. We define

$$B = \{f : \|f\|_B = \int_0^\infty \lambda_f(t)\,(1 + \log^+ \frac{1}{t\lambda_f(t)})\,dt < \infty\} .$$

It is not hard to see from definition (2) that

$$t\lambda_f(t)\,(1 + \log^+ 1/t\lambda_f(t)) \leqq \Lambda_f(t)$$

and, therefore, $\|f\|_B \leqq \|f\|_{B_\infty}$. Thus theorem 1 implies $B_\infty \subset B$. On the other hand, if $f \in B$ and f_d is its non-increasing rearrangement we have from remark 2 that $f_d \in B_\infty$ and $\|f_d\|_B = \|f_d\|_{B_\infty}$. Since B is rearrangement invariant we have $B = \widetilde{B}$.

In [5] it is shown that $\|f\|_B \leqq C_q M_q(f)$, where $C_q \sim q/(q-1)$, and we obtain as a consequence, that B is also the smallest rearrangement invariant class containing B_q. In particular, if $f \in B_q$ then $f_d \in B_\infty$ and $M_\infty(f_d) \leqq C_q M_q(f)$.

B plays, with respect to B_∞, the same role as $L \log^+ L$ does with respect to the class J. The following gives some information about this space B:

Proposition 6: (a) B is a quasi-Banach space, the quasi-norm being given by

$$\|f\|_B^0 = \int_0^\infty \lambda(t)\left(1 + \log \frac{\|f\|_1}{t\lambda(t)}\right) dt ,$$

and B is also a log convex space (see Kalton [2]).

(b) B contains $L \log^+ \log^+ L$. In particular, given f in the latter class, $f_d \in B_\infty$.

The proof can be found in [5]. Since $L \log^+ \log^+ L$ contains functions with a.e. divergent Fourier series, theorem A does not apply to the Class B.

References

1. R.A. Fefferman, A theory of entropy in Fourier analysis, Advances in Math. 30 (1978), pp. 171-201.

2. N.J. Kalton, Convexity, type and the three space problem, Studia Math., T. LXIX (1981), pp. 247-287.

3. Y. Meyer, M. Taibleson, G. Weiss, preprint.

4. C.A. Rogers, Hausdorff measures, Cambridge U. Press, 1970.

5. F. Soria, Ph.D. thesis, Washington University.

6. M.H. Taibleson and G. Weiss, Certain function spaces connected with a.e. convergence of Fourier series, Proceedings of the Conference on Harmonic Analysis, in Honor of Antoni Zygmund; W. Beckner, A.P. Calderon, R. Fefferman, P.W. Jones, editors, Wadsworth Publ. Co., Vol. I (1982), pp. 95-113.

Washington University
St. Louis, Missouri

Another Characterization of H^p, $0 < p < \infty$, with an Application to Interpolation

by

Stephen Semmes

I. Introduction

Suppose that $f(x)$ is an integrable function on $\mathbb{R}$ with Poisson integral $f(x,y)$ and radial maximal function $f^+(x) = \sup\{|f(x,y)| : y > 0\}$. By definition, f lies in the (real variable) Hardy space H^1 exactly if f^+ is integrable. If we define a weak type norm by

$$\|g(y)\|_{L^1_{wk}(dy)} = \sup_{\lambda>0} \lambda |\{y : |g(y)| > \lambda\}|$$

and define f_1 by

$$f_1(x) = \|y^{-1} f(x,y)\|_{L^1_{wk}(dy)},$$

then $f_1(x) \le f^+(x)$, and hence f_1 is integrable if f lies in H^1.

In Section II we prove the converse: if f_1 is integrable, then so is f^+. In fact, an analogous result holds for H^p, $0 < p < \infty$, and distributions f on R^n.

In Section III we prove an interpolation theorem using the ideas of Section II. An example of the situation that we consider is the following. Suppose that A^2 is the space of holomorphic functions in the upper half plane which are square integrable with respect to area measure. Let H^1_{Hol} denote the Hardy space of holomorphic functions in the upper half plane. In proving our interpolation theorem, we show that if f lies in H^1_{Hol}, then f is weakly integrable with respect to $y^{-1}dxdy$. Then we interpolate between $L^1_{wk}(y^{-1}dxdy)$ and $L^2(dxdy)$ by the real method of obtain information about the intermediate spaces between H^1 and A^2. These results are related to work of Janson and Peetre, and extend work by Peller and by Peetre and Svennsson.

In the final section we give an application of this interpolation theorem to show that the class of homeomorphisms which preserve BMO also preserve a family of Besov spaces.

The author wishes to thank Richard Rochberg for many valuable conversations, and acknowledges support from the National Science Foundation in the form of a graduate fellowship.

II. H^p spaces

Let $0<p<\infty$ be fixed, and suppose that f is a distribution on R^n with Poisson integral $f(x,y)$ defined on $R^{n+1}_+ = \{(x,y) \in R^n \times R : y>0\}$. For $\alpha>0$ and $x \in R^n$, let $\Gamma_\alpha(x) = \{(z,y) : |z-x| < \alpha y\}$. Define $f^+(x) = \sup\{|f(x,y)| : y>0\}$ and $f^*_\alpha(x) = \sup\{|f(u,v)| : (u,v) \in \Gamma_\alpha(x)\}$. We say that f lies in the Hardy space H^p if f^*_α lies in L^p, and we set $\|f\|_{H^p} = \|f^*_\alpha\|_{L^p}$. As is well known, this is essentially independent of $\alpha>0$, and is equivalent to the requirement that f^+ lie in L^p.

For a measure space (A,μ) and $0<p<\infty$ we define

$$\|g\|_{L^p_{wk}(A,\mu)} = \sup_{\lambda>0} \lambda\, \mu(\{a \in A : |g(a)| > \lambda\})^{1/p}$$

and we let $L^p_{wk}(A,\mu)$ be the space of functions for which the quasinorm is finite.

As in the introduction, let

$$f_p(x) = \|y^{-1/p} f(x,y)\|_{L^p_{wk}(dy)}$$

and

$$f_{p,\alpha}(x) = \|v^{-(n+1)/p} f(u,v)\|_{L^p_{wk}(\Gamma_\alpha(x),\, dudy)} .$$

Then it is easy to verify that $f_p(x) \leq C f^+(x)$ and $f_{p,\alpha}(x) \leq C f^*_\alpha(x)$.

This implies part of our basic result, which is the following.

Theorem 1. The following are equivalent.

(a) $f \in H^p$;

(b) $f_p \in L^p$;

(c) $f_{p,\alpha} \in L^p$, $0<\alpha<\infty$.

Furthermore, the appropriate quasinorms are equivalent.

To prove this, we need two preliminary lemmas. The first is based on an inequality of Kolmogorov, and a proof of it may be found in [6]. The second lemma is essentially due to Hardy and Littlewood; see [4] for a proof.

Lemma 1. Suppose that g lies in $L^p_{wk}(A,\mu)$. Then for $s<p$ and $E \subseteq A$,

$$\int_E |g|^s \, d\mu \le C_{s,p} |E|^{1-s/p} \|g\|^s_{L^p_{wk}} .$$

Lemma 2. Let U be harmonic on a ball $B \subseteq R^n$ with center x_0, and suppose that U is continuous on the closure of B. Then

$$|U(x_o)|^p \le C_p \frac{1}{|B|} \int_B |U(t)|^p dt .$$

To prove the theorem, notice that (a) implies both (b) and (c), by our earlier comments. Assume now that (c) holds. We shall show that $f^*_{\alpha/2}(x) \le C\, f_{p,\alpha}(x)$ for all $x \in R^n$, where C depends on p and α.

Let x in R^n and (u,v) in $\Gamma_{\alpha/2}(x)$ be given. Then there is a ball B with center (u,v) and radius comparable to v such that B is contained in $\Gamma_\alpha(x)$. Hence, by Lemma 2 and Lemma 1,

$$|f(u,v)|^q \le C \frac{1}{|B|} \int_B |f(w,y)|^q dwdy$$

$$\le C \frac{v^{(n+1)q/p}}{|B|} |B|^{1-q/p} f_{p,\alpha}(x)^q$$

$$= C\, f_{p,\alpha}(x)^q ,$$

for any $q<p$, where C does not depend on x. This implies the desired inequality.

To show that (b) implies (a), we shall use a technique in [4, p. 180] for controlling the nontangential maximal function by the radial one. It is enough to show that $\|f^*_\beta\|_p \le C\,\|f_p\|_p$, where $\beta>0$ is fixed.

Let x in R^n and (u,v) in $\Gamma_\beta(x)$ be given. Let Q be the cube centered at (u,v) with sides parallel to the axes and side length v, and let Q_0 be its projection onto R^n, i.e., the cube with center u and side length v. Let B be the ball which is inscribed in Q. Then, by Lemma 2 and Lemma 1,

$$|f(u,v)|^{p/2} \le C\,\frac{1}{|B|}\int_B |f(w,y)|^{p/2}\,dydw$$

$$\le C\,\frac{1}{|Q|}\int_Q |f(w,y)|^{p/2}\,dydw$$

$$\le C\,\frac{v^{1/2}}{|Q|}\int_{Q_0}\int_{v/2}^{3v/2} |y^{-1/p} f(w,y)|^{p/2}\,dydw$$

$$\le C\,\frac{v^{1/2}}{|Q|}\int_{Q_0} v^{1/2} f_p(w)^{p/2}\,dw$$

$$\le C\,\frac{1}{|Q_0|}\int_{Q_0} f_p(w)^{p/2}\,dw$$

$$\le C\,M\,(f_p(u)^{p/2})\,,$$

where Mg is the Hardy-Littlewood maximal function. Since $\|Mg\|_2 \leq C\|g\|_2$ (see, e.g., [16]), the desired inequality follows immediately. This completes the proof of Theorem 1.

III. Interpolation

In this section we restrict our attention to holomorphic functions in the upper half plane U. However, all of our techniques extend directly to the case of harmonic functions on R_+^{n+1}.

Let $A^{p,r}$ denote the Bergman space of functions holomorphic on U which lie in $L^p(y^{2r}dxdy)$, where $0<p<\infty$ and $-1/2<r<\infty$. Define the homogeneous Besov space B_s^p for s in R by

$$B_s^p = A^{p,-(ps+1)/2} \quad \text{if} \quad s<0$$

and

$$f \in B_s^p \Leftrightarrow Df \in B_{s-1}^p$$

for all s in R, where D denotes the operator of differentiation. (This latter space is sometimes denoted B_{pp}^s.) See [2], [10], [16], and [17] for information about these two families of spaces. Alternatively, many of the properties of the B_s^p may be derived from the results in [17] about the Bergman spaces. This is especially desirable for the case $0<p<1$, since in this case the equivalence of the various definitions is less clear.

For s real, $p>0$, define H_s^p to be the space of functions f such that $D^s f$ lies in H^p_{Hol}, where D^s is defined via the Fourier transform. These potential spaces are, in some sense, close to the Besov spaces (see references below). In particular, recall that $B_s^2 = H_s^2$.

For an appropriate pair A, B of Banach spaces, let $[A,B]_{\theta,p}$ denote the intermediate interpolation space, constructed by the real method, where $0<\theta<1$ and $0<p\leq\infty$. (See [2] for facts about interpolation.)

The following fact is due to Grisvard [5]:

$$(3.1) \qquad [B_{s_0}^{p_0}, B_{s_1}^{p_1}]_{\theta,p} = B_p^s$$

for $s = (1-\theta)\, s_0 + \theta s_1$ and $1/p = (1-\theta)/p_0 + \theta/p_1$. We are going to prove that B_s^p is still the intermediate space even if one or both of the endpoint spaces is replaced by the appropriate potential space, so long as $s_0 \neq s_1$.

<u>Theorem 2</u>. Let $0<p_0, p_1<\infty$, $p_0 \neq p_1$, $0<\theta<1$, $s_0, s_1 \in R$, and $s_0 \neq s_1$ Then

$$(3.2) \qquad [H_{s_0}^{p_0}, B_{s_1}^{p_1}]_{\theta,p} = B_s^p$$

$$(3.3) \qquad [B_{s_0}^{p_0}, H_{s_1}^{p_1}]_{\theta,p} = B_s^p$$

(3.4) $$[H^{p_0}_{s_0}, H^{p_1}_{s_1}]_{\theta,p} = B^p_s,$$

where $s = (1-\theta)s_0 + \theta s_1$ and $1/p = (1-\theta)/p_0 + \theta/p_1$.

This result has the following dualized version. Recall that BMO ("bounded mean oscillation") is the space of functions f on $\mathbb{R}$ for which $|I|^{-1}\int_I |f - f_I|\,dx$ is uniformly bounded over all intervals I, and where f_I is the mean of f on I. Then BMO A is the space of functions in BMO of analytic type, and BMO A is essentially the dual of H^1_{Hol}. (See [4].)

<u>Theorem 3</u>. Suppose $s_0 \in \mathbb{R}$, $0 < p_0 < \infty$, $s_0 \neq 0$, and $0 < \theta < 1$. Then

(3.5) $$[H^{p_0}_{s_0}, \text{BMO A}]_{\theta,p} = B^p_s$$

(3.6) $$[B^{p_0}_{s_0}, \text{BMO A}]_{\theta,p} = B^p_s,$$

where $s = (1-\theta)s_0$ and $1/p = (1-\theta)/p_0$.

Much of Theorem 3 (and hence, by a duality argument, also much of Theorem 2) is known. Peller ([12], [13]) proved (3.6) in the case $n = 1$, $s_0 = 1/p_0$, $p_0 \geq 1$ (using techniques that do not appear to generalize to $\mathbb{R}^{n+1}_+$, $n > 1$). Peetre and Svensson [11] obtained the result for all dimensions, $p_0 \geq 1$, and $s_0 > 0$, and, more recently, Janson and Peetre [7] showed that $s_0 > 0$ could be relaxed to $s_0 \neq 0$. Peetre [9] has since found a very simple proof of much of Theorem 2 using the theory of Triebel spaces.

We should point out that, as in [11], the conditions $s_0 \neq s$ and $s_0 \neq 0$ for the two theorems, respectively, are necessary. For example, we know [4] that $[H^2_{Hol}, \mathrm{BMOA}]_{\theta,p} = H^p_{Hol}$, but $H^p_{Hol} \neq B^p_0$ for $p \neq 2$.

Let us see first how Theorem 3 follows from Theorem 2. For $p_0 > 1$, this is immediate from the duality theorem for real interpolation, and the reflexitivity of the intermediate spaces. (Here we use the facts that $(H^p_s)^* = H^q_{-s}$ and $(B^p_s)^* = B^q_{-s}$, $1 < p < \infty$ and $p^{-1} + q^{-1} = 1$, and that $(H^1_{Hol})^* = \mathrm{BMOA}$, as noted above.) The rest of Theorem 3 follows from Theorem 2 and Wolff's 4-space reiteration theorem [18].

To prove Theorem 2, observe that it is enough to prove (3.2) for $p_0 \leq 2$. The rest follows from duality and Wolff's theorem. Also, one inclusion of (3.2) follows from (3.1). Indeed, if $p \leq 2$, then B^p_s is contained in H^p_s . This may be proven by adapting the proof of Theorem 5 in [16 , Chapter 5] to $0 < p \leq 2$, using the fact that the Littlewood-Paley g-function characterizes H^p for $0 < p < \infty$. (See [4].)

Because $[H^{p_0}_{s_0}, B^{p_1}_{s_1}]_{\theta,p}$ is a space of holomorphic functions in the upper half plane, we need only show that its elements have an appropriate derivative lying in a corresponding L^p space. To do this, we use the fact that such an estimate holds for elements of $B^{p_2}_{s_2}$ and that a certain wea type estimate holds for $H_{s_1}^{p_1}$. Then we combine these estimates by applying the results of [14] to obtain the other half of (3.2).

The weak type condition that we use is relative to a weighted Lebesgue on the upper half plane. From the point of view of Section II, a more natural condition might seem to be one of the form $L^p(L^p_{wk})$, relative to appropriate measures. However, this is not convenient for the arguments below because such a condition would require interpolation of vector-valued L^p spaces (with changing vector spaces), rather than just the Marcinkiewicz theorem and its variants. Also, we wish to avoid having to deal with spaces of functions which take values in non-normable spaces (such as L^p_{wk}, $p \leq 1$).

Choose $a > s_0, s_1$ so that $a - s_0$ is an integer, and set $r_i = p_i(a - s_i)$, $i = 0, 1$. We then have $D^a(B_{s_1}^{p_1}) \subseteq A^{p_1, (r_1 - 1)/2} \subseteq L^{p_1}(y^{r_1 - 1}\, dxdy)$. We now show that $D^a(H_{s_0}^{p_0})$ is contained in a weighted weak type space and can then obtain the desired inclusion by an application of the Marcinkiewicz theorem with change of weight as presented by Stein and Weiss in [14].

To use the theory of [14] we need a bit more notation. Let $d\mu = (y^{r_0} + y^{r_1})\, y^{-1}\, dxdy$, and let $\alpha_i = y^{r_i}(y^{r_0} + y^{r_1})^{-1}$ for $i = 0, 1$, so that $y^{r_i - 1}\, dxdy = \alpha_i d\mu$. With $K(x,y) = (y^{r_0 - r_1})^{1/(p_0 - p_1)}$ and $d\xi = y^{p_1 p_0 (s_1 - s_0)/(p_1 - p_0) - 1}\, dxdy$, we define $L_0^{p_0}$ to be the space of functions h defined on U for which

$$\lambda\, \xi\left(\{(x,y) : K(x,y)|h(x,y)| > \lambda\}\right)^{1/p_0}$$

is bounded, with the norm of h defined to be the least upper bound. By Theorem (2.9) of [14],

$$[L_0^{p_0}, L^{p_1}(y^{r_1-1}\, dxdy)]_{\theta,p} = L^p(d\mu_\theta),$$

where $1/p = (1-\theta)/p_0 + \theta/p$ and $d\mu_\theta = \alpha_0^{(1-\theta)p/p_0}\alpha_1^{\theta p/p_1}\, d\mu$ $= y^{p(a-s)-1}\, dxdy$, by direct calculation. To finish the proof, we now show that $D^a(H_{s_0}^{p_0}) \subseteq L_0^{p_0}$.

Suppose f lies in H^1 and n is nonnegative and, say, an integer. If we define $M_n f(x) = \sup\{y^n|f^{(n)}(x)| : y>0\}$ for x in $\mathbb{R}$, then $\|M_n f(x)\|_{L^p(\mathbb{R})} \leq C\|f\|_{H^p}$. (See Theorem 10 of [4].) Hence, if g is in $H_{s_0}^{p_0}$, then $|D^a g(x,y)| \leq y^{s_0-a} g_1(x)$, where g_1 lies in $L^{p_0}(\mathbb{R})$, with the proper norm control. It now remains to show that

$$\xi(\{(x,y) : K(y)\, y^{s_0-a} g_1(x) > \lambda\}) \leq \frac{C\|g_1\|_{p_0}}{\lambda^{p_0}}.$$

Write $K(y) = y^\gamma$. Observe that our hypothesis $s_0 \neq s$ implies that $\gamma \neq a - s_0$; assume that $\gamma > a - s_0$, the other case being handled similarly. Let $u = p_1 p_0(s_1 - s_0)/(p_1 - p_0)$, and note that

(3.7) $$\frac{1}{(\gamma + s_0 - a)}\, u = -p_0 .$$

Then

$$\{(x,y) : K(y)\, y^{s_0 - a}\, g_1(x) > \lambda\}$$

$$= \{(x,y) : y > (g_1(x)^{-1}\lambda)^{1/(\gamma + s_0 - a)}\},$$

so that

$$\xi\,(\{(x,y) : K(y)\, y^{s_0 - a}\, g_1(x) > \lambda\})$$

$$\leq \int_{-\infty}^{\infty} \int_{(g_1(x)^{-1}\lambda)^{1/(\gamma + s_0 - a)}} y^{u-1}\, dy dx$$

$$\leq C \int_{-\infty}^{\infty} (g_1(x)^{-1}\lambda)^{u/(\gamma + s_0 - a)}\, dx$$

$$\leq C\, \|g_1\|_{p_0}^{p_0} ,$$

where the convergence of the inner integral follows from (3.7) and the assumption that $\gamma > a - s_0$. (Notice that it would diverge if $\gamma = a - s_0$, i.e., if $s_0 = s_1$.) Thus $D^a(H_{s_0}^{p_0}) \subseteq L_0^{p_0}$, and Theorem 2 is proved.

IV. An extension of a theorem of Jones

Suppose that φ is an increasing homeomorphism of $\mathbb{R}$ onto itself, that φ is locally absolutely continuous, and φ' belongs to the Muckenhoupt class of A^{∞} weights. (See [3] about this class of weights.) A result of Jones [8] is that the operator V_{φ} defined by $V_{\varphi}(f) = f\circ\varphi$ preserves BMO. (In fact, he shows that the above conditions are also necessary.) Let $\mathrm{Re}(B^p_{1/p})$ be the space of real parts of the elements of $B^p_{1/p}$, and identify the elements of this space with their boundary value functions. (Hence, by a Sobolev-type inclusion theorem, $\mathrm{Re}(B^p_{1/p})$ is contained in VMO.) An application of Theorem 3 is the following.

Proposition 1. Under the conditions on φ above, V_{φ} preserves $\mathrm{Re}(B^p_{1/p})$, $1<p<\infty$.

To prove this, note that by duality and a change of variables, φ must preserve H^1_1 . Hence Theorem 3 above applies, and we conclude that V_{φ} preserves $B^p_{1/p}$, as desired.

We should point out that the assumption on φ will not, in general, be necessary. For if $p = 2$, then $B^p_{1/p}$ is the usual Dirichlet space, and it can be proven that $V_{\varphi}(B^2_{1/2}) \subseteq B^2_{1/2}$ if $d\varphi$ is a doubling measure. (Outline of proof: By a theorem of Beurling and Ahlfors (see [1]), φ extends to a quasiconformal mapping Φ of the upper half plane onto itself. If $f\in B^2_{1/2}$ and Pg denotes the Poisson integral of g, then

$$\int_U\int |\nabla P(f\circ\varphi)|^2\, dxdy \leq \int_U\int |\nabla((Pf)\circ\Phi)|^2\, dxdy$$

$$\leq C\int_U\int |\nabla Pf|^2\, dxdy,$$

where the first inequality follows from the Dirichlet principle, and the second follows from the quasi-invariance of the Dirichlet integral.)

This proposition is used in [19].

References

[1] L.V. Ahlfors, Lectures on Quasiconformal Mappings, Van Nostrand, 1966.

[2] J. Bergh and J. Löfström, Interpolation Spaces: An Introduction, Springer-Verlag, Berlin, Heidelberg, New York, 1976.

[3] R.R. Coifman and C. Fefferman, Weighted norm inequalities for maximal functions and singular integrals, Studia Math. 54 (1974), 241-250.

[4] C. Fefferman and E. Stein, H^p spaces of several variables, Acta. Math. 129 (1972), 136-193.

[5] P. Grisvard, Commutativité de deux foncteurs d'interpolation et applications, J. Math. Pures Appl. 45 (1966), 143-290.

[6] M. de Guzman, Real variable methods in Fourier analysis, North-Holland, New York, 1981.

[7] S. Janson and J. Peetre, to appear in Math. Scand.

[8] P. Jones, Homeomorphisms of the line which preserve BMO, to appear in Arkiv för Matematik.

[9] J. Peetre, personal communication, 1982.

[10] J. Peetre, New Thoughts on Besov Spaces, Duke University Mathematics Series 1, Mathematics Department, Duke University, Durham, 1976.

[11] J. Peetre and E. Svensson, On the generalized Hardy's inequality of McGhee, Pigno, and Smith and the problem of interpolation between BMO and a Besov space, preprint, 1982.

[12] V.V. Peller, Smooth Hankel operators and their applications (ideals γ_p, Besov classes, random processes), Dokl. Akad. Nauk. SSSR 252 (1980), 43-48. (Russian)

[13] V.V. Peller, Hankel operators of class $\mathfrak{S}^p$ and their applications (rational approximation, Gaussian processes, the majorant problem for operators), Mat. Sb. 113 (1980), 538-581. (Russian)

[14] E.M. Stein and G. Weiss, Interpolation of operators with change in measures, Trans. Amer. Math. Soc. 87 (1958), 159-172.

[15] E.M. Stein and G. Weiss, Introduction to Fourier Analysis on Euclidean Spaces, Princeton University Press, Princeton, N.J., 1971.

[16] E.M. Stein, Singular Integrals and Differentiability Properties of Functions, Princeton University Press, Princeton, N.J., 1970.

[17] R.R. Coifman and R. Rochberg, Representation theorems for holomorphic and harmonic functions in L^p, Asterisque 77 (1980), 11-66.

[18] T. Wolff, A note on interpolation spaces, Harmonic Analysis, Proceedings, Minneapolis, 1981, Lecture Notes in Mathematics 908, 199-2(Springer-Verlag, New York, 1982.

[19] S. Semmes, Ph.D. Dissertation, Washington University, St. Louis, 1983.

Current Address:
Institut Mittag-Leffler
Djursholm, Sweden

The Maximal Function on Weighted BMO

by Steven Bloom

I. Introduction

Recently, Bennett, Devore, and Sharpley [1] showed that the maximal function of a function with bounded mean oscilation, BMO, is itself in BMO.

Smoothness properties of functions are expressed by Lipschitz classes, or more generally, by weighted BMO. In this paper we address the question: What sort of weighted BMO classes are preserved by the Hardy-Littlewood maximal operator?

For notation, I will always denote an interval, with Lebesgue measure $|I|$, $w(I) = \int_I w\,dx$, and $I(w) = \frac{w(I)}{|I|}$. C will denote a universal constant, and may change at subsequent appearances. A weight w is in the class A_1 if $I(w) \leq C \operatorname{ess\,inf}_I w$ for all intervals I. $w \in A_p$ for $1<p<\infty$ if $I(w)I(w^{-1/(p-1)})^{p-1} \leq C$ for all I. w satisfies the Reverse Hölder Inequality if there exists a $p>1$ for which $I(w^p)^{1/p} \leq CI(w)$, for all I. w satisfies this Reverse Holder condition if and only if $w \in A_p$ for some p[2,4]. In either case, w must be a doubling measure, that is $w(2I) \leq Cw(I)$, where $2I$ is the interval concentric with I of twice the measure. A function f is in the weighted BMO class BMO_w if $\frac{1}{w(I)} \int_I |f-I(f)| \leq C$ for all I, with norm

$$\|f\|_{*,w} = \sup_I \frac{1}{w(I)} \int_I |f-I(f)| \ .$$

Finally the maximal function of f, denoted f^*, is defined by $f^*(x) = \sup\{I(|f|) : x \in I\}$.

In section II we introduce a variant of the classes A_p and develop their properties. In section III we examine the maximal function on weighted BMO.

II. <u>The Class</u> α_p

<u>Definition</u>: A weight w is in the class α_p if for each interval I and subset E of I,

$$\left[\frac{|E|}{|I|}\right]^p \leq C \frac{w(E)}{w(I)} \ .$$

<u>Theorem 1</u> i) $A_1 = \alpha_1$

ii) For $1 < p < \infty$,

$$A_p \underset{\neq}{\subset} \alpha_p \underset{\neq}{\subset} \bigcap_{q>p} A_q \ .$$

Theorem 1 has been derived independently and with roughly similar proof by Kerman and Torchinsky [3] in their study of restricted weak-type inequalities. Hence we omit our proof.

One other class of weights will also appear in the next section.

<u>Definition</u>: $w \in \text{weak } \alpha_p$ if for all intervals I and J with $I \subset J$,

$$\left[\frac{|I|}{|J|}\right]^p \leq C\frac{w(I)}{w(J)} .$$

For $p = 1$, this class is familiar. If $w \in \text{weak } \alpha_1$ and if x is any Lebesgue point, there exists an interval J containing x so that $w^*(x) \leq 2J(w)$. So for $I \subset J$ with $x \in I$, $w^*(x) \leq 2J(w) \leq 2CI(w)$, and since x is a Lebesgue point for w, we conclude $w^*(x) \leq Cw(x)$. Thus weak $\alpha_1 \subset A_1 = \alpha_1$. As the converse inclusion is trivial, we have weak $\alpha_1 = \alpha_1 = A_1$. The problem of characterizing weak α_p for $p > 1$ is still open.

III. <u>The Maximal Operator on Weighted BMO</u>

Our setting will be the unit circle T in the complex plane. The results do not extend nicely to higher dimensions. Some of the problems we will point out as we go along.

We are interested in a stronger class than BMO_w, functions of weighted bounded lower oscillation. A function f belongs to BLO_w if $I(f) - \text{ess}_I\inf f \leq CI(w)$ for all intervals I. Also we use RH to denote the class of weights which satisfy the Reverse Hölder Inequality.

<u>Theorem 2</u> i) If $w \in \text{weak } \alpha_p \cap RH$ for some $p < 2$, then for any $f \in BMO_w$, $f^* \in BLO_w$ with

(1) $$I(f^*) - \text{ess}_I\inf f^* \leq C\|f\|_{*,w} I(w) \quad \text{for all } I .$$

ii) Conversely, if (1) holds for all f, then $w \in \text{weak } \alpha_2 \cap RH$.

<u>Proof</u>: i) We may assume that $f \geq 0$. Fix an interval I. For $x \in I$, define

$$F_1(x) = \sup \{J(f) : x \in J, \quad J \subset 4I\}$$

$$F_2(x) = \sup \{J(f) : x \in J, \quad J \not\subset 4I\}.$$

On I, $f^*(x) = \max \{F_1(x), F_2(x)\}$. In the event that $|I| > \frac{\pi}{2}$, then $4I = T$ and F_2 is considered zero. Put

$$E_1 = \{x \in I : F_1(x) \geq F_2(x)\}$$

$$E_2 = I \sim E_1.$$

Let

$$\beta = \text{ess}_I \inf f^*.$$

We want to estimate

$$\int_I (f^* - \beta) = \sum_{i=1}^{2} \int_{E_i} (F_i - \beta).$$

We must show

(2) $$\int_{E_i} (F_i - \beta) \leq C_w(i) \quad \text{for } i=1 \text{ and } 2.$$

For $i=1$ we proceed as follows: For each $x \in I$, $4I(f) \leq f^*(x)$, so that $4I(f) \leq \beta$. We apply the Calderon-Zygmund decomposition to f on $4I$, yielding disjoint intervals $\{I_n\}$, with $I_n \subset J_n$, and satisfying

$$f \leq \beta \quad \text{a.e. on} \quad 4I \sim \bigcup_n I_n, \tag{3}$$

$$|J_n| = 2|I_n|, \quad \text{and} \tag{4}$$

$$J_n(f) \leq \beta < I_n(f). \tag{5}$$

Define good and bad functions g and b by

$$g = \sum_n J_n(f)\chi_{I_n} + f\chi_{4I \sim \bigcup_n I_n}$$

$$b = \sum_n (f-J_n(f))\chi_{I_n},$$

so that $b+g = f\chi_{4I}$. By (3) and (5), $\|g\|_\infty \leq \beta$.

Let $r>1$ be a Reverse Hölder exponent for w. For r sufficiently near 1 we will show that $b \in L^r$. First,

$$\|b\|_r^r = \sum_n \int_{I_n} |f-J_n(f)|^r \leq \sum_n \int_{J_n} |f-J_n(f)|^r.$$

Define

$$f^{\#,n}(x) = \sup\{J(|f-J(f)|) : x \in J \subset J_n\}$$

$$w^{*,n}(x) = \sup\{J(w) : x \in J \subset J_n\}.$$

Since $f \in BMO_w$, for $x \in J \subset J_n$, $J(|f-J(f)|) \leq CJ(w) \leq Cw^{*,n}(x)$, so that $f^{\#,n}(x) \leq Cw^{*,n}(x)$. By the Sharp Function Theorem of Fefferman and Stein,

$$\int_{J_n} |f-J_n(f)|^r \leq C\int_{J_n} (f^{\#,n})^r \leq C\int_{J_n} (w^{*,n})^r \leq C\int_{J_n} w^r .$$

Thus $\|b\|_r^r \leq C \sum_n \int_{J_n} w^r$. We choose r sufficiently near 1 so that $w^r \in RH$, which means, in particular, that w^r is a doubling measure [2], so that

$$\int_{J_n} w^r \leq C\int_{I_n} w^r .$$

So we conclude,

$$\|b\|_r^r \leq C \sum_n \int_{I_n} w^r \leq C\int_{4I} w^r \leq C\int_I w^r .$$

Now let $\frac{1}{r} + \frac{1}{r'} = 1$. By Hölder's Inequality,

$$\int_{E_1} b^* \leq |E_1|^{1/r'}[\int_{E_1} b^{*r}]^{1/r} \leq |I|^{1/r'}\|b^*\|_r$$

$$\leq C|I|^{1/r'}\|b\|_r \leq C|I|^{1/r'}[\int_I w^r]^{1/r}$$

$$= C|I|I(w^r)^{1/r} \leq Cw(I), \quad \text{by RH}.$$

Hence,

$$\int_{E_1} (F_1-\beta) = \int_{E_1} (f\chi_{4I})^* - \beta \le \int_{E_1} b^* + g^* - \beta$$

$$\le Cw(I) + (\|g^*\|_\infty - \beta)\ |E_1| \le Cw(I)$$

which is (2) for $i=1$.

For $i=2$, it will suffice to show

$$(6) \qquad F_2(x)-\beta \le CI(w) \quad \text{for all} \quad x \in E_2 .$$

So let $x \in E_2$ with $x \in J$ and $J \not\subset 4I$. If $I \subset J$, then $J(f) \le f^*(y)$ for all $y \in I$, or $J(f) \le \beta$, so that $J(f)-\beta \le 0 \le CI(w)$. Otherwise, $I=[a,b]$ and $J=[c,d]$, with $a<c\le b<d$, or a similar, symmetric situation. Put $\bar{J}=[c-|I|,d]$. Since $I \subset \bar{J}$, $\bar{J}(f) \le \beta$, and hence

$$J(f)-\beta \le J(f)-\bar{J}(f)$$

Define $J_0 = [c-|I|,c]$, and $J_k = [c,c+2^{k-1}|I|]$ for $k\ge 1$. Let n be the largest integer for which $J_n \subsetneq J$. Call $J_{n+1} = J$. Then

$$J(f) - \beta \le \frac{1}{|J|}\int_J f - \frac{1}{|\bar{J}|}\int_{\bar{J}} f = \frac{1}{|J|}\int_J f - \frac{1}{|\bar{J}|}\int_J f - \frac{1}{|\bar{J}|}\int_{J_0} f$$

$$= \frac{|I|}{|\bar{J}|}[J(f) - J_0(f)] \le 2^{-n}[J_1(f)-J_0(f)+J_{n+1}(f)-J_1(f)]$$

$$= 2^{-n}[J_1(f)-J_0(f)] + 2^{-n}\sum_{k=1}^{n}[J_{k+1}(f)-J_k(f)] .$$

Now let $R = J_0 \cup J_1$. Then

$$J_1(f)-J_0(f) = J_1(f)-R(f)+R(f)-J_0(f)$$

$$\leq \frac{1}{|J_1|}\int_{J_1} |f-R(f)| + \frac{1}{|J_0|}\int_{J_0} |f-R(f)|$$

$$\leq \frac{2}{|R|}\int_R |f-R(f)| \leq CR(w)$$

$$\leq CI(w)\ , \quad \text{as } w \text{ is a doubling measure,}$$

while

$$\sum_{k=1}^{n} J_{k+1}(f)-J_k(f) \leq \sum_{k=1}^{n} \frac{1}{|J_k|}\int_{J_k} |f-J_{k+1}(f)|$$

$$\leq 2 \sum_{k=1}^{n} \frac{1}{|J_{k+1}|}\int_{J_{k+1}} |f-J_{k+1}(f)|$$

$$\leq C \sum_{k=1}^{n} J_{k+1}(w) \leq Cn \max_k J_{k+1}(w) .$$

and so,

$$2^{-n} \Sigma\, [J_{k+1}(f)-J_k(f)] \leq C \frac{|I|}{|J|} \log \frac{|J|}{|I|} \max J_{k+1}(w) .$$

Now $x \log \frac{1}{x}$ is an increasing function for $x \leq \frac{1}{e}$, so that if k is the maximum term,

$$\frac{|I|}{|J|}\log\frac{|J|}{|I|} \le \frac{|I|}{|J_{k+1}|}\log\frac{|J_{k+1}|}{|I|} \quad .$$

and so,

$$J(f) - \beta \le C\left[I(w) + \frac{|I|}{|J_{k+1}|}\log\frac{|J_{k+1}|}{|I|}\,J_{k+1}(w)\right]$$

$$= C\left[I(w) + \left[\frac{|I|}{|J_{k+1}|}\right]^2\left[\log\frac{|J_{k+1}|}{|I|}\right]\frac{1}{|I|}\,w(J_{k+1})\right] .$$

But $w \in$ weak α_p for some $p<2$, so that

$$\left[\frac{|I|}{|J_{k+1}|}\right]^2 \log\frac{|J_{k+1}|}{|I|} \le \left[\frac{|I|}{|J_{k+1}|}\right]^p - C\frac{w(J_1)}{w(J_{k+1})} ,$$

and therefore

$$J(f)-\beta \le C[I(w)+J_1(w)] \le C[I(w)+2R(w)] \le CI(w) .$$

Taking supremums over J gives (6).

Before proceeding to the converse, we should remark about higher dimensions. Attempting this last argument in R^n does not lead to a sequence of cubes, but a sequence of rectangles with one side much shorter than the others. The problem, then, is that the weighted BMO hypothesis need not apply to such rectangles. If $w \in$ weak $\alpha_p \cap RH$ for some $p < \frac{n+1}{n}$, and if f satisfies such a stronger weighted BMO condition R^n, then either f^* is identically infinite, or belongs to BLO_w .

ii) First we prove that w is a doubling measure. For simplicity, we take $I = [0,2h]$ and $J = [2h,3h]$. It will suffice to show $w(J) \leq Cw(I)$. Let $f = w\chi_J$. Then for any interval R,

$$\frac{1}{|R|}\int_R |f-R(f)| \leq 2R(f) \leq 2R(w)$$

so that $f \in BMO_w$. Hence there exists a constant C independent of I so that $I(f^*) - \text{ess}_I\inf f^* \leq CI(w)$. Now

$$\text{ess}_I\inf f^* \leq f^*(0) \leq \frac{1}{|I|} w(J) ,$$

while

$$f^*(x) \geq \frac{1}{3h-x} w(J) .$$

Thus

$$I(f^*) - \text{ess}_I\inf f^* \geq \frac{w(J)}{|I|}\left[-1 + \int_0^{2h} \frac{1}{3h-x}\,dx\right]$$

$$= \frac{w(J)}{|I|}[\log 3 - 1]$$

and we conclude that

$$w(J) \leq \frac{C}{\log 3 - 1} w(I) .$$

Next we derive the Reverse Hölder Inequality. Take $I = [0,h]$ and put $\bar{I} = [0,2h]$. Let $f = w\chi_{\bar{I}}$. If $J \subset I$, we have

$$J(f^*) \le \text{ess}_J\inf f^* + CJ(w) \le (1+C)\ \text{ess}_J\inf f^* .$$

So the function f^* belongs to A_1 on the space I. In particular, $f^* \in RH$ on I, with constants which depend only on the A_1-constant for f^*, $1+C$, which is independent of I.[1]

So we have $r>1$ and a constant C for which $I(f^{*r})^{1/r} \le CI(f^*)$. Hence, $I(w^r)^{1/r} \le I(f^{*r})^{1/r} \le CI(f^*) \le 2C\bar{I}(f^*)$. Ah, but

$$\bar{I}(f^*) \le \text{ess}_I\inf f^* + C\bar{I}(w)$$

$$\le f^*(2h) + CI(w), \quad \text{since } w \text{ is a doubling measure,}$$

$$\le I(w) + CI(w)$$

so we conclude that $I(w^r)^{1/r} \le CI(w)$. An identical argument applies to any interval.

For the weak α_2 condition, if $I \subset J$ with $|J| \le 8|I|$, then

$$\left(\frac{|I|}{|J|}\right)^2 w(J) \le w(J) \le Cw(I) ,$$

so we must prove such a condition for $|I|$ small relative to $|J|$.

---

[1] Following the proof of Reverse Hölder in [2], we obtain the inequality $\frac{1}{\delta} - \frac{2(1+C)^{1+\delta}}{1+\delta} J(w^{1+\delta}) \le \frac{1}{\delta} J(w)^{1+\delta}$ for all $\delta > 0$ and $J \subset I$. Taking δ sufficiently close to zero so the left constant is positive gives the inequality, and that constant is independent of I.

We will study $I = [0,h]$ and $J = [0,\delta]$, $\delta \geq 4h$, and show

$$\left(\frac{h}{\delta}\right)^2 w(J) \leq Cw(I) . \tag{7}$$

A similar argument applies when I is a right-end segment of J. In general, if $I = [-h,h]$ and $J = [-\beta,\gamma]$, then

$$\left(\frac{h}{\beta}\right)^2 w([-\beta,0]) \leq Cw(I) , \quad \text{and}$$

$$\left(\frac{h}{\gamma}\right)^2 w([0,\gamma]) \leq Cw(I) ,$$

so that $h^2 w(J) \leq C(\beta^2+\gamma^2)w(I) \leq C(\beta+\gamma)^2 w(I)$ and the weak α_2 condition follows.

For (7), put $R = [\delta,2\delta]$ and set $f = w\chi_R$. Then

$$\text{ess}_I\inf f^* \leq f^*(0) = \sup_{r \in R} \frac{1}{r}\int_\delta^r w .$$

Since the integral is continuous, the supremum is attained, and there exists an $r \in R$ for which

$$f^*(0) = \frac{1}{r}\int_\delta^r w .$$

Now for $x \in I$,

$$f^*(x) \geq \frac{1}{r-x}\int_\delta^r w = \frac{r}{r-x} f^*(0) .$$

Thus

$$f^*(x) - \text{ess}_I\inf f^* \geq \frac{x}{r-x} f^*(0) \geq \frac{x}{2\delta} f^*(0) .$$

Hence,

$$I(f^*) - \text{ess}_I\inf f^* \geq \frac{h}{4\delta} f^*(0) \geq \frac{h}{8\delta^2} w(R) .$$

Since $\|f\|_{*,w} = 2$, we conclude that

$$\frac{h}{8\delta^2} w(R) \leq 2CI(w) .$$

On the other hand,

$$w(J) \leq w(J \cup R) \leq Cw(R) ,$$

so we obtain

$$\frac{h^2}{\delta^2} w(J) \leq Cw(I) ,$$

which is (7) .

References

1. C. Bennett, R.A. Devore, and R. Sharpley, Weak-L^∞ and BMO, Ann. of Math. 113(1981), 601-611.

2. R.R. Coifman and C. Fefferman, Weighted Norm Inequalities for Maximal Functions and Singular Integrals, Studia Math. 51(1974), 241-250.

3. R. A. Kerman and A. Torchinsky, Integral Inequalities with weights for the Hardy Maximal Function, To appear in Studia Math.

4. B. Muckenhoupt, The Equivalence of Two Conditions for Weight Functions, Studia Math. 49(1973/74), 101-106.

Siena College
Loudonville, New York

WEIGHTED HARDY SPACES AND THE LAPLACE TRANSFORM

by

J. J. Benedetto* and H. P. Heinig*

The classical Hardy spaces H^p, $1 \le p \le \infty$, are defined to consist of those functions f, holomorphic in the right half plane, with the property that $\mu_p(f;x)$ is uniformly bounded for $x > 0$, where

$$\mu_p(f;x) = \begin{cases} \left(\frac{1}{2\pi} \int_{-\infty}^{\infty} |f(x+iy)|^p dy\right)^{1/p}, & 1 \le p < \infty \\ \sup_{y \in \mathbb{R}} |f(x+iy)|, & p = \infty. \end{cases}$$

The relationship between these spaces and the Laplace transform has been known since the 1930's [12; 13; 15; 11; 7]. For example, the following representation theorem was formulated by Doetsch in 1936 [7], cf., Remark 2a in Section 3. *If* $F \in L^p(0,\infty)$, $1 < p \le 2$, *and* $f = L(F)$ *is the Laplace transform of* F, *that is*

$$L(F)(z) = \int_0^{\infty} e^{-zt} F(t)dt, \quad \text{Re } z > 0,$$

then $f \in H^{p'}$, *where* $\frac{1}{p} + \frac{1}{p'} = 1$; *and, conversely, if* $f \in H^p$, $1 < p \le 2$, *then there exists a function* $F \in L^{p'}(0,\infty)$ *such that* $f = L(F)$. In the case $p = 2$ this representation is the Paley-Wiener theorem [20, Theorem V] (1934). Rooney [21;22;23] has initiated and developed a program to extend this Laplace transform representation to some weighted H^p and L^p spaces. The proofs

---

* The authors were supported by NSF and NSERC grants, respectively.

of all these results are intimately related to generalizations of Plancherel's theorem.

Recently, several authors independently obtained weighted norm inequalities for the Fourier transform where the weights satisfy certain integrability conditions [10; 16; 18]. These results include the generalizations of Plancherel's theorem due to Titchmarsh [24, Theorem 74] (1924), Rooney [23], and others. In light of our previous remarks, it is natural to expect that Doetsch's Laplace transform representation theory and Rooney's weighted theory can be extended in the direction implied by these generalizations. It is the purpose of this paper to formulate and prove such an extension which contains all previous results.

We begin Section 1 by defining the weighted L^p spaces used throughout the paper. Theorem 1 is a general Plancherel theorem which provides the Fourier analysis of these spaces, cf., Remark 1d in Section 3. The weighted norm inequalities in this result are proven best possible in Theorem 2. These and similar inequalities are verified in [10; 16; 18]; the present proof is simpler than that in [10] and different than those in [16;18]. In Section 2 we define weighted H^p spaces corresponding to the weight condition introduced in Section 1. Then we use Theorem 1 to prove our main results (Theorem 3 and Theorem 4) which establish a Laplace transform representation theory relating these weighted H^p and L^p spaces. This weighted representation theory has several important applications, including a clarification of the uncertainty principle, and there is at least one major refinement to be made, viz., a weighted Laplace transform representation theory by means of Widder-Post inversion. In Section 3 we list a few pertinent examples and relevant historical remarks in order to put our definitions, results, and proofs in perspective.

Notation

The conjugate index of $p \in [1,\infty]$ is denoted by p' and defined by the rule $1/p + 1/p' = 1$, where $p' = \infty$, resp., 1, when $p = 1$, resp., ∞. Constants are denoted by C, and it may take different values at different occurrences. Also, if $S \subseteq \mathbb{R}$, the real line, is a Lebesgue measurable set then its Lebesgue measure is denoted by $|S|$; $L^1_{loc}(S)$ is the space of locally integrable functions defined on S.

For a given function (weight) $w \in L^1_{loc}$ and index $r \in [1,\infty)$, the weighted L^r space L^r_w is the set of locally integrable functions F on $\mathbb{R}$ for which

$$\|F\|_{r,w} = \|Fw\|_r = \left(\int_{-\infty}^{\infty} |w(t)F(t)|^r dt\right)^{1/r} < \infty.$$

The Fourier transform $\hat{F}$ of $F \in L^1$ is defined as

$$\hat{F}(y) = \int_{-\infty}^{\infty} e^{-iyt}F(t)dt.$$

1. Fourier transforms for weighted spaces

In order to define the Fourier transform in weighted L^p spaces and to establish boundedness of the Fourier operator we need the following known generalization of Hardy's inequality.

<u>Lemma 1</u> ([4]). Suppose that u and v are non-negative locally integrable functions defined on $(0,\infty)$ and that $1 \le p \le q \le \infty$.

a. There is $C > 0$ such that for every non-negative function $G \in L^1_{loc}(0,\infty)$,

$$(1.1) \qquad \left(\int_0^\infty \left(u(t) \int_0^t G(s)ds\right)^q dt\right)^{1/q} \le C\left(\int_0^\infty (v(t)G(t))^p dt\right)^{1/p}$$

if and only if

$$(1.2) \qquad \sup_{s>0}\left(\int_s^\infty u(t)^q dt\right)^{1/q}\left(\int_0^s v(t)^{-p'}dt\right)^{1/p'} < \infty.$$

b. Dually, there is $C > 0$ such that for every non-negative function $G \in L^1_{loc}(0,\infty)$,

$$(1.3) \qquad \left(\int_0^\infty \left(u(t)\int_t^\infty G(s)ds\right)^q dt\right)^{1/q} \le C\left(\int_0^\infty (v(t)G(t))^p dt\right)^{1/p}$$

if and only if

$$(1.4) \qquad \sup_{s>0}\left(\int_0^s u(t)^q dt\right)^{1/q}\left(\int_s^\infty v(t)^{-p'}dt\right)^{1/p'} < \infty.$$

In (1.1) - (1.4) the usual modifications are made for the cases $p = 1$ or $q = \infty$. With regard to Hardy's inequality we also mention the papers of Artola, Muckenhoupt, Talenti, and Tomasselli who obtained weighted Hardy estimates for $p = q$ in the late 1960's and early 1970's.

<u>Definition 1</u>. Let F be a Lebesgue measurable function on $\mathbb{R}$.

a. The <u>equimeasurable decreasing rearrangement</u> of $|F|$ is denoted by F^* and defined as $F^*(t) = \inf\{y>0: |\{x: |F(x)|>y\}| \le t\}$.

b. The <u>radially decreasing rearrangement</u> of $|F|$ is denoted by $F^{\circledast}$ and defined as $F^{\circledast}(t) = F^{*}(2t)$ if $t > 0$ and as $F^{\circledast}(t) = F^{\circledast}(-t)$ if $t < 0$. Observe that

$$\int_{-\infty}^{\infty} |F(t)|dt = 2\int_{0}^{\infty} F^{\circledast}(t)dt = \int_{-\infty}^{\infty} F^{\circledast}(|t|)dt.$$

We refer to [2] for properties of decreasing rearrangements of functions.

<u>Definition 2</u> a. Suppose that u and v are non-negative even functions on $\mathbb{R}$ such that u is decreasing and v is increasing for $t > 0$. The notation $(u,v) \in F(p,q)$, $1 < p \leq q < \infty$, signifies that

$$(1.5) \qquad \sup_{s>0}\left(\int_{0}^{1/(2s)} u(t)^q dt\right)^{1/q}\left(\int_{0}^{s/2} v(t)^{-p'}dt\right)^{1/p'} < \infty,$$

cf., (1.2) and (1.4).

b. In case $p = 1$ and $1 \leq q < \infty$ the notation $(u,v) \in F(1,q)$ signifies that

$$\sup_{s>0}\left[\left(\int_{0}^{1/(2s)} u(t)^q dt\right)^{1/q}\left(\sup_{t\in(0,s/2]} (1/v(t))\right)\right] < \infty,$$

and, in particular, $v(0\pm) > 0$. In case $p = q = 2$ the notation $(u,v) \in F(2,2)$ signifies not only (1.5) but also

$$(1.6) \qquad \sup_{s>0}\left(\int_{1/(2s)}^{\infty} u(t)^2 t^{-1}dt\right)^{1/2}\left(\int_{s/2}^{\infty} v(t)^{-2}t^{-1}dt\right)^{1/2} < \infty,$$

cf., <u>Remark 1d</u> (<u>Section 3</u>).

c. Note that for $1 < p \leq q < \infty$ one has $(u,v) \in F(p,q)$ if and only if $(1/v,1/u) \in F(q',p')$.

The following result is elementary in the case $p = 1$ and it is instructive to see the idea of proof unencumbered by Hardy's inequality and decreasing rearrangements, e.g., Proposition 1 (Section 3).

Theorem 1. Suppose that $(u,v) \in F(p,q)$, $1 \le p \le q < \infty$, and $F \in L^p_v$.

a. If $\lim_{n\to\infty}\|F_n - F\|_{p,v} = 0$ for a sequence $\{F_n\}$ of simple functions then $\{\hat{F}_n\}$ converges in L^q_u to a function $\hat{F} \in L^q_u$. $\hat{F}$ is independent of the sequence $\{F_n\}$ and it is called the Fourier transform of F.

b. $\hat{F}$ has the pointwise representation

$$\hat{F}(y) = \frac{d}{dy}\int_{-\infty}^{\infty} \frac{1-e^{-iyt}}{it} F(t)dt \quad \text{a.e.}$$

c. There is $C > 0$ such that

$$\forall F \in L^p_v, \quad \|\hat{F}\|_{q,u} \le C\|F\|_{p,v}. \tag{1.7}$$

d. If $G \in L^{q'}_{1/u}$ and $q > 1$ then Parseval's formula

$$\int_{-\infty}^{\infty} \hat{F}(y)G(y)dy = \int_{-\infty}^{\infty} F(t)\hat{G}(t)dt \tag{1.8}$$

is valid.

Proof. i. We prove the result in the following way. First, we verify (1.7) for simple functions F; this is the subject matter of parts ii-viii. An essential estimate in terms of rearrangements is formulated in part ii, and parts iii-vii are devoted to its proof. In part viii we use this estimate to complete (1.7) for simple functions. It is then straightforward to prove $\underline{c}$ for all $F \in L_v^p$. This, as well as the verification of $\underline{a}$, is accomplished in part $\underline{ix}$. We conclude with routine calculations to obtain $\underline{d}$ and $\underline{b}$ in parts $\underline{x}$ and $\underline{xi}$ of the proof.

ii. Since the Fourier transform maps L^1 into L^∞ and L^2 onto L^2, we can invoke a result of Calderón [5] which implies the existence of a constant $C > 0$ such that

$$\hat{F}^*(y) \le C\left(\int_0^{1/y} F^*(s)ds + y^{-1/2}\int_{1/y}^{\infty} s^{-1/2}F^*(s)ds\right),\ y > 0 \tag{1.9}$$

for all simple functions F. Replacing y by $2y$ and s by $2s$, (1.9) can be written in terms of radially decreasing rearrangements as

$$\hat{F}^{\circledast}(y) \le 2C\left(\int_0^{1/(4y)} F^{\circledast}(s)ds + y^{-1/2}\int_{1/(4y)}^{\infty} s^{-1/2}F^{\circledast}(s)ds\right),\ y > 0.$$

An application of Minkowski's inequality yields the inequality,

$$\begin{aligned}\left(\int_0^\infty (u(y)\hat{F}^{\circledast}(y))^q dy\right)^{1/q} &\le C\Big[\left(\int_0^\infty \left(u(y)\int_0^{1/(4y)} F^{\circledast}(s)ds\right)^q dy\right)^{1/q} \\ &+ \left(\int_0^\infty \left(u(y)y^{-1/2}\int_{1/(4y)}^\infty s^{-1/2}F^{\circledast}(s)ds\right)^q dy\right)^{1/q}\Big] \equiv C(J_0 + J_1),\end{aligned} \tag{1.10}$$

for $q \in [1,\infty)$.

We shall now estimate J_0 and J_1 as defined in (1.10).

iii. We first prove

$$J_0 \le C\left(\int_0^\infty (v(t)F^{\circledast}(t))^p dt\right)^{1/p},\quad 1 \le p \le q < \infty \tag{1.11}$$

for every simple function F, where C is independent of F.

We begin by verifying the equivalence of (1.11) and

$$\text{(1.12)} \qquad \sup_{s>0}\Big(\int_0^{1/(4s)} u(t)^q dt\Big)^{1/q}\Big(\int_0^s v(t)^{-p'}dt\Big)^{1/p'} < \infty,$$

where the usual adjustment is made for $p = 1$. In fact, if we let $t = \frac{1}{4y}$ in the definition of J_0 we obtain

$$J_0^q = \frac{1}{4}\int_0^\infty \Big(u\big(\tfrac{1}{4t}\big)t^{-\frac{2}{q}}\int_0^t F^{\circledast}(s)ds\Big)^q dt,$$

and so part a of Lemma 1 allows us to assert (1.11) if and only if

$$\text{(1.13)} \qquad \sup_{s>0}\Big(\int_s^\infty u\big(\tfrac{1}{4t}\big)^q\, t^{-2}dt\Big)^{1/q}\Big(\int_0^s v(t)^{-p'}dt\Big)^{1/p'} < \infty.$$

The change of variable $y = \frac{1}{4t}$ in (1.13) yields the equivalence of (1.12) and (1.13).

The hypothesis $(u,v) \in F(p,q)$ is obviously the same as (1.12) and so we have verified (1.11).

iv. We now note that the estimate,

$$\text{(1.14)} \qquad J_1 \le C\Big(\int_0^\infty (v(t)F^{\circledast}(t))^p dt\Big)^{1/p}, \quad 1 \le p \le q < \infty,$$

is equivalent to the condition,

$$\text{(1.15)} \qquad \sup_{s>0}\Big(\int_{1/(4s)}^\infty u(t)^q t^{-q/2}dt\Big)^{1/q}\Big(\int_s^\infty (v(t)t^{1/2})^{-p'}dt\Big)^{1/p'} < \infty,$$

where F is simple and C is independent of F.

To see this, we first let $t = \frac{1}{4y}$ to calculate

$$J_1^q = \frac{1}{4}\int_0^\infty \Big(u\big(\tfrac{1}{4t}\big)\big(\tfrac{1}{4t}\big)^{-1/2}\int_t^\infty s^{-1/2}F^{\circledast}(s)ds\Big)^{1/q} t^{-2}dt$$

$$= 2^{q-2}\int_0^\infty \Big(u\big(\tfrac{1}{4t}\big)t^{\frac{1}{2}-\frac{2}{q}}\int_t^\infty s^{-1/2}F^{\circledast}(s)ds\Big)^q dt.$$

Then, setting $U(t) = u(\frac{1}{4t})t^{\frac{1}{2}-\frac{2}{q}}$ and $V(t) = t^{1/2}v(t)$, an application of part b of Lemma 1 yields the equivalence of the estimate,

$$(1.16) \quad \left(\int_0^\infty (U(t) \int_t^\infty s^{-1/2}F^{\circledast}(s)ds)^q dt\right)^{1/q} \le C\left(\int_0^\infty (V(t)t^{-1/2}F^{\circledast}(t))^p\right)^{1/p},$$

and the condition,

$$(1.17) \qquad \sup_{s>0}\left(\int_0^s U(t)^q dt\right)^{1/q}\left(\int_s^\infty V(y)^{-p'}dy\right)^{1/p'} < \infty,$$

where C is independent of F. Letting $t = \frac{1}{4y}$, (1.17) becomes

$$\sup_{s>0}\left(\frac{1}{4}\int_{\frac{1}{4s}}^\infty u(y)^q\left(\frac{1}{4y}\right)^{\frac{q}{2}-2}y^{-2}dy\right)\left(\int_s^\infty (v(y)y^{1/2})^{-p'}dy\right)^{1/p'} < \infty,$$

and, clearly, this is (1.15). The definitions of U and V give the equivalence of (1.14) and (1.16). Combining all these facts yields the equivalence of (1.14) and (1.15).

In parts v-vii we shall show that (1.15) is a consequence of (1.5), or (1.6) in case $p = q = 2$, for various values of p and q.

v. Consider the case $1 < p < 2$ and $p \le q < \infty$. Since u is decreasing, (1.5) implies

$$\forall s > 0, \quad u\left(\frac{1}{2s}\right)\left(\frac{1}{2s}\right)^{1/q} \le C\left(\int_0^{s/2} v(t)^{-p'}dt\right)^{-1/p'},$$

where C is independent of $s > 0$. Letting $y = \frac{1}{2s}$, this inequality becomes

$$\forall y > 0, \quad u(y) \le Cy^{-1/q}\left(\int_0^{1/(4y)} v(t)^{-p'}dt\right)^{-1/p'};$$

and substituting this bound for u into (1.15) we see that (1.15) is bounded by

$$C \sup_{s>0}\left(\int_{1/(4s)}^{\infty} y^{-(1+\frac{q}{2})}\left(\int_0^{1/(4y)} v(t)^{-p'}dt\right)^{-\frac{q}{p'}}dy\right)^{\frac{1}{q}}\left(\int_s^{\infty}(v(y)y^{\frac{1}{2}})^{-p'}dy\right)^{\frac{1}{p'}}$$

$$\le C \sup_{s>0} v(s)^{-1}s^{-\frac{1}{2}+\frac{1}{p'}}\left(\int_{1/(4s)}^{\infty} y^{-(1+\frac{q}{2})}y^{\frac{q}{p'}}v(\tfrac{1}{4y})^q dy\right)^{1/q}$$

$$\le C \sup_{s>0} v(s)^{-1}s^{-\frac{1}{2}+\frac{1}{p'}}v(s)s^{-(\frac{q}{p'}-\frac{q}{2})\frac{1}{q}} = C,$$

where the first inequality depends on the facts that v increases and $p' > 2$, and where the second inequality depends only on the fact v increases and hence $v(\frac{1}{4y})$ is a decreasing function of y.

Thus, for $1 < p < 2$, we have verified (1.15) by means of (1.5). The calculation for $p = 1$ is made in a similar manner.

vi. Next, we consider the case $2 < q < \infty$. Since v is increasing, (1.5) implies

$$\forall s > 0, \quad v(\tfrac{s}{2})^{-1}(\tfrac{s}{2})^{1/p'} \le C\left(\int_0^{1/(2s)} u(t)^q dt\right)^{-1/q},$$

where C is independent of $s > 0$. Letting $y = s/2$ this inequality becomes

$$\forall s > 0, \quad v(y)^{-1} \le Cy^{-1/p'}\left(\int_0^{1/(4y)} u(t)^q dt\right)^{-1/q};$$

and substituting this bound for v^{-1} into (1.15) we see that (1.15) is bounded by

$$C \sup_{s>0}\left(\int_{1/(4s)}^{\infty} u(y)^q y^{-\frac{q}{2}} dy\right)^{1/q}\left(\int_s^{\infty} y^{-(1+\frac{p'}{2})}\left(\int_0^{1/(4y)} u(t)^q dt\right)^{-\frac{p'}{q}} dy\right)^{1/p'}$$

$$\leq C \sup_{s>0} u(\tfrac{1}{4s}) s^{\frac{1}{2}-\frac{1}{q}}\left(\int_s^{\infty} y^{-(1+\frac{p'}{2})} y^{\frac{p'}{q}} u(\tfrac{1}{4y})^{-p'} dy\right)^{1/p'}$$

$$\leq C \sup_{s>0} u(\tfrac{1}{4s}) s^{\frac{1}{2}-\frac{1}{q}} u(\tfrac{1}{4s})^{-1} s^{-(\frac{p'}{2}-\frac{p'}{q})\frac{1}{p'}} = C,$$

where the first inequality depends on the facts that u decreases and $q > 2$, and where the second inequality depends only on the fact that u decreases and hence $u(\frac{1}{4y})^{-1}$ is a decreasing function of y.

Thus, for $q > 2$, we have verified (1.15) by means of (1.5).

vii. Finally, we consider $p = q = 2$. In this case, (1.15) is the same as (1.6) and so the result follows immediately from the definition of $(u,v) \in F(2,2)$.

viii. Combining the results of parts ii-vii, we have proved

$$(1.18) \qquad \left(\int_0^{\infty}(u(t)\hat{F}^{\theta}(t))^q dt\right)^{1/q} \leq C\left(\int_0^{\infty}(v(t)F^{\theta}(t))^p dt\right)^{1/p}$$

for all simple functions F, where C is independent of F and $1 \leq p \leq q < \infty$. In fact, the left hand side of (1.18) is bounded by $J_0 + J_1$ from (1.10), J_0 is bounded by the right hand side of (1.18) from (1.11), and J_1 is bounded by the right hand side of (1.18) from the equivalence of (1.14) and (1.15) and the verification of (1.15) for various values of p and q in parts v - vii.

The norm inequality (1.7) for indices $1 \leq p \leq q < \infty$ and simple functions F is a consequence of (1.18) and the following calculation :

$$\|\hat{F}\|_{q,u} \leq C(\int_{-\infty}^{\infty} |u(t)\hat{F}^{\circledast}(|t|)|^q dt)^{1/q}$$

$$\leq C(\int_0^{\infty} (u(t)\hat{F}^{\circledast}(t))^q dt)^{1/q} \leq C(\int_0^{\infty} (v(t)F^{\circledast}(t))^p dt)^{1/p}$$

$$= C(\int_{-\infty}^{\infty} |v(|t|)F^*(|t|)|^p dt)^{1/p} \leq C\|F\|_{p,v},$$

where we have applied the continuous analogue of [8, Theorem 368] to obtain the first and last inequalities.

ix. L_v^p is a Banach space and the simple functions are dense in L_v^p. Since (1.7) is valid for simple functions, the mapping $F \mapsto \hat{F}$ is a continuous transformation from a dense subspace of L_v^p to L_u^q. Consequently, the transformation has a unique continuous extension to all of L_v^p. This extension is also denoted by $\hat{F}$ and, hence, the proof of part c as well as part a is complete.

x. If F and G are simple functions then Parseval's formula (1.8) is true by an easy calculation. Also, if $F \in L_v^p$ and $G \in L_{1/u}^{q'}$ then we compute

$$(1.19) \quad |\int_{-\infty}^{\infty} \hat{F}(y)G(y)dy| \leq \|\hat{F}\|_{q,u}\|G\|_{q',1/u} \leq C\|F\|_{p,v}\|G\|_{q',1/u}$$

and

$$(1.20) \quad |\int_{-\infty}^{\infty} F(t)\hat{G}(t)dt| \leq \|F\|_{p,v}\|\hat{G}\|_{p',1/v} \leq C\|F\|_{p,v}\|G\|_{q',1/u},$$

where both sets of inequalities use part c and where the second set makes use of Definition 2c, i.e., $(1/v,1/u) \in F(q',p')$.

Given $F \in L^p_v$ and $G \in L^{q'}_{1/u}$, we choose sequences $\{F_n\}$ and $\{G_n\}$ of simple functions for which $\lim\|F_n - F\|_{p,v} = \lim\|G_n - G\|_{q',1/u} = 0$. It is at this point that the restriction $q > 1$ is required. Then, (1.8) is a consequence of Parseval's formula for simple functions, the estimates (1.19) and (1.20), and the identity $\hat{F}G - F\hat{G} = \hat{F}(G-G_n) + G_n(\hat{F}-\hat{F}_n) + (\hat{F}_n G_n - F_n \hat{G}_n) + F_n(\hat{G}_n - \hat{G}) + \hat{G}(F_n - F)$.

xi. It remains to prove part b. Define

$$G_x(y) = \begin{cases} 1 & \text{if} \quad 0 < y < x, \\ -1 & \text{if} \quad x < y < 0, \\ 0 & \text{otherwise,} \end{cases}$$

so that we have $\hat{G}_x(t) = (1-e^{-ixt})/(it)$. We also make the estimate,

$$\|G_x\|^{q'}_{q',1/u} = \int_0^x u(y)^{-q'}dy \le xu(x)^{-q'} < \infty,$$

which follows since $u^{-q'}$ is increasing, where we have chosen any $q > p$ (and thus $q \ne 1$) in order to implement part d in the following way:

$$\int_0^x \hat{F}(y)dy = \int_{-\infty}^{\infty} \hat{F}(y)G_x(y)dy = \int_{-\infty}^{\infty} F(t)\hat{G}_x(t)dt$$

$$= \int_{-\infty}^{\infty} \frac{1-e^{-ixt}}{it} F(t)dt.$$

The proof of part b is complete.

qed

At the expense of a more complicated proof, condition (1.6) for the $p = q = 2$ case of Theorem 1 can be replaced by the condition that $t^{\varepsilon}u(t)$ is decreasing on $(0,\infty)$ for each $\varepsilon > 0$.

We now prove that Theorem 1 is sharp.

Theorem 2 ([10]). Suppose that u and v are non-negative even functions on $\mathbb{R}$ such that u is decreasing and v is increasing for $t > 0$. If (1.7) is valid for $1 < p,q < \infty$ then (1.5) holds.

Proof. Define the function F as

$$F(t) = \begin{cases} v(|t|)^{-p'} & \text{if } 0 < |t| \le s/2, \\ 0 & \text{otherwise.} \end{cases}$$

Then F and its Fourier transform $\hat{F}$ are radial. In fact, we have

$$\hat{F}(y) = 2\int_0^{s/2} \cos(ty)v(t)^{-p'}dt.$$

Because (1.7) can be written as

$$\left(\int_0^{\infty} |u(t)\hat{F}(t)|^q dt\right)^{1/q} \le C\left(\int_0^{s/2} v(t)^{-p'}dt\right)^{1/p},$$

we obtain the inequality

$$\left(\int_0^{2/s} u(y)^q \left|\int_0^{s/2} \cos(ty)v(t)^{-p'}dt\right|^q dy\right)^{1/q} \le C\left(\int_0^{s/2} v(t)^{-p'}dt\right)^{1/p}.$$

This inequality in turn implies the condition,

$$(1.21) \qquad \left(\int_0^{2/s} u(y)^q dy\right)^{1/q}\left(\int_0^{s/2} v(t)^{-p'}dt\right)^{1/p'} \le C,$$

since $\cos(ty) \ge 1/2$ for the range of t and y considered.

Also, the fact that u is non-negative yields

$$\int_0^{2/s} u(y)^q dy \geq \int_0^{1/(2s)} u(y)^q dy.$$

The result follows by substituting this last inequality into the left hand side of (1.21).

qed

2. Laplace transforms for weighted spaces

We now define weighted Hardy spaces, and, using Theorem 1, establish the Laplace transform representation of these spaces by means of weighted L^p spaces; this is accomplished in Theorem 3 and Theorem 4.

Definition 3. Suppose that $0 < p < \infty$, $z = x + iy$, and that w is a non-negative radial function on $\mathbb{R}^2\backslash\{0\}$ which as a function of $|z|$ is continuous a.e. on $(0,\infty)$. The function f belongs to the weighted Hardy space H_w^p if f is holomorphic in the right half plane $\operatorname{Re} z > 0$, the limit $\lim_{x\to 0+} f(x+iy)$ exists a.e., and $\mu_{p,w}(f;x)$ is uniformly bounded for $x > 0$, where

$$\mu_{p,w}(f;x) = \left(\frac{1}{2\pi}\int_{-\infty}^{\infty} |w(x+iy)f(x+iy)|^p dy\right)^{1/p}$$

Lemma 2. If $f \in H_w^p$, $0 < p < \infty$, and $f(iy) \equiv \lim_{x\to 0+} f(x+iy)$ then

$$w(y)f(iy) \in L^p.$$

Proof. The continuity hypothesis on w implies that

$$\lim_{x\to 0+} |w(x+iy)f(x+iy)| = |w(y)f(iy)| \quad \text{a.e.};$$

and hence we can apply Fatou's lemma and the definition of H_w^p to obtain

$$\int_{-\infty}^{\infty} |w(y)f(iy)|^p dy \le \underline{\lim}_{x \to 0+} \int_{-\infty}^{\infty} |w(x+iy)f(x+iy)|^p dy < \infty.$$

qed

The following lemma is proved along the lines of the result in [24, p. 125].

Lemma 3. Suppose that v is a non-negative, increasing function defined on $(0,\infty)$ which is extended radially to $\mathbb{R}^2\setminus\{0\}$, and suppose that f is holomorphic in the right half plane $\mathrm{Re}\, z > 0$. For $1 \le p < \infty$ and $0 < x_1 < x_2$, assume

$$\sup_{x \in [x_1,x_2]} \int_{-\infty}^{\infty} |v(x+iy)f(x+iy)|^p dy < \infty. \tag{2.1}$$

If $0 < \delta < (x_2-x_1)/2$ then one obtains

$$f(x+iy) = o(1/v(y/2)), \quad y \to \pm\infty$$

uniformly for $x \in [x_1+\delta, x_2-\delta]$.

Proof. For $x_1 + \delta \le x \le x_2 - \delta$, any $y, z = x + iy$, and $0 < \rho < \delta$ the analyticity of f yields

$$f(z) = \frac{1}{2\pi} \int_0^{2\pi} f(z+\rho e^{i\theta})d\theta$$

and so

$$\frac{\delta^2}{2} v(y/2)|f(z)| = \left|\frac{1}{2\pi} \int_0^{\delta} \rho d\rho \int_0^{2\pi} v(y/2)f(z+\rho e^{i\theta})d\theta\right|.$$

Consequently, using Hölder's inequality for $p > 1$, we obtain

(2.2) $$\frac{\delta^2}{2} v(y/2)|f(z)| \le C\left(\int_0^\delta\int_0^{2\pi} |v(y/2)f(z+\rho e^{i\theta})|^p \rho\, d\theta\, d\rho\right)^{1/p}, 1 \le p < \infty$$

where $C = C(\delta,p') = (\int_0^\delta\int_0^{2\pi} \rho\, d\theta\, d\rho)^{1/p'}/(2\pi)$ for $p > 1$ and $C = C(\delta,\infty) = 1/(2\pi)$ for $p = 1$.

We now take $|y| \ge x_2 - x_1$. Thus, we have $1/2 \ge \rho/|y|$ and $|y||1 + \frac{\rho \sin\theta}{y}| \ge |y|/2$, and hence we compute

$$|z+\rho e^{i\theta}| = [(x+\rho\cos\theta)^2 + (y+\rho\sin\theta)^2]^{1/2}$$

$$= |y|\, [(\frac{x+\rho\cos\theta}{y})^2 + (1 + \frac{\rho\sin\theta}{y})^2]^{1/2} \ge |y|/2$$

for all $|y| \ge x_2 - x_1$. Since v is radial and increasing on $(0,\infty)$ we use this inequality to show that the right hand side of (2.2) is dominated by

$$C\left(\int_0^\delta\int_0^{2\pi} |v(z+\rho e^{i\theta})f(z+\rho e^{i\theta})|^p \rho\, d\theta\, d\rho\right)^{1/p}.$$

The disc with center $z = x + iy$ and radius δ is contained in the rectangle $[x_1,x_2] \times [y-\delta,y+\delta]$ and so this last integral is bounded by

(2.3) $$C\left(\int_{x_1}^{x_2} d\tau \int_{y-\delta}^{y+\delta} |v(\tau+i\gamma)f(\tau+i\gamma)|^p d\gamma\right)^{1/p}.$$

Because of (2.1) the inner integral of (2.3) is bounded by a constant C which is independent of $\tau \in [x_1,x_2]$. Since C is integrable on $[-x_1,x_2]$ we can apply the dominated convergence theorem to (2.3) taking the limit as $y \to \pm\infty$. The limit is clearly 0, so that we obtain our result by combining this limit with the left hand side of (2.2) which is bounded by (2.3).

<u>qed</u>

Lemma 2 and Lemma 3 are used to prove the following result.

Lemma 4. Suppose that $f \in H_v^p$, $1 < p < \infty$, where the radial function v is increasing on $(0,\infty)$ as a function of $|z|$, $z = x + iy$. If $(1/v(|z|))^{p'} \in L^1_{loc}$ then

$$\frac{1}{2\pi}\int_{-\infty}^{\infty} \frac{f(\tau+i\gamma)}{z-(\tau+i\gamma)}\,d\gamma = \begin{cases} f(z) & \text{if } 0 \le \tau < x, \\ 0 & \text{if } x < \tau \text{ and } \tau \ge 0, \end{cases}$$

for each fixed $z = x + iy$.

Proof. i. We first consider the case $0 < \tau < x$.

Choosing R and ρ for which $x < \rho$ and $|y| < R$ we obtain by analyticity that

$$f(z) = \frac{1}{2\pi i}\int_C \frac{f(\zeta)}{z-\zeta}\,d\zeta,$$

where C is the clockwise oriented rectangle determined by the vertices $\tau \pm iR$, $\rho \pm iR$.

Along the upper side of the rectangle the above integral is

$$-\frac{1}{2\pi i}\int_\tau^\rho \frac{f(\xi+iR)d\xi}{z-(\xi+iR)}.$$

Take $0 < \varepsilon < 1$ and apply Lemma 3. Thus, there is $R_0 > |y|$ such that $|f(\xi+iR)| < \varepsilon(1/v(R/2))$ for all $R \ge R_0$ uniformly in $\xi \in [\tau,\rho]$. Consequently, we compute

$$\left|\frac{1}{2\pi i}\int_\tau^\rho \frac{f(\xi+iR)d\xi}{z-(\xi+iR)}\right| \le \frac{\varepsilon}{2\pi v(R_0/2)}\int_\tau^\rho \frac{d\xi}{((x-\xi)^2+(y-R)^2)^{1/2}}$$

$$\le \frac{\varepsilon(\rho-\tau)}{2\pi v(R_0/2)|y-R|}\ ;$$

and hence we have

$$\lim_{R\to\infty} \frac{1}{2\pi i} \int_{\tau}^{\rho} \frac{f(\xi+iR)d\xi}{z-(\xi+iR)} = 0.$$

A similar calculation and answer is obtained for the lower side of the rectangle. Therefore, we can write $f(z)$ as

$$f(z) = \frac{1}{2\pi} \int_{-\infty}^{\infty} \frac{f(\tau+i\gamma)d\gamma}{z-(\tau+i\gamma)} - \frac{1}{2\pi} \int_{-\infty}^{\infty} \frac{f(\rho+i\gamma)d\gamma}{z-(\rho+i\gamma)} .$$

Our final task for this part of the proof is to verify

$$\text{(2.4)} \qquad \lim_{\rho\to\infty} \frac{1}{2\pi} \int_{-\infty}^{\infty} \frac{f(\rho+i\gamma)d\gamma}{z-(\rho+i\gamma)} = 0.$$

By Hölder's inequality and the hypotheses, $f \in H_v^p$ and v increasing, we can make the estimate

$$\left|\frac{1}{2\pi} \int_{-\infty}^{\infty} \frac{f(\rho+i\gamma)d\gamma}{z-(\rho+i\gamma)}\right| = \frac{1}{2\pi}\left| \int_{-\infty}^{\infty} \frac{v(\rho+i\gamma)f(\rho+i\gamma)v(\rho+i\gamma)^{-1}}{z-(\rho+i\gamma)} d\gamma\right|$$

$$\le \mu_{p,v}(f;\rho)\left(\frac{1}{2\pi} \int_{-\infty}^{\infty} \frac{v(\rho+i\gamma)^{-p'}d\gamma}{((x-\rho)^2+(y-\gamma)^2)^{p'/2}}\right)^{1/p'}$$

$$\le Cv(\rho_0)^{-1}\left(\int_{-\infty}^{\infty} \frac{d\gamma}{((x-\rho)^2+(y-\gamma)^2)^{p'/2}}\right)^{1/p'},$$

for all $\rho > \rho_0 > x$, where ρ_0 is fixed. For any such ρ we make the substitution $\lambda = (\gamma-y)/(\rho-x)$, noting that $\rho - x > 0$. Then the last term is dominated by

$$Cv(p_0)^{-1}\left(\int_{-\infty}^{\infty} \frac{(\rho-x)d\lambda}{(\rho-x)^{p'}(1+\lambda^2)^{p'/2}}\right)^{1/p'} \le Cv(p_0)^{-1}(\rho-x)^{\frac{1}{p'}-1}$$

where the right hand side follows since $p' > 1$. Also, the fact

$p' > 1$ allows us to assert that the right hand side tends to 0 as $\rho \to \infty$. Therefore, we obtain (2.4).

ii. We next check the case $\tau > x$ and $\tau > 0$. Taking $\rho > \tau$, $R > |y|$, and the contour C as in part i we have

$$\frac{1}{2\pi i}\int_C \frac{f(\zeta)}{z-\zeta}\,d\zeta = 0$$

by analyticity and the fact that z is outside C. The estimates from part i of this contour integral clearly yield the result.

iii. It remains to prove the case $\tau = 0$. We shall show

$$\lim_{\xi\to 0}\frac{1}{2\pi}\int_{-\infty}^{\infty}\frac{f(\xi+i\gamma)}{z-(\xi+i\gamma)}\,d\gamma = \frac{1}{2\pi}\int_{-\infty}^{\infty}\frac{f(i\gamma)}{z-i\gamma}\,d\gamma, \tag{2.5}$$

a finite limit, recalling from Lemma 2 that $v(\gamma)f(i\gamma) \in L^p$. (2.5) yields the result since, by part i and part ii, the left hand side of (2.5) is $f(z)$ for all $0 < \xi < x$ and is 0 for all $x < \xi$, $0 < \xi$.

Define $g(\gamma) = g_z(\gamma) = [(x/2)^2 + (y-\gamma)^2]^{-1/2}/v(\gamma)$.

Since $(1/v)^{p'} \in L^1_{loc}$ and $p' > 1$ we have $g \in L^{p'}$; in fact, using the hypothesis that v is radial and increasing on $(0,\infty)$, we compute

$$\int_{-\infty}^{\infty}|g(\gamma)|^{p'}d\gamma \le \left(\frac{2}{x}\right)^{p'}\int_{-a}^{a} v(\gamma)^{-p'}d\gamma + v(a)^{-p'}\int_{|\gamma|\ge a}\frac{d\gamma}{|y-\gamma|^{p'}},$$

which is finite for $a > |y|$. This hypothesis on v also allows us to make the estimate

$$
\begin{aligned}
&|v(\xi+i\gamma)(z-(\xi+i\gamma))g(\gamma)|^{-1} \le |v(\gamma)(z-(\xi+i\gamma))g(\gamma)|^{-1} \\
(2.6)\qquad &= \left(\frac{(x-\xi)^2+(y-\gamma)^2}{(\frac{x}{2})^2+(y-\gamma)^2}\right)^{-1/2} \le 1
\end{aligned}
$$

for all $\xi \le x/2$ and for all γ.

Setting $H_\xi(\gamma) = f(\xi+i\gamma)/[g(\gamma)(z-(\xi+i\gamma))]$ and using (2.6) and the hypothesis $f \in H_v^p$ we see that

$$
\begin{aligned}
&\int_{-\infty}^{\infty} |H_\xi(\gamma)|^p d\gamma = \int_{-\infty}^{\infty} \left|\frac{v(\xi+i\gamma)f(\xi+i\gamma)}{v(\xi+i\gamma)(z-(\xi+i\gamma))g(\gamma)}\right|^p d\gamma \\
(2.7)\qquad &\le 2\pi \sup_{\xi>0} \mu_{p,v}(f;\xi)^p \le C,
\end{aligned}
$$

a bound independent of $\xi \in (0,x/2]$. Since $p > 1$, we know that L^p is the dual of $L^{p'}$, and (2.7) allows us to assert the weak * compactness of the weak * closure of the set $\{H_\xi : \xi \in (0,x/2]\} \subseteq L^p$. Thus, there is a sequence $\{\xi_n\}$ and a function $H_0 \in L^p$ such that $\lim \xi_n = 0$ and

$$
(2.8)\qquad \lim_{n\to\infty} \int_{-\infty}^{\infty} H_{\xi_n}(\gamma)h(\gamma)d\gamma = \int_{-\infty}^{\infty} H_0(\gamma)h(\gamma)d\gamma
$$

for each $h \in L^{p'}$. Taking $h = g$, (2.8) becomes

$$
\lim_{n\to\infty} \frac{1}{2\pi} \int_{-\infty}^{\infty} \frac{f(\xi_n+i\gamma)}{z-(\xi_n+i\gamma)} d\gamma = \frac{1}{2\pi} \int_{-\infty}^{\infty} H_0(\gamma)g(\gamma)d\gamma .
$$

(2.5) will be obtained and the proof completed once we show

$$
(2.9)\qquad \int_{-\infty}^{\infty} H_0(\gamma)g(\gamma)d\gamma = \int_{-\infty}^{\infty} \frac{f(i\gamma)}{z-i\gamma} d\gamma ,
$$

where we know the left hand side is finite. By Lemma 2 we know

$\lim_{n\to\infty} f(\xi_n+i\gamma) = f(i\gamma)$ a.e. and hence $\lim H_{\xi_n}(\gamma) =$ $f(i\gamma)/[g(\gamma)(z-i\gamma)] \equiv H_1(\gamma)$ a.e. Since $p > 1$ and $\|H_{\xi_n}\|_p \leq C$ we conclude by a basic result in real variable [1, Theorem 6.10] (which combines Egorov, Fatou, Hölder, and Minkowski) that $\lim H_{\xi_n} = H_1$ in the weak topology on L^p. We saw however that $\lim H_{\xi_n} = H_0$ in the weak * topology which, by reflexivity, is the weak topology in L^p, $p > 1$. Hence, $H_0 = H_1$ a.e. and so (2.9) is obtained.

qed

The proof of the previous result required the hypothesis $p > 1$. For example, it is trivial to find examples of sequences $\{H_n\} \subsetneq L^1$ for which $\|H_n\|_1 \leq C$ and $\lim H_n = H$ a.e. but for which $\lim H_n \neq H$ weakly. For the case of H_v^1 there are no problems because of the following lemma.

Lemma 5. Suppose that $f \in H_v^1$, where the radial function v is increasing on $(0,\infty)$ as a function of $|z|$, $z = x + iy$. If $1/v(|z|)$ is bounded on $(0,a)$ for each $a > 0$ then $H_v^1 \subsetneq H^1$ and the conclusion of Lemma 4 is valid.

Proof. Suppose $f \in H_v^1$ and note that $v(0\pm) > 0$ by the boundedness hypothesis. The assertion $f \in H^1$ is immediate since

$$v(0\pm) \int_{-\infty}^{\infty} |f(\tau+i\gamma)|d\gamma \leq \int_{-\infty}^{\infty} |v(\tau+i\gamma)f(\tau+i\gamma)|d\gamma.$$

A similar but simpler proof than Lemma 4 yields the result for $0 < \tau < x$.

The case $\tau = 0$ is a consequence of a result due to Hille and Tamarkin for H^p spaces [15, Theorem 2.1]. For the sake of completeness we give the following proof: since $f(i\gamma) \in L^1$ and

$$\lim_{\xi \to 0} \int_{-\infty}^{\infty} |f(\xi+i\gamma) - f(i\gamma)| d\gamma = 0$$

the right hand side of the estimate,

$$\left|\int_{-\infty} \left[\frac{f(\xi+i\gamma)}{z-(\xi+i\gamma)} - \frac{f(i\gamma)}{z-i\gamma}\right] d\gamma\right|$$

$$\leq \int_{-\infty}^{\infty} |f(\xi+i\gamma) - f(i\gamma)| \frac{1}{|z-(\xi+i\gamma)|} d\gamma$$

$$+ \int_{-\infty}^{\infty} |f(i\gamma)| \left|\frac{1}{z-(\xi+i\gamma)} - \frac{1}{z-i\gamma}\right| d\gamma,$$

tends to 0 as $\xi \to 0$.

qed

The existence of the Laplace transform in the following result does not require Theorem 1 although we do use Lemma 1.

Theorem 3. Suppose that $(u,v) \in F(p,q)$, $1 \leq p \leq q < \infty$.

a. If $F \in L_v^p(0,\infty)$ then $L(F) = f$ exists as a holomorphic function for $\operatorname{Re} z > 0$, $z = x + iy$.

b. There is a constant $C > 0$ such that for all $F \in L_v^p(0,\infty)$

$$(2.10) \quad \forall x > 0, \ \mu_{q,u}(f;x) \leq C\left(\int_0^{\infty} e^{-xtp} |v(t)F(t)|^p dt\right)^{1/p},$$

where $L(F) = f$. In particular, if $\lim_{x \to 0+} f(x+iy)$ exists a.e. then $L(F) = f \in H_u^q$ and $\hat{F}(y) = L(F)(iy)$ for each $F \in L_v^p(0,\infty)$.

Proof. i. For $z = x + iy$, $x > 0$, we have the formal inequality

$$(2.11) \quad |L(F)(z)| \leq \int_0^{1/x} |F(t)| dt + (ex)^{-1} \int_{1/x}^{\infty} \frac{|F(t)|}{t} dt.$$

In fact, the estimate of the first integral in (2.11) follows since $e^a \le 1$ for $a \le 0$ and that of the second integral is a consequence of the inequality $e \le e^a/a$ for $a \ge 1$, a fact which is easily checked since e^a/a is increasing on $[1,\infty)$.

We shall show that $u(x)L(F)(x)$ exists a.e. on $(0,\infty)$ by using (2.11) to verify

$$(2.12)\qquad \left(\int_0^\infty |u(x)L(F)(x)|^q dx\right)^{1/q} \le C\|F\|_{p,v},$$

where C is independent of $F \in L_v^p(0,\infty)$. Since u is continuous except at possibly countably many points, the existence of $uL(F)$ a.e. yields the existence of $L(F)(x)$ a.e. on $(0,\infty)$. By elementary properties of the Laplace transform we conclude that $L(F)$ exists and is holomorphic for $\operatorname{Re} z > 0$, e.g., [25]; and, hence, part a will be proved. Thus the existence of $L(F)$ follows from (2.12), an inequality we verify in part ii.

The fact that C is independent of F in (2.12) only plays a role in (2.10) which we prove in part iii.

ii. From (2.11) we make the estimate

$$\left(\int_0^\infty |u(x)L(F)(x)|^q dx\right)^{1/q} \le \left(\int_0^\infty |u(x)\int_0^{1/x} |F(t)|dt|^q dx\right)^{1/q}$$

$$+ e^{-1}\left(\int_0^\infty \left|\frac{u(x)}{x}\int_{1/x}^\infty \frac{|F(t)|}{t}\,dt\right|^q dx\right)^{1/q} \equiv I_0 + e^{-1}I_1$$

by means of Minkowski's inequality. We shall verify $I_0 + e^{-1}I_1 \le C\|F\|_{p,v}$; and therefore the formal inequality (2.11), as well as (2.12), will be obtained.

Using part a of Lemma 1 and making the change of variable $y = 1/x$ in I_0, we see that I_0 is dominated by $\|F\|_{p,v}$ if and only if there is $C > 0$ such that

$$(2.13) \quad \forall s > 0, \left(\int_s^\infty u(1/y)^q \frac{dy}{y^2}\right)^{1/q}\left(\int_0^s v(y)^{-p'}dy\right)^{1/p'} \le C,$$

where the second integral in (2.13) must be modified in case $p = 1$. The change of variable $t = 1/(4y)$ in the first integral of (2.13) and the fact that u is decreasing on $(0,\infty)$ show that (2.13) is a consequence of our hypothesis $(u,v) \in F(p,q)$. Hence, $I_0 \le C\|F\|_{p,v}$.

The calculation, $I_1 \le C\|F\|_{p,v}$, is similar. In fact, using part b of Lemma 1, we see that I_1 is dominated by $\|F\|_{p,v}$ if and only if

$$(2.14) \quad \forall s > 0, \left(\int_0^s u(1/y)^q y^{q-2} dy\right)^{1/q}\left(\int_s^\infty (yv(y))^{-p'}dy\right)^{1/p'} \le C,$$

where the second integral in (2.14) must be changed in case $p = 1$ to $\sup\{|yv(y)|^{-1} : y \ge s\}$. This equivalence requires several routine substitutions besides Lemma 1.

Thus, part a will be complete once we verify (2.14). To this end observe that $(u,v) \in F(p,q)$ and the fact that u is decreasing on $(0,\infty)$ imply

$$u(1/y)^q \le Cy\left(\int_0^{y/4} v(t)^{-p'}dt\right)^{-q/p'}, \quad p > 1$$

adjusted to $u(1/y)^q \le Cyv(0\pm)^q$ in case $p = 1$. Substituting this into the left hand side of (2.14) yields the estimates

$$\left(\int_0^s y^{q-1}\left(\int_0^{y/4} v(t)^{-p'}dt\right)^{-q/p'}dy\right)^{1/q}\left(\int_s^\infty (yv(y))^{-p'}dy\right)^{1/p'}$$

$$\le Cv(s/4)\left(\int_0^s y^{q-1-q/p'}dy\right)^{1/q} v(s)^{-1}s^{-1+1/p'}$$

$$\le Cv(s/4)v(s)^{-1} \le C,$$

and (2.14) is obtained. The case $p = 1$ is analogous but simpler.

iii. It remains to prove (2.10). If we define $G_x(t) = e^{-xt}F(t)$ for $x > 0$ and $F \in L^p_v(0,\infty)$ then $G_x \in L^p_v$, since F vanishes on $(-\infty,0)$, and $\hat{G}_x(y) = f(x+iy)$ where $f = L(F)$ is well-defined by part <u>a</u>. <u>Theorem 1</u> is applicable and hence there is $C > 0$ such that

(2.15) $\quad \forall x > 0$ and $\forall F \in L^p_v(0,\infty)$, $\|\hat{G}_x\|_{q,u} \le C\|G_x\|_{p,v}$.

Therefore, since $u(y) \ge u(x+iy)$, we compute

$$\mu_{q,u}(f;x) = \left(\frac{1}{2\pi}\int_{-\infty}^{\infty}|u(x+iy)f(x+iy)|^q dy\right)^{1/q}$$

$$\le C\left(\int_{-\infty}^{\infty}|u(y)f(x+iy)|^q dy\right)^{1/q} = C\|\hat{G}_x\|_{q,u} \le C\|G_x\|_{p,v},$$

and this is the right hand side of (2.10). C is independent of F because of (2.15) and the theorem is proved. <u>qed</u>

<u>Theorem 4</u>. Suppose that $(u,v) \in F(p,q)$, where $1 \le p \le q < \infty$ and $q > 1$, and that the weight function u satisfies the condition,

(2.16) $$\forall \alpha > 0, \int_1^\infty |e^{-\alpha t}u(t)^{-1}|^{q'}dt < \infty.$$

a. If $f \in H_v^p$ then there is $F \in L_u^q(0,\infty)$ for which $f(z) = L(F)(z)$, Re $z > 0$.

b. There is a constant $C > 0$ such that for all $f \in H_v^p$

$$(2.17) \qquad \forall x > 0, \ \left(\int_0^\infty e^{-xtq}|u(t)F(t)|^q dt\right)^{1/q} \le C\mu_{p,v}(f;x),$$

where $L(F) = f$ and F is well-defined by part a.

Proof. i. For each $x \ge 0$, define $F_x(y) = f(x+iy)$. Since v is increasing we obtain the inequality,

$$\forall x > 0, \ \|F_x\|_{p,v} \le \left(\int_{-\infty}^\infty |v(x+iy)f(x+iy)|^p dy\right)^{1/p}.$$

Consequently, the hypothesis $f \in H_v^p$ yields the fact $F_x \in L_v^p$, for each $x \ge 0$, where the case $x = 0$ requires Fatou's lemma for its verification. Hence, Theorem 1 applies, and F_x has a Fourier transform $\hat{F}_x \in L_u^q$ for each $x \ge 0$.

ii. Since $(u,v) \in F(p,q)$ we know that $1/v(|z|) \in L_{loc}^{p'}$ and, consequently, the hypothesis $f \in H_v^p$ allows us to apply Lemma 4, resp., Lemma 5. As such we have

$$(2.18) \qquad \frac{1}{2\pi}\int_{-\infty}^\infty \frac{f(\tau+i\gamma)}{z-(\tau+i\gamma)}\, d\gamma = \begin{cases} f(z) & \text{if } 0 \le \tau < x, \\ 0 & \text{if } x < \tau \text{ and } \tau \ge 0, \end{cases}$$

for each fixed $z = x + iy$. Also, we introduce the notation

$$\check{H}(t) = \frac{1}{2\pi}\int_{-\infty}^\infty H(\gamma)e^{i\gamma t}d\gamma.$$

iii. We shall first use (2.18) to prove $\check{F}_\tau \in L_u^q$ vanishes a.e. on $(-\infty,0)$, that is, we shall show $\check{F}_\tau \in L_u^q(0,\infty)$ for each $\tau > 0$.

To this end, pick $\tau \geq 0$ and define

$$G_{z,\tau}(t) = \begin{cases} e^{(z+\tau)t} & \text{if} \quad t < 0, \\ 0 & \text{if} \quad t > 0, \end{cases}$$

where $z = x + iy$ and $x > \tau$. Note that $\tau \geq 0 > -x$. $G_{z,\tau}$ is an element of L^1 since $x + \tau > 0$, and $2\pi\check{G}_{z,\tau}(\gamma) = 1/(z+\tau+i\gamma)$. We compute

$$\int_0^\infty e^{-zt}(e^{-t\tau}\check{F}_\tau(-t))dt = \int_{-\infty}^0 e^{(z+\tau)t}\check{F}_\tau(t)dt =$$

$$(2.19) \quad \int_{-\infty}^\infty G_{z,\tau}(t)\check{F}_\tau(t)dt = \int_{-\infty}^\infty \check{G}_{z,\tau}(\gamma)F_\tau(\gamma)d\gamma =$$

$$-\frac{1}{2\pi}\int_{-\infty}^\infty \frac{f(\tau+i\gamma)}{(-z)-(\tau+i\gamma)}\,d\gamma = 0.$$

The last equality is a consequence of (2.18) since $-x < \tau$. The third equality is Parseval's and requires that $G_{z,\tau} \in L^{q'}_{1/u}$ because of <u>Theorem 1d</u> and the facts $F_\tau \in L^p_v$ and $\check{F}_\tau \in L^q_u$. To see that $G_{z,\tau}$ is an element of $L^{q'}_{1/u}$ we use (2.16), noting there is no integrability problem at the origin since $1/u$ increases on $(0,\infty)$. If $\tau \geq 0$ is fixed then (2.16) allows us to assert that $G_{z,\tau} \in L^{q'}_{1/u}$ since $x + \tau > 0$. Thus, (2.19) is true for all $z = x + iy$ with $x > \tau$. Therefore, by the uniqueness property of the Laplace transform, we see that $\check{F}_\tau(-t) = 0$ for all $t \in (0,\infty)$, and hence

$$(2.20) \qquad \forall\tau \geq 0, \quad \operatorname{supp} \check{F}_\tau \subseteq [0,\infty).$$

iv. Our next step is to verify the formula,

(2.21) $\forall\tau \geq 0$ and $\forall x > \tau$, $f(z) = \int_0^\infty e^{-zt}(e^{\tau t}\check{F}_\tau(t))dt$,

by means of (2.18), where $z = x + iy$.

Since $\tau - x < 0$, the function,

$$H_{z,\tau}(t) = \begin{cases} 0 & \text{if } t < 0 \\ e^{(-z+\tau)t} & \text{if } t > 0, \end{cases}$$

is an element of L^1 and $2\pi\check{H}_{z,\tau}(\gamma) = -1/(-z+\tau+i\gamma)$. We compute

$$(2.22) \quad \begin{aligned} &\int_0^\infty e^{-zt}(e^{\tau t}\check{F}_\tau(t))dt = \int_{-\infty}^\infty H_{z,\tau}(t)\check{F}_\tau(t)dt = \\ &\int_{-\infty}^\infty \check{H}_{z,\tau}(\gamma)F_\tau(\gamma)d\gamma = -\frac{1}{2\pi}\int_{-\infty}^\infty \frac{f(\tau+i\gamma)}{(-z+\tau+i\gamma)}\,d\gamma = f(z). \end{aligned}$$

The last equality is a consequence of (2.18) since $x > \tau$. The second equality is Parseval's and requires that $H_{z,\tau} \in L^{q'}_{1/u}$ because of Theorem 1d and the facts $F_\tau \in L^p_v$ and $\check{F}_\tau \in L^q_u$. To see that $H_{z,\tau}$ is an element of $L^{q'}_{1/u}$ we again use (2.16), noting once again that there is no integrability problem at the origin since $1/u$ increases on $(0,\infty)$. If $\tau \geq 0$ is fixed then (2.16) allows us to assert that $H_{z,\tau} \in L^{q'}_{1/u}$ since $x > \tau$. Thus, (2.22) is true for all $z = x + iy$ with $x > \tau$, and so (2.21) is obtained.

v. Set $F(t) = \check{F}_0(t)$. We have proved $F \in L^q_u$ has the properties, $\operatorname{supp} F \subseteq [0,\infty)$ and $L(F)(z) = f(z)$ for all $z = x + iy$ with $x > 0$, by means of (2.20) and (2.21) respectively. Consequently, part a is complete.

vi. To establish part b we observe

$$\forall x \geq 0, \; F(t) \;=\; e^{xt}\check{F}_x(t) \quad \text{a.e. on } \mathbb{R};$$

this follows from (2.21) and the uniqueness property of the Laplace transform. Therefore, for each $x > 0$, we compute

$$\left(\int_0^\infty e^{-xtq}|u(t)F(t)|^q dt\right)^{1/q} = \left(\int_0^\infty |u(t)\check{F}_x(t)|^q dt\right)^{1/q}$$

$$= \frac{1}{2\pi}\left(\int_0^\infty |u(-t)\hat{F}_x(-t)|^q dt\right)^{1/q} = \frac{1}{2\pi}\left(\int_{-\infty}^0 |u(t)\hat{F}_x(t)|^q dt\right)^{1/q}$$

$$\leq C\|F_x\|_{p,v} \leq C\left(\int_{-\infty}^\infty |v(|y|)f(x+iy)|^p dy\right)^{1/p} \leq C\mu_{p,v}(f;x),$$

by means of Theorem 1, where C is independent of $f \in H_v^p$.

qed

Condition (2.16) can be weakened, to asserting the existence of $\alpha > 0$ for which the integral is bounded, at the expense of obtaining a smaller half-plane of convergence in Theorem 4a.

3. Examples and remarks

Remark 1. Theorem 1 and Plancherel's theorem.

It is easy to check that $L^p \subseteq S'$, the space of tempered distributions, for each $p \in [1,\infty]$; and thus the Fourier transform $\hat{F}$ exists in S' for each $F \in L^p$. The following observations deal with the existence of $\hat{F}$ as an element of a specified function space.

a. Titchmarsh [24, Theorem 74] (1924) established the existence of $\hat{F} \in L^{p'}$ for $F \in L^p$, $1 \leq p \leq 2$; and Hardy and Littlewood [24, Theorem 80 and 81] (1927) proved an analogous result in weighted L^p spaces for specific weights and various values of p including $p > 2$. Rooney [23, Theorem 2.1] generalized these theorems, and Theorem 1 extends Rooney's result; this latter fact follows when we choose the weights $u(t) = |t|^{\frac{1}{r'} - \frac{1}{q}}$ and $v(t) = |t|^{\frac{1}{r} - \frac{1}{p}}$, where $1 < p < \infty$, $1 < r \le \min(p,p')$, and $p \le q \le r'$, cf., [9].

b. One expects that $\hat{F} \in L^1_{loc}$ for $F \in L^p_v$ if v increases quickly enough. The following elementary facts quantify this point of view.

i. If $v(0\pm) > 0$ and $p \in [1,\infty)$ then $L^p_v \subsetneq L^p$.

ii. If $p \in [1,\infty)$, $\alpha > 1 - \frac{1}{p}$, and $v_\alpha(t) = (1+|t|)^\alpha$ then $L^p_{v_\alpha} \subsetneq L^1$.

iii. Given $p \in [1,\infty)$, $\alpha > 1 - \frac{1}{p}$, $q \in [p,\infty)$, and any $\beta > 1/q$. Assume p or q is not 2. If $u_\beta(t) = 1/(1+|t|)^\beta$ then $(u_\beta, v_\alpha) \in F(p,q)$ and thus $(L^p_{v_\alpha})\hat{} \subsetneq L^q_{u_\beta}$.

iv. If $p \in (2,\infty)$ and $\alpha < \frac{2}{p'} - 1$ then

$$\forall u \text{ and } \forall q \in [p,\infty),\ (u,v_\alpha) \notin F(p,q). \tag{3.1}$$

Of course, (3.1) does not preclude the possibility that $\hat{F} \in L^1_{loc}$ exists for many elements $F \in L^p_{v_\alpha}$.

v. The weights v_α and u_β do not deal with behavior at the origin. We can deal with the origin by taking $1 < p \le q < \infty$ and setting $r = \frac{1}{p} + \frac{1}{q} - 1$, $v_{0,\alpha}(t) = |t|^\alpha$ and $u_{0,\beta}(t) = |t|^{-\beta}$, where $\alpha, \beta > 0$. Then we have

$\forall\alpha \geq |r|, \exists\beta = \alpha + r$ such that $(u_{0,\beta}, v_{0,\alpha}) \in F(p,q)$.

c. We now consider the L^p case, i.e., $v = 1$. If $p \leq q = \infty$ then $(u,1) \in F(p,\infty)$ can only occur for $u = 0$. Given $1 \leq p \leq q < \infty$ and let $u \in L^q$ have the property that $u(0\pm) < \infty$. If $q \leq p'$ then $(u,1) \in F(p,q)$; the case $p = 1$ does not require the condition $u(0\pm) < \infty$. The following are easy to check.

i. If $q \leq p'$, so that $p \in [1,2]$, then $(L^p)\hat{} \subseteq L^q_u$.

ii. If $p' < q$ then $(u,1) \notin F(p,q)$. This material is related to the representation theory of $(L^p)\hat{}$, e.g., [6].

d. The proof of Theorem 1 is based on a weak type (2,2) result; and, hence, we cannot expect a strong type conclusion. In particular, the proof of Theorem 1 does not reduce to a proof of the Plancherel theorem (or the theorems of Titchmarsh and Hardy and Littlewood) when we take $p = q = 2$ and $u = v = 1$; and this is also the situation for those results which Theorem 1 generalizes.

On the other hand, the Plancherel theorem for weighted L^2 spaces has been proved independently of (1.6), e.g., [16; 18]; these proofs do not depend on weak type methods and are more difficult than the proof of Theorem 1.

For the sake of completeness we provide an outline of our own proof of the Plancherel theorem for weighted L^2 spaces. Our hypothesis is (1.5) for $p = q = 2$ and not (1.6). The result we invoke that goes beyond weak type arguments is Stein's inequality,

$$\left(\int_{-\infty}^{\infty} |\hat{F}(y)u_0(y)^{\theta}u_1(y)^{1-\theta}|^q dy\right)^{1/q} \le C\left(\int_{-\infty}^{\infty} |F(t)v_0(t)^{\theta}v_1(t)^{1-\theta}|^p dt\right)^{1/p},$$

for simple functions (E. Stein, "Interpolation of linear operators" Trans. Amer. Math. Soc. 83 (1956) 485); this inequality requires the two inequalities listed below as hypotheses. We begin our proof by noting that if $(u,v) \in F(p,q)$, $1 < p \le q < \infty$, then the factorizations

$$u = u_0^{1/q}\, u_1^{1/q'} \quad \text{and} \quad v = v_0^{1/p} v_1^{1/p'}$$

are valid for some $(u_0, v_0) \in F(1,1)$ and $(u_1, v_1) \in F(\infty,\infty)$. Next, using the methods of Theorem 1, we can verify the inequalities,

$$\|\hat{F}\|_{1,u_0} \le C\|F\|_{1,v_0}$$

and

$$\|\hat{F}\|_{\infty,u_1} \le C\|F\|_{\infty,v_1},$$

for simple functions. These verifications do not involve Plancherel's theorem, e.g., Proposition 1. We then apply Stein's result for $p = q = 2$ and $\theta = 1/2$. This yields the fundamental inequality,

$$\|\hat{F}\|_{2,u} \le C\|F\|_{2,v},$$

for simple functions F; and routine arguments allow us to extend this inequality to all of L_v^2.

The following result is Theorem 1 for $p = 1$.

Proposition 1. Suppose that $(u,v) \in F(1,q)$, $q \ne 1$, where $u(t) > 0$ for all t. There is $C > 0$ such that

$$\forall F \in L^1_v, \quad \|\hat{F}\|_{q,u} \le C\|F\|_{1,v}.$$

Proof. i. Since $(u,v) \in F(1,q)$ and $u > 0$ we have $\sup\{|1/v(t)| : t \in (0,s/2]\} < \infty$ for each $s > 0$. Thus, $v(0\pm) > 0$ and, hence, we obtain $L^1_v \subseteq L^1$ (Remark 1bi) and $u \in L^q$.

ii. If F and G are simple functions then $\int F(t)\hat{G}(t)dt = \int \hat{F}(y)G(y)dy$. Therefore, for a fixed simple function F, we compute

$$\|\hat{F}\|_{q,u} = \sup\left|\int \hat{F}(y)G(y)dy\right| = \sup\left|\int Fv\hat{G}(1/v)\right|$$

$$\le (1/v(0\pm))\|F\|_{1,v} \sup \sup_y |\hat{G}(y)| \le (1/v(0\pm))\|F\|_{1,v} \sup\|G\|_1$$

$$\le (\|u\|_q/v(0\pm))\|F\|_{1,v},$$

where each supremum is taken over the simple functions G for which $\|G\|_{q',1/u} \le 1$ and where the last inequality follows since

$$\|G\|_1 = \int |u(t)G(t)(1/u(t))|dt \le \|u\|_q\|G(1/u)\|_{q'} \le \|u\|_q.$$

The first equality in the estimate of $\|\hat{F}\|_{q,u}$ shows that we must take $q > 1$.

qed

Remark 2. Laplace transform representation of H^p.

a. In 1933 Paley and Wiener [19; 20, Theorem XII] proved the following important theorem: *given* $\phi \ge 0, \phi \in L^2\backslash\{0\}$; *there is a function* $F \in L^2$ *which vanishes on* $(-\infty,0)$ *and which satisfies* $|\hat{F}| = \phi$ *if and only if*

$$(3.2) \qquad \int_{-\infty}^{\infty} \frac{|\log \phi(y)|}{1+y^2}\, dy < \infty.$$

Condition (3.2) has played a basic role in several areas including the profound work of Beurling and Malliavin [3; 17]. Hille and Tamarkin [12, Theorem 5] (1933) generalized the above theorem for the L^p case, $1 < p \leq 2$, and in the process proved the following facts: i. if $f \in H^p$ then $\operatorname{supp} F \subseteq [0,\infty)$, where $\hat{F}(y) = f(iy)$ (actually $f(iy) \in L^p$ and so $F \in L^{p'}$ is the Fourier transform of $f(iy)$); ii. if $f \in H^p$, $\hat{F}(y) = f(iy)$, and $F \in L^{p'}(0,\infty)$ then the Poisson integral P of f has the representation

$$P(f(is),z) = \int_0^{\infty} e^{-zt}F(t)dt, \quad \operatorname{Re} z > 0.$$

These two results are Lemma 4.1 and Lemma 4.2 of [15] (Hille-Tamarkin, 1935), and Lemma 4.1 was quoted by Doetsch in his proof of the Laplace transform representation of H^p, $1 < p \leq 2$, which we stated in the Introduction. Clearly, Lemma 4.1, Lemma 4.2, and the well-known fact $P(f(is),z) = f(x+iy)$ (where $f \in H^p$, $z = x + iy$, and $x > 0$) yield an immediate proof of this representation.

The analogue of Lemma 4.1 in this paper is found in part iii of the proof of Theorem 4.

b. Hille and Tamarkin made fundamental contributions to Laplace transform theory in the early 1930's [13; 14]; and [11] (1934) provides an interesting survey. They did not explicitly formulate the Laplace transform representation of H^p, $1 < p \leq 2$, although they developed basic techniques for this range of p as well as $p > 2$.

c. Rooney [22] formulated an extension of H^p spaces to which our weighted results Theorem 3 and Theorem 4 can be generalized. Naturally,. Rooney [23] earlier noted the possibility of such a generalization for the weights he considered, cf., Remark 1a.

d. Our proof of Theorem 1 can be adjusted to eliminate the monotonicity hypothesis on the weights u. At this point monotonicity of both u and v plays an essential role for the results of Section 2.

Acknowledgement. A fundamental problem, which we have not yet solved and which was also pointed out to us by P. G. Rooney, is to omit the hypothesis,

$$\lim_{x \to 0+} f(x+iy) = f(iy) \quad \text{a.e.},$$

from Definition 3 for as many weights as possible. We hasten to point out that this limit does exist for most weights including power weights. A salubrious meeting with W. Schneider, who outlined the subtleties involved in this problem, leaves us cautiously optimistic.

Bibliography

1. J. Benedetto, Real variable and integration B.G. Teubner Verl., Stuttgart, 1976.

2. J. Bergh and J. Löfstrom, Interpolation spaces, an introduction Springer Verl., N.Y., 1976.

3. A. Beurling and P. Malliavin, "On Fourier transforms of measures with compact suport" Acta Math. 107 (1962) 291-309.

4. J. Bradley, "Hardy inequalities with mixed norms" Can. Math. Bull. 21 (1978) 405-408.

5. A. P. Calderón, "Spaces between L^1 and L^∞" Studia Math. 26 (1966) 273-299.

6. J. L. B. Cooper, "Fourier transforms and inversion formulae for L^p functions" Proc. London Math. Soc. 14 (1964) 271-298.

7. G. Doetsch, "Bedingungen für die Darstellbarkeit einer Funktion als Laplace-Integral und eine Umkehrformel für die Laplace-Transformation" Math. Zeit. 42 (1937) 263-286.

8. G. H. Hardy, J. E. Littlewood, and G. Pólya, Inequalities Cambridge University Press, 1952.

9. H. Heinig, Ph.D. thesis, U. of Toronto, 1965.

10. H. Heinig, "Weighted norm inequalities for classes of operators" (preprint).

11. E. Hille, "On Laplace integrals" Ått. Skand. Mathematikerkong, Stockholm (1934) 216-227.

12. E. Hille and J. D. Tamarkin, "On a theorem of Paley and Wiener" Ann. of Math. 34 (1933) 606-614.

13. E. Hille and J. D. Tamarkin, "On moment functions" Proc. Nat. Acad. Sci. 19 (1933) 902-908.

14. E. Hille and J. D. Tamarkin, "On the theory of Laplace integrals, I and II" Proc. Nat. Acad. Sci. 19 (1933) 908-912 and 20 (1934) 140-144.

15. E. Hille and J. D. Tamarkin, "On the absolute integrability of Fourier transforms" Fund. Math. 25 (1935) 329-352.

16. W. Jurkat and G. Sampson, "On rearrangement and weight inequalities for the Fourier transform" (preprint).

17. P. Malliavin, "On the multiplier theorem for Fourier transforms of measures with compact support" Ark. Mat. 17 (1979) 69-81.

18. B. Muckenhoupt, "Weighted norm inequalities for the Fourier transform" Trans. Amer. Math. Soc. (to appear).

19. R. E. A. C. Paley and N. Wiener, "Notes on the theory and application of Fourier transforms" Trans. Amer. Math. Soc. 35 (1933) 348-355.

20. R. E. A. C. Paley and N. Wiener, Fourier transforms in the complex domain AMS Colloquium Publications 19 (1934).

21. P. G. Rooney, "On some properties of functions analytic in a half-plane" Can. J. Math. 11 (1959) 432-439.

22. P. G. Rooney, "A generalization of the Hardy spaces" Can. J. Math. 16 (1964) 358-369.

23. P. G. Rooney, "Generalized H^p spaces and Laplace transforms" Abstract Spaces and Approximation, Proc. Conf. Oberwolfach (July 18-27, 1968) 258-269.

24. E. C. Titchmarsh, Introduction to the theory of Fourier integrals, second edition, Oxford University Press, 1948.

25. D. V. Widder, The Laplace transform Princeton University Press, 1946.

J. Benedetto
U. of Maryland
College Park, MD 20742
USA

H. Heinig
McMaster University
Hamilton, Ontario L8S-4K1
Canada

VECTOR VALUED INEQUALITIES OF MARCINKIEWICZ-ZYGMUND AND GROTHENDIECK TYPE FOR TOEPLITZ FORMS

by
Mischa Cotlar and Cora Sadosky

ABSTRACT. The generalized Bochner-Herglotz theorem for generalized Toeplitz kernels (GTKs) [10] contains as special cases the solutions of several classical moment problems that, in turn, contain the germs of Grothendieck's theory of bilinear forms. In this paper some Grothendickian properties of the GTKs are studied, through the consideration of matrix-valued Hilbertian forms. Generalizations for GTKs of the Bochner-Eberlein-Horn theorems and of the vector-valued Marcinkiewicz-Zygmund and Grothendieck inequalities are given. Some applications to vector-valued weighted norm inequalities for the Hilbert transform and to Toeplitz and Hankel operators are outlined.

INTRODUCTION. Many moment problems in Fourier Analysis can be similarly stated: Given a family F of finite measures in the circle T and a set of integers $\Gamma \subset Z$, characterize the class $[F,\Gamma]$ of all sequences $s(n)$, $n \in \Gamma$, such that $s(n) = \hat{\mu}(n)$ for all $n \in \Gamma$ and for some $\mu \in F$ (where $\hat{\mu}$ is the Fourier transform of μ).

Some classical examples are:

(a) $\Gamma = Z$, F = the set of all positive finite measures in T. The Bochner-Herglotz theorem characterizes this class $[F,\Gamma]$ as those sequences $s(n)$ for which the Toeplitz kernel $K(m,n) = s(m-n)$ is positive definite (see Section 1).

(b) $\Gamma = Z$, F = the set of all complex (or real) finite measures in T. The Bochner-Eberlein and R. Horn's theorems characterize this class $[F,\Gamma]$ in terms of "majorized" Toeplitz kernels (see Section 1).

(c) $\Gamma = \{\pm 2^n, n = 1,2,\dots\}$, $F = \{d\mu = f dt, f \in L^\infty\} \sim L^\infty$. Here $[F,\Gamma] = \ell^2(\Gamma)$ (see [26] and [6]).

Problems (a), (b) and (c) are related to the Grothendieck Fundamental Theorem. In fact, J. Gilbert pointed out ([15],[16]) that the characterizations (a),(b) contain the germs of Grothendieck's theory of Hilbertian and integral forms, and R. Blei [6] showed that Grothendieck's inequality is a consequence of (a refinement of) (c). Moreover, Grothendieck's inequality is a complement to previous inequalities of Marcinkiewicz and Zygmund [19] which arose in the study of Fourier coefficients.

On the other hand, there are other examples of $[F,\Gamma]$ moment problems:

(d) $\Gamma = Z_1 = \{0,1,2,\dots\}$, $F = \{d\mu = f dt,\ f \in L^\infty\} \sim L^\infty$. Here the Nehari theorem characterizes this class $[F,\Gamma]$ as those sequences $s(n)$ for which the Hankel kernel $H(m,n) = s(m+n)$ defines a bounded operator in ℓ^2_+ (see [21] and [1]).

(e) $\Gamma = Z$, $F = H^{pq}_M$, the class of all positive finite measures $\mu \geq 0$ in T satisfying

$$\left(\int |Hf|^q d\mu\right)^{1/q} \leq M\left(\int |f|^p d\mu\right)^{1/q}, \quad f \in C(T) \tag{1}$$

where Hf is the Hilbert transform of f. Here the class $[F,\Gamma]$ was characterized in terms of positive definite generalized Toeplitz kernels (GKTs) in [10] and [2] for the case $p = q = 2$, and in terms of "majorized" GTKs in [11] and [12] for the cases $p = 2$, $q > 2$ and $p = q$. (The definition of GTK is given in Section 1. See also Section 3.)

Furthermore, there is a generalized Bochner-Herglotz representation of positive definite and majorized GTKs which furnishes characterizations of the Helson-Szegö type of the classes H^{pq}_M, as well as several other results in prediction theory and Carleson measures (cfr. [2],[11],[12],[24]). This general representation contains as special cases the results of (a), (d), (e) and, as shown in Section 1 of this paper, also of (b).

Since (a), (b), (d) and (e) are special cases of the theory of GTKs, it is natural to ask whether the GTKs also have some Grothendieck type properties. In this paper, which is a continuation of [12] and [24], we show that this is the case and that

(1) the properties (a), (b) for GTKs can also be expressed in terms of (vector valued) Hilbertian forms;

(2) the generalized Bochner-Herglotz theorem furnishes inequalities of the Marcinkiewicz-Zygmund and Grothendieck types for the cases of GTKs.

Section 1 contains a summary of the basic definitions and results of [2], [12] and [24] and an extension of Bochner-Eberlein and Horn's theorems for GTKs.

In Section 2 the notion of GTKs is considered from the viewpoint of Hilbertian forms.

In Section 3 some inequalities of Marcinkiewicz-Zygmund and Grothendieck type are derived for GTKs and some applications are briefly mentioned. The complete study of those and other applications will be the subject of a forthcoming paper.

BASIC NOTATIONS. We shall work in the unit circle $T \sim [0,2\pi]$ and its dual group Z. Let $Z_1 = \{n \in Z : n \geq 0\}$, $Z_2 = \{n \in Z : n < 0\}$, $P = \{f(t) = \Sigma \hat{f}(n)e^{int}$, $\hat{f}(n) = 0$ for $|n| > n_0$ some fixed integer$\}$ the set of all trigonometric polynomials, $P_1 = \{f \in P : \hat{f}(n) = 0 \text{ for } n \in Z_2\}$, $P_2 = \{f \in P : \hat{f}(n) = 0 \text{ for } n \in Z_1$, dt the Lebesgue measure in T, $L^p = L^p(T,dt)$, $H^1 = H^1(T) = \{f \in L^1 : \hat{f}(n) = 0$ for $n \in Z_2\}$.

If $\lambda = \lambda(n) \in \ell^1(Z)$, then $\hat{\lambda}(t) = \Sigma_{n \in Z} \lambda(n) e^{int}$. Therefore, $f \in P$ and $\lambda(n) = \hat{f}(n)$ imply $f(t) = \hat{\lambda}(t)$.

If μ is a measure in T, then $\hat{\mu}(n) = \mu(e^{-int}) = \int e^{-int} d\mu$. By the theorem of F. and M. Riesz,

$$\hat{\mu}(n) = 0, \forall n \in Z_2 \Rightarrow d\mu = h dt, \ h \in H^1. \tag{2}$$

1. INTEGRAL REPRESENTATION OF POSITIVE DEFINITE GTKs

By a kernel we mean a bounded function $K : Z \times Z \to C$ satisfying $K(m,n) = \overline{K(n,m)}$, $\forall (m,n) \in Z \times Z$.

To each kernel corresponds a sesquilinear continuous bilinear form $b = b_K$ in $\ell^1(Z)$, given by

$$b(\lambda,\gamma) = K[\lambda,\gamma] = \sum_{m,n \in Z} K(m,n)\lambda(m)\overline{\gamma(n)}, \ \forall \lambda \in \ell^1(Z), \ \gamma \in \ell^1(Z). \tag{3}$$

The kernel K is <u>positive definite</u>, p.d., if $b = b_K$ is positive, i.e., if $b(\lambda,\lambda) \geq 0$, $\forall \lambda \in \ell^1$.

To each bounded sequence $s = s(n)$, $n \in Z$, with $s(-n) = \overline{s(n)}$, corresponds a kernel K given by $K(m,n) = s(m-n)$, $(m,n) \in Z \times Z$. Such kernels are called <u>Toeplitz kernels</u> and the following classical characterizations hold.

1.1. (<u>Theorem of Bochner-Herglotz</u> [7]). A sequence $s = s(n)$, $n \in Z$, satisfies $s(n) = \hat{\mu}(n)$ for all $n \in Z$ and for some positive finite measure μ in T iff the Toeplitz kernel $K(m,n) = s(m-n)$ is p.d.

1.2. (<u>Theorems of Bochner-Eberlein</u> [8] <u>and Horn</u> [18]). A sequence $s = s(n)$, $n \in Z$, satisfies $s(n) = \hat{\mu}(n)$ for all $n \in Z$ and for some real finite measure in T iff $s(m-n) = K_1(m,n) - K_2(m,n)$, K_1 and K_2 p.d. Toeplitz kernels, and iff

$$\Big|\sum_{n \in Z} \lambda(n)s(n)\Big| \leq c\|\hat{\lambda}\|_\infty, \ \forall \lambda \in \ell^1(Z) \tag{4}$$

and iff

$$\Big|\sum_{m,n\in Z} s(m+n)\lambda(m)\lambda(n)\Big| \leq \sum_{m,n\in Z} u(m-n)\lambda(m)\overline{\lambda(n)},\ \forall\ \lambda \in \ell^1(Z) \tag{4a}$$

and for some Toeplitz kernel $u(m-n)$.

<u>Remark 1.1</u>. Another variant of this theorem in terms of Schur multipliers is due to G. Bennett [5].

The study of the classes $H_M^{2,2}$ and $[H_M^{2,2},Z]$ (see (1)) leads to the following two closely related notions of p.d. generalized Toeplitz kernels and weakly positive matrix measures. Since $Z \times Z = \bigcup_{\alpha,\beta=1,2} Z_\alpha \times Z_\beta$, in order to define a kernel K it is enough to give its values in each quadrant $Z_\alpha \times Z_\beta$, $\alpha,\beta = 1,2$. We say that K is a <u>generalized Toeplitz kernel</u>, <u>GTK</u>, if there are four sequences $s_{\alpha\beta} = s_{\alpha\beta}(n)$, $\alpha,\beta = 1,2$, $s_{21}(-n) = \overline{s_{12}(n)}$, $n \in Z$, such that

$$K(m,n) = s_{\alpha\beta}(m-n),\ (m,n) \in Z_\alpha \times Z_\beta. \tag{5}$$

In this case we write $K \sim S = (s_{\alpha\beta})$ and call S a defining matrix of sequences for K. Two such defining matrices of sequences, S and S', may give rise to the same kernel K. In fact,

$$K \sim S \text{ and } K \sim S' \iff \begin{array}{l} s'_{\alpha\alpha} = s_{\alpha\alpha},\ \alpha = 1,2 \\ s'_{12}(n) = s_{12}(n) \text{ only for } n > 0. \end{array} \tag{5a}$$

The basic example of a GTK is as follows. Consider 2×2 matrices $M = (\mu_{\alpha\beta})$, $\alpha,\beta = 1,2$, having as elements $\mu_{\alpha\beta}$ (complex) finite measures in T, such that $\mu_{21} = \overline{\mu_{12}}$. Set

$$M[f_1,f_2] = \sum_{\alpha,\beta=1,2} \int f_\alpha \bar{f}_\beta d\mu_{\alpha\beta},\forall (f_1,f_2) \in C(T) \times C(T). \tag{6}$$

To each such $M = (\mu_{\alpha\beta})$ we associate the GTK $K = K_M$ defined by $S = (s_{\alpha\beta})$, $s_{\alpha\beta} = \hat{\mu}_{\alpha\beta}$, $\alpha,\beta = 1,2$. In this case we write $K \sim \hat{M} = (\hat{\mu}_{\alpha\beta})$. It is easy to see that, in this case, $(f_1,f_2) \in P_1 \times P_2$ and $\lambda(n) = (f_1 + f_2)\hat{}(n)$ imply

$$K[\lambda,\lambda] = M[f_1,f_2]. \tag{7}$$

In particular, if $K \sim \hat{M}$, then K is p.d. iff

$$M[f_1,f_2] \geq 0, \ \forall\ (f_1,f_2) \in P_1 \times P_2 \tag{7a}$$

This suggests the following definitions:

$$M \succ 0 \quad \text{if} \quad M[f_1,f_2] \geq 0, \ \forall\ (f_1,f_2) \in P_1 \times P_2 \tag{7b}$$

In this case M is called <u>weakly positive</u>, and M is called <u>positive</u>

$$M \geq 0 \quad \text{if} \quad M[f_1,f_2] \geq 0, \ \forall\ (f_1,f_2) \in P \times P \tag{7c}$$

The equivalence

$$M \geq 0 \quad \text{iff} \ (\mu_{\alpha\beta}(\Delta)) \text{ is p.d. for all Borel sets } \Delta \subset T \tag{7d}$$

is immediate.

By (5a) and (2), two matrix measures, M and M', give rise to the same K,

$$K \sim M \ \text{and}\ K \sim M' \quad \text{iff} \quad \mu'_{\alpha\alpha} = \mu_{\alpha\alpha},\ \alpha = 1,2,$$

$$\mu'_{12} = \mu_{12} + h\,dt,\ h \in H^1(T). \tag{7e}$$

In this case we write $M \equiv M'$.

The following result characterizes the p.d. GTKs in terms of positive matrix measures.

1.3. (Generalized Bochner Theorem [10]). A GTK K is p.d. iff $K \sim \hat{M}$ for some positive matrix measure $M \geq 0$.

In view of the equivalence (7a), this theorem can be restated as:

1.4. (Lifting Property of Weakly Positive Matrix Measures [10]). If $M = (\mu_{\alpha\beta})$, $\alpha,\beta = 1,2$, $\mu_{21} = \overline{\mu_{12}}$ is a 2 × 2 matrix of measures, then the following conditions are equivalent:

(i) $M \succ 0$. (ii) $M \equiv M'$ for some $M' \geq 0$. (iii) There exists $h \in H^1(T)$ such that $M + \begin{pmatrix} 0 & hdt \\ \bar{h}dt & 0 \end{pmatrix} \geq 0$. (iv) $\mu_{11} \geq 0$, $\mu_{22} \geq 0$ and $|\int_\Delta (d\mu_{12} + hdt)|^2 \leq \mu_{11}(\Delta) \cdot \mu_{22}(\Delta)$, $\forall\, \Delta \subset T$ Borel set and for some $h \in H^1(T)$.

The characterizations corresponding to problems (a) and (d) of the Introduction are special cases of Theorem 1.3 (cfr. [10] and [3] respectively). Furthermore, theorems 1.3 and 1.4 yield refinements of the Helson-Szegö theorem and of other results in prediction theory and for Carleson measures (cfr. [10] and [2]).

If $C^2 = C \times C$ is the 2-dimensional Hilbert space and $E = L(C^2)$ are the linear operators acting on C^2, then every 2 × 2 matrix measure $M = (\mu_{\alpha\beta})$ can be considered as a measure $M(\Delta) = (\mu_{\alpha\beta}(\Delta))$ with values in E.

If $K \sim S = (s_{\alpha\beta})$ is a GTK, then s_{11}, s_{22}, as well as $s_{12}(n) = s_{21}(-n)$ for $n > 0$, are well determined by K. Therefore, if each $\lambda_{\alpha\beta} \in \ell^1(Z)$, $\alpha,\beta = 1,2$, and $\lambda_{12}(n) = \overline{\lambda_{21}(-n)}$ has support in Z_2, then the number

$$K[(\lambda_{\alpha\beta})] = \sum_{\alpha,\beta=1,2} \sum_{n\in Z} s_{\alpha\beta}(n)\lambda_{\alpha\beta}(n) \tag{8}$$

depends only on K and $(\lambda_{\alpha\beta})$.

These remarks lead to the following new result.

1.5. <u>(Generalized Theorems of Bochner-Eberlein and Horn)</u>. If $K \sim S = (s_{\alpha\beta})$ is a GTK the following conditions are equivalent:

(i) $K \sim \hat{M} = (\hat{\mu}_{\alpha\beta})$, where $\mu_{21} = \overline{\mu_{12}}$ and $M(\Delta) = (\mu_{\alpha\beta}(\Delta))$ is an E-valued measure of bounded variation, $E = L(C^2)$.

(ii) $K = K_1 - K_2$, with K_1, K_2 p.d. GTKs.

(iii) $|K[(\lambda_{\alpha\beta})]| \leq c \sup_t \{\sup_{\alpha,\beta} |\hat{\lambda}_{\alpha\beta}(t)|\}$ whenever $\lambda_{\alpha\beta} \in \ell^1(Z)$, $\alpha,\beta = 1,2$, and $\lambda_{12}(n) = \overline{\lambda_{21}(-n)}$ has support in Z_2.

(iv) There exists a GTK K' such that

$$|\sum_{m,n} K(m,-n)\lambda(m)\lambda(n)| \leq \sum_{m,n} K'(m,n)\lambda(m)\overline{\lambda(n)}, \forall \lambda \in \ell^1(Z). \quad (9)$$

<u>Remark 1.2</u>. Corollary 1 of [24] provides yet another equivalent characterization.

<u>Proof</u>. <u>(i) and (ii) ard equivalent</u>: immediate consequence of Theorem 1.3. <u>(i) and (iii) are equivalent</u>: Let $P(E) = \{(f_{\alpha\beta}) : f_{\alpha\beta} \in P, \alpha,\beta = 1,2\}$ be the set of E-valued trigonometric polynomials, and $P_*(E) = \{(f_{\alpha\beta}) \in P(E) : f_{21} = \overline{f_{12}} \in P_2\}$. Then, if $(f_{\alpha\beta}) \in P_*(E)$, $\hat{f}_{\alpha\beta} = \lambda_{\alpha\beta} \in \ell^1(Z)$, $\alpha,\beta = 1,2$, with $\lambda_{12}(n) = \overline{\lambda_{21}(-n)}$ having support in Z_2. Then (8) defines a linear functional ℓ in $P_*(E)$ given by $\ell((f_{\alpha\beta})) = K[(\hat{f}_{\alpha\beta})]$, and condition (iii) insures that ℓ extends to a bounded linear functional in $P(E) \subset C(E)$. By the Riesz representation theorem for C(E) (cfr. [14]), such ℓ is given by an E-valued measure M, and (i) follows from (7). Conversely, the same argument applies. <u>(iv) implies (iii)</u>: Condition (9) implies that K' is p.d. and, by Theorem 1.3, $K' \sim \hat{M}$ with $M \geq 0$. Let ℓ be the linear functional in $P_*(E)$ defined above. If $f_1 \in P_1$, $f_2 \in P_2$, $\lambda(n) = f_1(n)$ for $n \in Z_1$ and $\lambda(n) = f_2(n)$ for

$n \in Z_2$, $f_{\alpha\beta} = f_\alpha \overline{f_\beta}$, $\alpha,\beta = 1,2$, then $(f_{\alpha\beta}) \in P_*(E)$ and $\ell((f_{\alpha\beta})) = \Sigma_{m,n} K(m,-n)\lambda(m)\lambda(n)$. By (9),

$$|\ell((f_{\alpha\beta}))| \leq \sum_{m,n} K'(m,n)\lambda(m)\overline{\lambda(n)} = \sum_{\alpha,\beta} \int f_\alpha \overline{f_\beta} d\mu_{\alpha\beta}$$

$$\leq c \sup_t\{|f_1(t)|^2 + |f_2(t)|^2\}.$$

Thus, if $\ell_{\alpha\beta}$ is the linear functional defined in P by $\ell_{\alpha\beta}(f) = \Sigma_n s_{\alpha\beta}(n)\hat{f}(n)$, $\alpha,\beta = 1,2$, so that $\ell((f_{\alpha\beta})) = \Sigma_{\alpha,\beta}\ell_{\alpha\beta}(f_{\alpha\beta})$, then

$$|\sum_{\alpha,\beta} \ell_{\alpha\beta}(f_\alpha \overline{f_\beta})| \leq c \sup_t\{|f_1(t)|^2 + |f_2(t)|^2\}$$

whenever $(f_1,f_2) \in P_1 \times P_2$. Taking $f_2 \in P_2$, $f_1 = \overline{f_2} \in P_1$, we get, since $\ell_{21} = \overline{\ell_{12}}$, $\ell_{11}(|f_1|^2) \leq c_1\|f_1\|_\infty^2$, $\ell_{22}(|f_1|^2) \leq c_1\|f_1\|_\infty^2$, $|\ell_{12}(f_1^2)| \leq c_1\|f_1\|_\infty^2$ whenever $f_1 \in P_1$, $\overline{f_1} \in P_2$. Since every $f \in P$ is of the form $f = e^{int}f_1$ for some $n < 0$ and $f_1 \in P_1$, and since $4f_1 = (f_1 + 1)^2 - (f_1 - 1)^2$, it follows easily from the preceding inequalities that $|\ell_{\alpha\beta}(f)| \leq c\|f\|_\infty$ $\forall f \in P$, $\alpha,\beta = 1,2$. Therefore, $\ell((f_{\alpha\beta})) = \Sigma_{\alpha,\beta}\ell_{\alpha\beta}(f_{\alpha\beta}) = K[(\hat{f}_{\alpha\beta})] = K[(\lambda_{\alpha\beta})]$ satisfies (iii). <u>(ii) implies (iv)</u>: If $K = K_1 - K_2$, K_1,K_2 p.d. GTKs, (9) is satisfied taking $K' = K_1 + K_2$. Q.E.D.

<u>Remark 1.3</u>. G. Bennet's characterization, mentioned in Remark 1.1, can also be extended to the case of GTKs, but will be considered elsewhere.

2. GROTHENDIECKIAN INTERPRETATIONS

A sesquilinear form b will be called <u>Toeplitz</u> (respectively, <u>generalized Toeplitz</u>) if $b = b_K$ with K a Toeplitz kernel (respectively, a GTK). The Toeplitz kernels and the GTKs will be considered now from the point of view of sesquilinear forms.

Theorems 1.1 and 1.2 lead to the notion of Hilbertian and integral forms as follows (cfr. [15]). If K is a p.d. Toeplitz kernel and if $b = b_K$ is given by (3), then by Theorem 1.1 there is a measure $\mu \geq 0$ such that $K(m,n) = \int e^{-imt} e^{int} d\mu$, and $b(\lambda,\gamma) = \int \hat{\lambda}(t)\overline{\hat{\gamma}(t)} d\mu$. If $H = L^2(\mu)$ and if $\Phi : \ell^1(Z) \to H$ is given by $\ell^1 \ni \lambda \mapsto \Phi_\lambda = \hat{\lambda}(t) \in H$, then we can write

$$b(\lambda,\gamma) = (\Phi_\lambda, \Phi_\gamma)_H, \ \forall\, \lambda \in \ell^1(Z),\ \gamma \in \ell^1(Z). \tag{10}$$

If $K = K_1 - K_2$, with K_1, K_2 p.d. Toeplitz kernels, and $b = b_K$, then it is easy to deduce from (10) that there exists a Hilbert space H and two bounded linear operators $\Phi : \ell^1 \to H$, $\Psi : \ell^1 \to H$ such that

$$b(\lambda,\gamma) = (\Phi_\lambda, \Phi_\gamma)_H, \ \forall\, \lambda \in \ell^1(Z),\ \gamma \in \ell^1(Z). \tag{10a}$$

(In fact, by (10), $b_\alpha = b_{K_\alpha} = (\Phi^\alpha_\lambda, \Phi^\alpha_\gamma)_{H_\alpha}$, $\alpha = 1,2$, and setting $H = H_1 \oplus H_2$, $\Phi_\lambda = \Phi^1_\gamma \oplus (-\Phi^2_\lambda)$, $\Psi_\gamma = \Psi^1_\gamma \oplus \Psi^2_\gamma$, we get (10a), since $\Phi^\alpha_\lambda, \Psi^\beta_\gamma$ act in orthogonal subspaces.)

An arbitrary sesquilinear form b is said to be <u>Hilbertian</u> if it admits a representation of the form (10a); b is said to be <u>positively Hilbertian</u> if it admits a representation of the form (10). The following general property is true: <u>b is positively Hilbertian iff $b \geq 0$, and b is Hilbertian iff it is a linear combination of positive forms.</u>

As pointed out by Gilbert, Theorem 1.2 leads to the following characterization of Hilbertian forms: the bilinear form b is Hilbertian iff there exist two positive sesquilinear forms P, Q, and a constant A, such that

$$|b(\lambda,\gamma)| \leq A(P(\lambda,\lambda))^{1/2}(Q(\gamma,\gamma))^{1/2}, \ \forall\, \lambda,\gamma \tag{10b}$$

If (10a) holds for b with $H = L^2(\mu)$ and $\Phi_\lambda, \Psi_\gamma \in L^\infty$, then b is said to be <u>integral</u>. From (10b) it follows easily that every integral form is Hilbertian.

Conversely, the invariant Hilbertian forms are integral. $K(m,n) = s(m-n)$ is a Toeplitz kernel if it satisfies the invariance property $K(m+1,n+1) = K(m,n)$ or, equivalently, if $b = b_K$ satisfies $b(\tau\lambda,\tau\gamma) = b(\lambda,\gamma)$, where $(\tau\lambda)(n) = \lambda(n+1)$, $\forall$ n. Let us say that b is an <u>invariant Hilbertian form</u> if the operators Φ, Ψ in (10a) have the additional invariance property

$$(\Phi_{\tau\lambda},\Phi_{\tau\gamma}) = (\Phi_\lambda,\Phi_\gamma),(\Psi_{\tau\lambda},\Psi_{\tau\gamma}) = (\Psi_\lambda,\Psi_\gamma),(\Phi_{\tau\lambda},\Psi_{\tau\gamma}) = (\Phi_\lambda,\Psi_\gamma) \tag{10c}$$

Thus, b is a p.d. Toeplitz form iff it is an invariant positively Hilbertian form.

Similarly, K is a GTK iff it satisfies the generalized invariance property: $K(m+1,n+1) = K(m,n)$ whenever $(m+1,n+1)$ and (m,n) belong to the same quadrant $Z_\alpha \times Z_\beta$, $\alpha,\beta = 1,2$. Or, equivalently, if the associated form $b = b_K$ satisfies the generalized invariance property. Accordingly, let us say that b is a <u>generalized invariant Hilbertian form</u> if (10c) holds whenever the supports of $\tau\lambda$ and λ are contained in the same halfline Z_α, and the supports of $\tau\gamma$ and γ are contained in the same Z_β, $\alpha,\beta = 1,2$. Then b is a p.d. generalized Toeplitz form iff it is a generalized invariant positively Hilbertian form. And b is a linear combination of p.d. generalized Toeplitz forms if it is a generalized invariant Hilbertian form.

All these properties are consequences of the above-mentioned general property of positive sesquilinear forms.

Let us see now how Theorems 1.3 and 1.5 lead to another Grothendieckian interpretation. Let $C^2 = C \times C$ be the two-dimensional Hilbert space, so that every 2×2 matrix can be considered as an element of $L(C^2)$, that is, a linear operator in C^2. If $K \sim (s_{\alpha\beta})$ is a GTK, $K_{\alpha\beta}(m,n) = s_{\alpha\beta}(m-n)$ are the four corresponding Toeplitz kernels, and $b = b_K$, $b_{\alpha\beta} = b_{K_{\alpha\beta}}$ are the sesquilinear forms associated to these kernels, then K is determined by the matrix $(K_{\alpha\beta})$ and b by the matrix $(b_{\alpha\beta})$, through the formula

$$b(\lambda,\gamma) = b_{\alpha\beta}(\lambda,\gamma) \text{ whenever supp } \lambda \subset Z_\alpha, \text{ supp } \gamma \subset Z_\beta. \tag{11}$$

Now $(K_{\alpha\beta})$ can be considered as a kernel with values in $L(C^2)$, and $(b_{\alpha\beta})$ as a sesquilinear form with values in $L(C^2)$. Thus the notion of GTK leads to the consideration of the set $\mathcal{B}(\ell^1;L(C^2))$ of all $L(C^2)$-valued sesquilinear forms $B(\lambda,\gamma)$ defined for $\lambda \in \ell^1$, $\gamma \in \ell^1$. We shall first discuss some concepts concerning such forms. Observe that there is a one-to-one correspondence between the set $\mathcal{B}(\ell^1;L(C^2))$ and the set $\mathcal{B}(\ell^1 \times \ell^1)$ of all scalar sesquilinear forms $b(\Lambda,\Gamma)$ defined for $\Lambda = (\lambda_1,\lambda_2) \in \ell^1 \times \ell^1$, $\Gamma = (\gamma_1,\gamma_2) \in \ell^1 \times \ell^1$. In fact, to each $B(\lambda,\gamma) \in \mathcal{B}(\ell^1;L(C^2))$ corresponds the form $b(\Lambda,\Gamma) \in \mathcal{B}(\ell^1 \times \ell^1)$ given by

$$b(\Lambda,\Gamma) = b((\lambda_1,\lambda_2),(\gamma_1,\gamma_2)) = \sum_{\alpha,\beta=1,2} b_{\alpha\beta}(\lambda_\alpha,\lambda_\beta) \tag{11a}$$

where $B(\lambda,\gamma) = (b_{\alpha\beta}(\lambda,\gamma))$, $\lambda_1,\lambda_2,\gamma_1,\gamma_2 \in \ell^1(Z)$.

Thus, each 2×2 matrix $(b_{\alpha\beta}(\lambda,\gamma))$ can be considered either as an element $B(\lambda,\gamma) \in \mathcal{B}(\ell^1;L(C^2))$, or as element $b(\Lambda,\Gamma) \in \mathcal{B}(\ell^1 \times \ell^1)$. According to the usual definition, an element $b \in \mathcal{B}(\ell^1 \times \ell^1)$ is positive if $b(\Lambda,\Lambda) \geq 0$, $\forall\, \Lambda \in \ell^1 \times \ell^1$, while $B \in \mathcal{B}(\ell^1;L(C^2))$ is said to be positive if the numerical matrix $B(\lambda,\lambda) = ((b_{\alpha\beta}(\lambda,\lambda))$ is p.d., $\forall\, \lambda \in \ell^1$. Both definitions agree through the above-mentioned correspondence $B \leftrightarrow b$.

$B = (b_{\alpha\beta})$ is invariant or Toeplitz if each $b_{\alpha\beta}$ is so, and b is Toeplitz if $b(\tau\Lambda,\tau\Gamma) = b(\Lambda,\Gamma)$, where $\tau\Lambda = (\tau\lambda_1,\tau\lambda_2)$ if $\Lambda = (\lambda_1,\lambda_2)$.

Observe that if H is a Hilbert space and $C = \{c1\}$ is the one-dimensional space, then to give an element $\xi \in H$ is the same as to give an operator $V_\xi : C \to H$, $V_\xi(c1) = c\xi$. If $\xi \mapsto V_\xi$, $\eta \mapsto V_\eta$ then $(\xi,\eta)_H = V_\eta^* V_\xi (= V_\eta^* V_\xi(1))$. According to this remark we give the following definition: A matrix form $B = (b_{\alpha\beta}) \in \mathcal{B}(\ell^1; L(C^2))$ is <u>Hilbertian</u> if there exist two bounded linear operators $U : \ell^1(Z) \to L(C^2;H)$, $V : \ell^1(Z) \to L(C^2;H)$ (H is a Hilbert space) assigning to each $\lambda \in \ell^1$ the operators $U : C^2 \to H$, $V_\lambda : C^2 \to H$ such that

$$B(\lambda,\gamma) = (b_{\alpha\beta}(\lambda,\gamma)) = V_\gamma^* U_\lambda . \tag{12}$$

On the other hand, a form $b(\Lambda,\Gamma) \in \mathcal{B}(\ell^1 \times \ell^1)$ will be said to be Hilbertian if it is Hilbertian in the usual way, that is, if there exist two operators $\ell^1 \times \ell^1 \ni \Lambda \mapsto \Phi_\Lambda \in H$ and $\ell^1 \times \ell^1 \ni \Gamma \mapsto \Psi_\Gamma \in H$, such that

$$b(\Lambda,\Gamma) = (\Phi_\Lambda, \Psi_\Gamma)_H, \quad \forall\ \Lambda,\Gamma \in \ell^1 \times \ell^1 \tag{12a}$$

It is not hard to verify that <u>both definitions, (12) and (12a), agree through the correspondence $B \leftrightarrow b$</u>. In fact, setting $u_1(\lambda) = U_\lambda \binom{1}{0}$, $u_2(\lambda) = U_\lambda \binom{0}{1}$, $\binom{1}{0} \in C^2$, $\binom{0}{1} \in C^2$, the generators of C^2, then $U_\lambda \binom{a}{b} = au_1(\lambda) + bu_2(\lambda)$, and

$$\begin{aligned} \left\langle B(\lambda,\gamma) \binom{c_1}{c_2}, \binom{d_1}{d_2} \right\rangle &= \sum_{\alpha,\beta=1,2} b_{\alpha\beta}(\lambda,\gamma) c_\alpha \overline{d_\beta} \\ &= \sum_{\alpha,\beta=1,2} (c_\alpha u_\lambda^\alpha, d_\beta u_\gamma^\beta) \\ &= \sum_{\alpha,\beta=1,2} ((V_\gamma^\beta)^* U_\lambda^\alpha c_\alpha, d_\beta). \end{aligned}$$

Therefore, $b_{\alpha\beta}(\lambda,\gamma) = ((V_\gamma^\beta)^* U_\lambda^\alpha)\, 1$. Since (11a) gives

$$b(\Lambda,\Gamma) = \sum_{\alpha,\beta=1,2} b_{\alpha\beta}(\lambda_\alpha,\gamma_\beta) = \sum_{\alpha,\beta=1,2} (V^\beta_{\gamma_\beta})^* U^\alpha_{\lambda_\alpha}$$

we get

$$b(\Lambda,\Gamma) = (\Phi_\Lambda,\Psi_\Gamma) \text{ with } \Phi_\Lambda = U\binom{\lambda_1}{\lambda_2} = u^1_{\lambda_1} + u^2_{\lambda_2} \tag{12b}$$

This proves that (12) and (12a) agree, and (12b) gives the relation between Φ_Λ and U_λ.

From the equivalence of both types of definitions we get the following property (cfr. Remark 2.2 below):

2.1. A form $B \in \mathcal{B}(\ell^1;L(C^2))$ is positive iff it is positively Hilbertian, hence a $L(C^2)$-valued kernel $(K_{\alpha\beta})$ is p.d. if $B = (b_{K_{\alpha\beta}})$ is positively Hilbertian. Similarly, $B \in \mathcal{B}(\ell^1;L(C^2))$ is a linear combination of positive forms if it is Hilbertian.

A form $B = (b_{\alpha\beta}) \in \mathcal{B}(\ell^1;L(C^2))$ will be called an <u>integral form</u> if there exist a 2×2 matrix measure $M = (\mu_{\alpha\beta}) \geq 0$ and two applications $\ell^1(Z) \ni \lambda \mapsto P_\lambda \in L^\infty$, $\ell^1(Z) \ni \lambda \mapsto Q_\lambda$ such that

$$B(\lambda,\gamma) = (b_{\alpha\beta}(\lambda,\gamma)) = (\int P_\lambda \overline{Q_\gamma}\, d\mu_{\alpha\beta})_{\alpha,\beta=1,2} \tag{13}$$

The integral forms $b(\Lambda,\Gamma) \in \mathcal{B}(\ell^1 \times \ell^1)$ are defined in the usual way, and again the definitions agree. The invariant Hilbertian forms are integral.

If we set $\ell^1_1 = \{\lambda \in \ell^1(Z) : \text{supp } \lambda \subset Z_1\}$, $\ell^1_2 = \{\lambda \in \ell^1(Z) : \text{supp } \lambda \subset Z_2\}$, then $\ell^1(Z) = \ell^1_1 \oplus \ell^1_2$, and $\ell^1(Z)$ can be identified with the subspace $\ell^1_1 \times \ell^1_2$ of $\ell^1 \times \ell^1$. Therefore, if $b \in \mathcal{B}(\ell^1 \times \ell^1)$, then the restriction of $b(\Lambda,\Gamma)$ to $\ell^1_1 \times \ell^1_2$ can be identified with a form $b(\lambda,\gamma)$, $\lambda,\gamma \in \ell^1(Z)$. In this case we write

$$b = \text{proj } \mathcal{b}, \ \mathcal{b} = \text{a lifting of b to } \ell^1 \times \ell^1. \tag{14}$$

If $b = b_K$, $b_{\alpha\beta} = b_{K_{\alpha\beta}}$, $B = (b_{\alpha\beta})$ and $\mathcal{b} \leftrightarrow B$, then

$$b = \text{proj } \mathcal{b} \text{ iff } K(m,n) = K_{\alpha\beta}(m,n) \text{ for } (m,n) \in Z_\alpha \times Z_\beta, \quad \alpha,\beta = 1,2. \tag{14a}$$

After these considerations we can give the Grothendieckian interpretation of Theorems 1.3 and 1.5. By (14a), K(m,n) is a GTK iff there exists a Toeplitz (invariant) form $\mathcal{b} \in \mathcal{B}(\ell^1 \times \ell^1)$ such that $b_K = \text{proj } \mathcal{b}$. That is, K is a GTK if b_K can be lifted to a Toeplitz form. But the Toeplitz lifting $\mathcal{b}(\Lambda,\Gamma)$, as well as the Toeplitz lifting $(K_{\alpha\beta}(m,n)) = (s_{\alpha\beta}(m-n))$ of K, is not unique. Therefore, in dealing with GTKs, it is usual to encounter problems of the following type: If $b = b_K$ (K a GTK) satisfies a certain property (P), does there exist some Toeplitz lifting of b satisfying the same property? For instance, if K is a GTK and $b = b_K$ is positive (i.e., K is p.d.), does there exist a Toeplitz lifting $B(\lambda,\gamma) \leftrightarrow \mathcal{b}(\Lambda,\Gamma)$, of b, such that B is also positive? Theorem 1.3 is just an affirmative answer to this last question. Thus Theorem 1.3 expresses a lifting property: If $b = b_K \geq 0$ then there exists a Toeplitz lifting $\mathcal{b} \geq 0$. (Cfr. this lifting formulation of 1.3 with the lifting property of 1.4, which is a restatement of 1.3.)

Similarly, the Grothendieckian properties can be lifted: If $b = b_K$ is invariant Hilbertian (K a GTK), then there exists an invariant Hilbertian Toeplitz lifting $\mathcal{b}$ of b. But of course not every Toeplitz lifting of b will have such properties.

If for $B \leftrightarrow \mathcal{b}$, $B' \leftrightarrow \mathcal{b}'$ we write $B \sim B'$ whenever $\text{proj } \mathcal{b} = \text{proj } \mathcal{b}'$, then we may say that a GTK, or a generalized Toeplitz form, coincides with a class of forms $B \in \mathcal{B}(\ell^1;L(C^2))$ (or $\mathcal{b}$ $\mathcal{B}(\ell^1 \times \ell^1)$) modulo the equivalence relation $\sim$. The study of the GKTs is thus the same as the study of the Toeplitz forms $B \in \mathcal{B}(\ell^1;L(C^2)) \sim \mathcal{B}(\ell^1 \times \ell^1)$ modulo the equivalence relation $\sim$. Therefore, the sense of the basic properties of GTKs, such as 1.3 and 1.5, is that positive or Hilbertian forms $b = b_K$, satisfying certain properties, can be lifted to forms in $\ell^1 \times \ell^1$ with the same properties.

In Section 3 we shall see that if $b = b_K$ satisfies a majorization property $b \leq \sigma$, then there is also a Toeplitz lifting $\mathcal{b}$ to $\ell^1 \times \ell^1$ satisfying the same majorization property, and moreover in a stronger vectorial form. This will give some inequalities of Marcinkiewicz-Zygmund type for GTKs.

Remark 2.1. Theorem 1.5 can be interpreted as the Gilbert characterization (10b) for the forms $\mathcal{b} \in \mathcal{B}(\ell^1 \times \ell^1)$.

Remark 2.2. The considerations of this section apply also when $B = (b_{\alpha\beta})$ takes values in $L(E)$, E any Hilbert space. The property 2.1, that $B(\lambda,\gamma)$ is p.d. iff it is positively Hilbertian, was proved by A. Weron [25], for general $L(E)$-valued forms, from the different point of view of stationary processes (cfr. also [20]).

Remark 2.3. The considerations of this paper can be extended to kernels $K(x,y)$, $(x,y) \in R \times R$, following the idea indicated in [2], but such extension is far from immediate and will be considered elsewhere. Some of the results on GTKs can be carried to the two-dimensional case of T^2 and Z^2, by considering instead of the partition $Z = Z_1 \cup Z_2$, partitions of Z^2 into N disjoint cones (cfr. [13]). In this case the liftings obtained are to $L(E)$-valued forms $B(\lambda,\gamma)$, $\dim E = N$.

3. VECTOR TRANSFERENCE OF MAJORIZED GTKs

The Grothendieck Fundamental Theorem asserts that in the case, considered here, of bilinear forms in $\ell^1(Z)$ (or in general L^1 spaces), all Hilbertian forms are integral. Another version of this theorem is

3.1. (Grothendieck's Fundamental Inequality [17]). If the kernel K(m,n) satisfies

$$\left|\sum_{m,n} K(m,n)\lambda(m)\gamma(n)\right| \leq \sup_m|\lambda(m)|\sup_n|\gamma(n)| \tag{15}$$

for every pair of sequences λ,γ of finite support, then

$$\left|\sum_{m,n} K(m,n)\langle\xi(m),\eta(n)\rangle\right| \leq K_G \sup_m\|\xi(m)\| \sup_n\|\eta(n)\| \tag{15a}$$

for every pair of H-valued sequences $\xi(m),\eta(n)$, of finite support, where H is a Hilbert space and K_G is the Grothendieck universal constant.

Roughly speaking, Theorem 3.1 says that the scalar inequality (12) can be transferred to H-valued sequences ξ,η. Equivalently, if the linear operator $T : C_0 \to L^1$ satisfies

$$\int|Tf| \leq \|T\| \sup_t |f(t)| \tag{15b}$$

then

$$\int\|TF\| \leq M_G\|T\| \sup_t \|F(t)\|$$

for $H \cong \ell^2$-valued functions $F \in L^1(H)$. This result complements an earlier theorem:

3.2. (Theorem of Marcinkiewicz-Zygmund [19]). If a linear operator $T : L^p \to L^q$, $1 < p,q < \infty$, satisfies

$$(\int |Tf|^q)^{1/q} \leq \|T\| (\int |f|^p)^{1/p} \tag{15c}$$

then the same inequality holds for ℓ^2-valued functions $F \in L^p(\ell^2)$, but with constant $M_{pq}\|T\|$ (if $p = q$, $M = 1$). Moreover, this is true also for ℓ^γ-valued functions $F \in L^p(\ell^\gamma)$, provided $p < \gamma$, $q < \gamma$, $0 < \gamma < 2$.

Of course, in the case of a particular K, or a particular T, there might be other variants of these inequalities.

In the preceding section we showed that the property $b_K \geq 0$ (or b_K a linear combination of such forms), for K a GTK, can be lifted to $\ell^1 \times \ell^1$, and that this fact has a Grothendieckian interpretation. We shall see now that the property $b_K \leq \sigma$ defined below can also be lifted, and that this fact furnishes vector transference properties for GTKs.

Let σ be a seminorm in $C(T)$ satisfying conditions

(σ_1) $|f| \leq |g|$ implies $\sigma(f) \leq \sigma(g)$

(σ_2) $\sigma(f) \leq c\|f\|_\infty$

Additional conditions that may be required are that σ be defined in L^∞ and

(σ_3) $f_n \downarrow f$ implies $\sigma(f_n) \to \sigma(f)$

(σ_4) $f = 0$ a.e. implies $\sigma(f) = 0$

In this paper we are interested mainly in the case when $\sigma(f) = (\int |f|^p d\mu)^{1/p}$, $\mu \geq 0$, $1 \leq p < \infty$ (cfr. [12] and [24]).

If K is a GTK and $M = (\mu_{\alpha\beta})$ a 2×2 matrix measure, such that $\mu_{21} = \overline{\mu_{12}}$, then we write (cfr. the basic notations (7b) and (7c)):

$$K \leq \sigma \quad \text{if} \quad K[\lambda,\lambda] \leq \sigma(|\hat{\lambda}|^2), \ \forall\, \lambda \in \ell^1(Z) \tag{16}$$

$$M \prec \sigma \quad \text{if} \quad M[f_1,f_2] \leq \sigma(|f_1 + f_2|^2), \ \forall\ (f_1,f_2) \in P_1 \times P_2 \tag{16a}$$

$$M \leq \sigma \quad \text{if} \quad M[f_1,f_2] \leq \sigma(|f_1 + f_2|^2), \ \forall\ (f_1,f_2) \in P \times P \tag{16b}$$

$$K \ll \sigma \quad \text{if} \quad K[\lambda,\lambda] \leq \sigma(|\hat{\lambda}|^2) \text{ only for } \hat{\lambda} \text{ real} \quad \text{i.e.: } \lambda(-n) = \overline{\lambda(n)} \tag{16c}$$

$$M \prec\!\prec \sigma \quad \text{if} \quad \text{(16a) holds only for } f_1 + f_2 \text{ real} \quad \text{i.e.: } f_1 = \overline{f_2} \tag{16d}$$

For simplicity only matrix measures M satisfying $\mu_{11} = \mu_{22}$, and GTKs $K \sim (s_{\alpha\beta})$ satisfying $s_{11} = s_{22}$, will be considered here. While this is enough for most applications, it can be seen in [24] that most results hold in the general case, with a change in constants.

It was proved in [12] (cfr. also [24]) that the above majorization properties can be lifted. More precisely,

3.3. (<u>Lifting properties of majorized matrix measures</u> [12]). Let $M = (\mu_{\alpha\beta})$, $\mu_{21} = \overline{\mu_{21}}$, $\mu_{11} = \mu_{22}$, be a 2×2 matrix of Radon measures in T, and σ a seminorm in C(T) satisfying conditions (σ_1) and (σ_2).

(a) If $\mu_{11} = \mu_{22} = 0$ then $M \prec \sigma$ iff $M \equiv N$ for some $N \leq \sigma$.

(b) If $M \prec \sigma$ then $M \equiv N$ for some $N \leq 2\sigma$, either if $\mu_{11} \geq 0$ or if σ satisfies (σ_3).

(c) If $\mu_{11} = \mu_{22} = 0$ then $M \prec\!\prec \sigma$ implies $M \equiv N$ for some $N \leq 4\sigma$.

(d) If σ satisfies (σ_1), (σ_2), (σ_3) and (σ_4), $M \prec \sigma$ implies $d\mu_{\alpha\beta} = w_{\alpha\beta}(t)dt$, $w_{\alpha\beta} \in L^1(T)$, $\alpha,\beta = 1,2$.

3.4. (<u>Generalized Bochner Theorem for Majorized GTK</u> [12]). Let $K \sim (s_{\alpha\beta})$ be a GTK, $s_{11} = s_{22}$, N a 2×2 matrix of Radon measures in T and σ a seminorm satisfying conditions (σ_1) and (σ_2).

(a) If $s_{11} = s_{22} = 0$ then $K \leq \sigma$ iff $K \sim N^{\wedge}$ for some $N \leq \sigma$.

(b) $K \leq \sigma$ implies $K \sim N^{\wedge}$ for some $N \leq 2\sigma$, either if s_{11} is a positive definite sequence or if σ satisfies (σ_3).

(c) If $s_{11} = s_{22} = 0$ then $K << \sigma$ implies $K \sim N^{\wedge}$ for some $N \leq 4\sigma$.

Remark 3.1. Theorems 3.3 and 3.4 extend to the cases $\mu_{11} \neq \mu_{22}$, $s_{11} \neq s_{22}$, as shown in [24].

In view of 3.3(d), only matrix measures M with elements $d\mu_{\alpha\beta} = w_{\alpha\beta}dt$, $w_{11} = w_{22}$, and GTKs $K \sim (s_{\alpha\beta})$, with $s_{\alpha\beta}(n) = \hat{w}_{\alpha\beta}(n)$, $w_{\alpha\beta} \in L^1(T)$, will be considered here. We write $W = (w_{\alpha\beta})$ instead of M (see Remark 3.2 below).

The next theorem contains the main result about vector transference.

3.5. (Vector Transference for Majorized Matrix Measures). Let $W = (w_{\alpha\beta}dt)$, $\alpha,\beta = 1,2$, be a matrix measure with $w_{11} = w_{22}$, $w_{21} = \overline{w_{12}}$, and B be a Banach space, B' its dual.

(I) $W \prec \sigma$ iff $\exists h \in H^1(T)$ such that the following three conditions are satisfied:

$$\mathrm{Re}\ (w_{12} - w_{11} + h) \geq 0 \tag{17}$$

$$\int |f|^2 w_{11} dt \leq 2\sigma(|f|^2),\ \forall f \in P \tag{17a}$$

and

$$\mathrm{Re} \int \langle g_1(t), g_2(t) \rangle (w_{12} - w_{11} + h)dt - \int \|g_2\|^2_{B'}\ \mathrm{Re}\ (w_{12} - w_{11} + h)dt \leq c\sigma(\|g_1\|^2_B) \tag{17b}$$

whenever $g_1 \in P_B$ is a B-valued trig. polynomial and $g_2 \in P_{B'}$ is a B'-valued trig. polynomial. The constant c is less than or equal to 4.

(II) If $\mathcal{B} = \ell^2$ = Hilbert space, then condition (17b) can be replaced by any of the two following inequalities:

$$\sum_{\alpha,\beta=1,2} \int \langle g_\alpha, g_\beta \rangle w_{\alpha\beta} dt + 2\mathrm{Re} \int \langle g_1, g_1 \rangle h dt \qquad (18)$$

$$\leq c\sigma(\|g_1 + g_2\|^2), \ \forall\ (g_1, g_2) \in P_{\mathcal{B}}$$

or

$$\sum_{\alpha,\beta=1,2} \int \langle g_\alpha, g_\beta \rangle w_{\alpha\beta} dt \leq c\sigma(\|g_1 + g_2\|^2), \qquad (18a)$$

$$\forall\ (g_1, g_2) \in (P_1)_{\mathcal{B}} \times (P_2)_{\mathcal{B}}$$

(III) Similar properties hold for $W \prec\!\prec \sigma$ if $w_{11} = w_{22} = 0$.

<u>Remark 3.2.</u> Since $W \prec \sigma$ means that $\forall\ (f_1, f_2) \in P_1 \times P_2$,

$$\int f_1 \overline{f_1} w_{11} dt + \int f_2 \overline{f_2} w_{22} dt + \int f_1 \overline{f_2} w_{12} dt + \int f_2 \overline{f_1} w_{21} dt \leq \sigma(|f_1 + f_2|^2),$$

(18a) is a <u>vector transference</u> of the condition $W \prec \sigma$, and (18) is a lifting of (18a) to $P_{\mathcal{B}} \times P_{\mathcal{B}}$.

<u>Remark 3.3.</u> Theorem 3.5 extends to general matrix measures M with $\mu_{11} \neq \mu_{22}$, using the argument given in [24] for the proof of Theorem 3.3.

<u>Proof.</u> To avoid obscuring the point of the proof we shall consider only the case $w_{11} = w_{22} \geq 0$. The case of an arbitrary real w_{11} can be treated by the same argument given in the proof of Theorem 3.3 (cfr. [24]). $W \prec \sigma$ means

$$W[f_1, f_2] = \sum_{\alpha,\beta=1,2} \int f_\alpha \overline{f_\beta} w_{\alpha\beta} dt \leq \sigma(|f_1 + f_2|^2) \qquad (19)$$

$$\forall\ (f_1, f_2) \in P_1 \times P_2.$$

By Theorem 3.3(b), there exists $h \in H^1(T)$ such that (since $w_{11} = w_{22}$)

$$\int(|f_1|^2+|f_2|^2)w_{11}dt + 2\mathrm{Re}\int f_1\overline{f_2}(w_{12}+h)dt \le 2\sigma(|f_1+f_2|)^2 \quad (19a)$$

$$\forall\ (f_1,f_2)\in P\times P,$$

or equivalently,

$$\int|f_1+f_2|^2w_{11}dt + 2\mathrm{Re}\int f_1\overline{f_2}(w_{12}-w_{11}+h)dt \le 2\sigma(|f_1+f_2|^2) \quad (19b)$$

$$\forall\ (f_1,f_2)\in P\times P.$$

Letting $f_2 = 0$ in (19b) we get

$$\int|f|^2w_{11}dt \le 2\sigma(|f|^2),\ \forall\ f\in P \quad (17a)$$

and therefore

$$\mathrm{Re}\int f_1\overline{f_2}(w_{12}-w_{11}+h)dt \le \sigma(|f_1+f_2|^2),\ \forall\ (f_1,f_2)\in P\times P \quad (19c)$$

Inequalities (17a) and (19c) together are clearly equivalent (but for a multiplicative constant ≤ 4) to (19a) and, a fortiori, to (19).

Letting $g_1 = f_1 + f_2$, $g_2 = f_2$, (19c) becomes

$$-\int g_2\overline{g_2}\ \mathrm{Re}\ (w_{12}-w_{11}+h)dt + \mathrm{Re}\int g_1\overline{g_2}(w_{12}-w_{11}+h)dt$$
$$\le \sigma(|g_1|^2),\ \forall\ (g_1,g_2)\in P\times P. \quad (20)$$

Replacing in (20) g_α by $\lambda_\alpha g_\alpha$, $\lambda_\alpha\in C$, $\alpha = 1,2$, we get a positive quadratic form in λ_1,λ_2. Therefore, the inequality (20) is equivalent to the two conditions that express that the determinant of the quadratic form is nonnegative:

$$\mathrm{Re}\ (w_{12} - w_{11} + h) \ge 0 \quad (17)$$

$$\left|\int g_1\overline{g_2}(w_{12}-w_{11}+h)dt\right|^2 \le \sigma(|g_1|^2)\int|g_2|^2\ \mathrm{Re}\ (w_{12}-w_{11}+h)dt \quad (20a)$$

$$\forall\ (g_1,g_2)\in P\times P.$$

Now (20a) is equivalent to

$$\int |g|^2 \frac{|w_{12} - w_{11} + h|^2}{\mathrm{Re}\ (w_{12} - w_{11} + h)} \leq \sigma(|g|^2),\ \forall\ g \in \mathcal{P} \tag{20b}$$

Thus, (19) or (19a) are equivalent (but for a multiplicative constant ≤ 4) to (17) + (17a) + (20a).

The validity of (17a) for all $f \in \mathcal{P}$ implies its validity for all $f \in C(T)$, since $\mathcal{P}$ is dense in $C(T)$ and σ satisfies conditions (σ_1) and (σ_2). Thus (17a) remains true replacing $|f|^2$ by $\|f\|_{\mathcal{B}}^2$ (a continuous function of t if $f \in C(T;\mathcal{B})$ is a $\mathcal{B}$-valued function). Therefore,

$$\int \|f\|_{\mathcal{B}}^2 w_{11} dt \leq 2\sigma(\|f\|_{\mathcal{B}}^2),\ \forall\ f \in C(T;\mathcal{B}) \tag{21}$$

and (17a) and (21) are equivalent. By (20a),

$$\left|\int |g_1||g_2|(w_{12} - w_{11} + h)dt\right|^2 \leq \sigma(|g_1|^2)\int |g_1|^2\ \mathrm{Re}\ (w_{12} - w_{11} + h)dt$$

$$\forall\ (g_1,g_2) \in \mathcal{P} \times \mathcal{P}.$$

Furthermore, if $g_1 \in \mathcal{P}_{\mathcal{B}}$ and $g_2 \in \mathcal{P}_{\mathcal{B}'}$, then

$$\left|\int \langle g_1(t),g_2(t)\rangle (w_{12} - w_{11} + h)dt\right|^2 \leq \left(\int \|g_1\|_{\mathcal{B}}\|g_2\|_{\mathcal{B}'}(w_{12} - w_{11} + h)dt\right)^2$$

Therefore, since $\mathcal{P}$ is dense in $C(T)$, we get

$$\left|\int \langle g_1,g_2\rangle (w_{12} - w_{11} + h)dt\right|^2 \leq \sigma(\|g_1\|_{\mathcal{B}}^2)\int \|g_2\|_{\mathcal{B}'}^2\ \mathrm{Re}\ (w_{12} - w_{11} + h)dt \tag{21a}$$

$$\forall\ (g_1,g_2) \in \mathcal{P}_{\mathcal{B}} \times \mathcal{P}_{\mathcal{B}'},$$

and (20a) and (21a) are equivalent.

Inequalities (21) and (21a) are equivalent to inequality (17b), since replacing g_1,g_2 by $\lambda_1 g_1,\lambda_2 g_2$, $\lambda_1,\lambda_2 \in C$, in (17b) we get a positive quadratic form in λ_1, λ_2 and (21),(21a) express that the determinant of this form is ≥ 0. These equivalences amount to the equivalence (but for a multiplicative constant ≤ 4) of (19) and (17) + (17a) + (17b), which is part (I).

Inequality (17b) is the vectorial transference of (20), and if $\mathcal{B}$ is a Hilbert space, we can go from this vectorial inequality back to the vectorial form of (19c) and then to vectorial (19a) and (19), which are part (II).

The proof of part (III) is similar, based on Theorem 3.3(c). Q.E.D.

Remark 3.3. The vector form of inequality (19c) is obtained from (21a) through the substitution $f_1 = g_1 - g_2$, $f_2 = g_2$. When $g_1 \in P_{\mathcal{B}}$, $g_2 \in P_{\mathcal{B}'}$ this substitution makes no sense in general, but it does if, for instance, $\mathcal{B}$ is a Hilbert space, $\mathcal{B}' = \mathcal{B}$. In the case of a general Banach space $\mathcal{B}$ we have to stop at inequality (21a), which is equivalent to (19a) in the Hilbert space or scalar cases. (See Remark 3.5 below for $\mathcal{B} = \ell^r$ and other cases.) Thus, in the case of a general Banach space $\mathcal{B}$, we do not transfer vectorially the initial inequality (19) (or (19a)), but the inequality (20), obtained by a formal change, into (21a). More precisely, for $(g_1, g_2) \in P_1 \times P_2$, (20) becomes

$$- \int g_2 \overline{g_2} \operatorname{Re} (w_{12} - w_{11}) dt + \operatorname{Re} \int g_1 \overline{g_2} (w_{12} - w_{11}) dt \leq \sigma(|g_1|^2) \tag{22}$$

or

$$V[g_1, g_2] = \sum_{\alpha,\beta=1,2} \int g_\alpha \overline{g_\beta} v_{\alpha\beta} dt \leq \sigma(|g_1|^2) \tag{22a}$$

where $V = (v_{\alpha\beta})$, with

$$v_{11} = 0,\ v_{22} = -\operatorname{Re} (w_{12} - w_{11}),\ v_{12} = w_{12} - w_{11} = \overline{v_{21}}. \tag{22b}$$

Thus, <u>instead of inequality</u>

$$W[f_1,f_2] \le \sigma(|f_1+f_2|^2) \tag{17}$$

<u>we transfer vectorially inequality (22a) for V as in (22b)</u>.

The case of majorized GTKs, $K \le \sigma$, is reduced by Theorem 3.4 to that of majorized matrix measures. Therefore, from Theorem 3.5 follows

3.6. <u>(Vector transference for majorized GTKs)</u>. Let $K \sim (s_{\alpha\beta})$ be a GTK, $s_{\alpha\beta} = \hat{w}_{\alpha\beta}$, $w_{\alpha\beta} \in L^1(T)$, $\alpha,\beta = 1,2$, $s_{11} = s_{22}$, and let H be a Hilbert space. If

$$K[\lambda,\lambda] \equiv \sum_{m,n} K(m,n)\lambda(m)\overline{\lambda(n)} \le \sigma(|\hat{\lambda}|^2) \tag{23}$$

$\forall$ λ of finite support, then

$$K[\xi,\xi] \equiv \sum_{m,n} K(m,n)\langle\xi(m),\xi(n)\rangle_H \le c\sigma(\|\hat{\xi}\|_H^2) \tag{23a}$$

$\forall$ $\xi(n)$, H-valued sequence of finite support, $\hat{\xi} = \Sigma\xi(n)e^{int}$ P_H, and c a constant ≤ 8.

<u>Remark 3.4</u>. This result is a version for GTKs of Grothendieck's Fundamental Inequality. If K is in addition p.d., then (23a) yields a corresponding inequality for $K[\xi,\eta]$ more similar to (15a).

In the case of a general Banach space B, from Theorem 3.5 and Remark 3.3 follows the corollary

<u>3.7</u>. Let $K \sim (s_{\alpha\beta})$ be a GTK, $s_{\alpha\beta} = \hat{w}_{\alpha\beta}$, $w_{\alpha\beta} \in L^1(T)$, $\alpha,\beta = 1,2$, $s_{11} = s_{22}$, and let B be a Banach space. Write $K = K_W$. If

$$K_W[\lambda,\lambda] \le \sigma(|\hat{\lambda}|^2) \tag{23}$$

$\forall\ \lambda$ of finite support, then

$$K_V[\xi,\xi] \leq c\sigma(\|\hat{\xi}_1\|_B^2) \tag{24}$$

$\forall\ \xi(n)$, B-valued sequence of finite support, $\hat{\xi}_1 = \Sigma_{n\geq 0}\xi(n)e^{int} \in (P_1)_B$, where K_V is the GTK defined by the matrix measure V given in (22b) and $c \leq 8$.

Remark 3.5. The sequences of finite support are dense in all ℓ^r, $1 \leq r < \infty$, so that if $B = \ell^r$ then B and B' have a dense subspace in common where the scalar product $<\xi,\eta>$ makes sense, and so does the change of variables referred to in Remark 3.3. Therefore, in such a case it is possible to go back from (21a) to (19c). Thus, we get

Corollary 3.8. Part (II) of Theorem 3.5 still holds if $B = \ell^r$, $1 \leq r < \infty$, i.e., in such a case inequalities (19), (19a) can be transferred vectorially to $(P_1)_B \times (P_2)_{B'}$ or $P_B \times P_{B'}$. This is also true for all spaces B such that there is a vector space contained in $B \cap B'$ and dense in B and B'.

More generally, this is true whenever the formal substitution $g_1 = f_1 + f_2$, $g_2 = f_2$ can be given sense. The same remarks apply to Theorem 3.6.

As the following examples show, several boundedness conditions of the form $\|Tf\| \leq c\|f\|$ can be expressed as $M \prec \sigma$ for appropriate M and σ, so that Theorem 3.5 gives vector transferences of these conditions. (For decisive results concerning vector transference in L^p spaces, cfr. [22] and [23].)

Example 1. Let $(\rho dt, wdt) \in H_M^{2,p}$, $2 \leq p < \infty$ (see (1)), that is

$$(\int|Hf|^2\rho dt)^{1/2} \leq M(\int|f|^p wdt)^{1/p},\ \forall\ f \in P. \tag{25}$$

where H is the Hilbert transform operator. Since every $f \in P$ can be written as $f = f_1 + f_2$, $(f_1, f_2) \in P_1 \times P_2$ and $Hf = -i(f_1 - f_2)$, (25) can be rewritten as $M \prec \sigma$, where $M = (\mu_{\alpha\beta})$ is given by $d\mu_{\alpha\beta} = w_{\alpha\beta}dt$, $w_{11} = w_{22} = \rho$, $w_{12} = w_{21} = -\rho$, and $\sigma(f) = (\int |f|^{p/2} w dt)^{2/p}$. By Theorem 3.5 and Corollary 3.8 it follows that <u>(25) can be transferred to functions $f \in L^p(B)$, B an ℓ^r-space, $1 \le r < \infty$, as well as for B an Orlicz, Lorentz, etc. space</u>. For $B = H$, a Hilbert space, this conclusion follows from the Marcinkiewicz-Zygmund theorem 3.2; for $B = \ell^r$ this is a result of Benedek-Calderón-Panzone [4] and Córdoba-Fefferman [9], but seems to be new in the other cases. Moreover, we have also the generalization for the case of arbitrary Banach spaces B given by (17b) (cfr. Remark 3.3). These conclusions extend to the cases when $(\rho dr, w dt) \in H_M^{p,p}$ or $H_M^{p,q}$ by using the argument given in [11], and to the case when $L^p(wdt), L^q(\rho dt)$ are replaced by Orlicz or Lorentz spaces, by using the arguments of [12]. As indicated in Remark 3.2, ρdt and wdt can be replaced by more general measures. Finally, while the theorems of Marcinkiewicz-Zygmund or Córdoba-Fefferman have the norm $\|T\|$ increased to $c_{pq}(B)\|T\|$, in Theorem 3.5 $\|T\|$ increases to $c\|T\|$ with $c \le 4$.

<u>Example 2</u>. Let us write $(\mu, \nu) \in H_M^{2,p}(N)$ if

$$(\int |Hf|^2 d\mu)^{1/2} \le M(\int |f|^p d\nu)^{1/p}$$

only for $f \in P$ satisfying the vanishing moment condition $\hat{f}(n) = 0$ for $|n| \le N$. Such classes appear (for $p = 2$) in prediction theory. As shown in [2] the condition $(\mu, \nu) \in H_M^{2,2}(N)$ can also be expressed in the form $M \prec \sigma$, for suitable M and $\sigma = \|\cdot\|_{L^1(\nu)}$ (cfr. [24]). The conclusions of Example 1 apply also to those cases, that is, vector transferences from $H_M^{p,q}(N)$ to Banach spaces-valued functions hold. Let us remark that even the cases $B = \ell^r$ are not covered by the Marcinkiewicz-Zygmund theorem or the subsequent generalizations.

Example 3. A measure $\rho \geq 0$ in T gives a bounded linear functional in the real H^1 space if $\int \mathrm{Re}\, F d\rho \leq c\int |F| dt$, $\forall\, F \in H^1$. This condition is easily seen to be equivalent to

$$2\mathrm{Re} \int f_1 \overline{f_2} d\rho \leq c\int |f_1 + f_2|^2 dt \equiv \sigma(|f_1 + f_2|^2),\ \forall\ (f_1,f_2) \in P_1 \times P_2. \tag{26}$$

This suggests the consideration of measures $d\rho$ satisfying

$$2\mathrm{Re} \int f_1 \overline{f_2} d\rho \leq c(\int |f_1 + f_2|^{2p} dt)^{1/p},\ 1 \leq p < \infty, \tag{26a}$$

or, more generally,

$$2\mathrm{Re} \int f_1 \overline{f_2} d\rho \leq \sigma(|f_1 + f_2|^2) \tag{26b}$$

for a general σ (a kind of p-dual or σ-dual of Re H^1).

Conditions (26a) or (26b) are of the form $M \prec \sigma$ for $M = (\mu_{\alpha\beta})$, $\mu_{11} = \mu_{22} = 0$, $\mu_{12} = \mu_{21} = 2\rho$. Therefore, as in Example 1, (26a) and (26b) can be transferred to functions $f \in L^p(B)$ if B is a ℓ^r, Lorentz or other such Banach spaces. In the case of ρ satisfying (26), (17b) gives, for a general Banach space B, that

$$2\int (\mathrm{Re} \langle f_1,f_2\rangle + \|f_2\|^2_{B'}) d\rho + \int (\mathrm{Re}\langle f_1,f_2\rangle - \|f_2\|^2_{B'}) h dt$$
$$- \int \mathrm{Im}\langle f_1,f_2\rangle\, \mathrm{Im}\, h dt \leq c\int \|f_1\|^2_B dt$$

for some $h \in H^1(T)$ and for all $f_1 \in P_B$, $f_2 \in P_{B'}$.

Example 4. It was shown in [12] that Theorem 3.3 contains as a special case Nehari's theorem that solves problem (d) of the Introduction, providing generalizations of this theorem to L^p and other spaces. Therefore, the vector transference of Theorem 3.5 can be applied also to Nehari's theorem, that is, to Hankel operators. Similar conclusions apply to the inequality (26) with H replaced by the Poisson operator, and more generally by a so-called quasimultiplicative operator (cfr. [12] and [24], Section 6). All these applications will be developed in a forthcoming paper.

Finally, let us remark that Theorem 3.4 applied to the case of Example 1 gives a necessary and sufficient condition for a sequence $\lambda(n)$ to satisfy $\lambda(n) = \hat{\mu}(n)$ for some $\mu \in H_M^{p,q}$. Therefore, Theorem 3.6, with its vectorial characterization, provides a much stronger necessary condition for such a case, in terms of $\mathcal{B}$-valued sequences $\xi(n)$.

Mischa Cotlar
Universidad Central de Venezuela

Cora Sadosky
Howard University

REFERENCES

1. V. M. Adamjan, D. Z. Arov, and M. G. Krein, Infinite Hankel matrices and problems of Carathéodory and Fejér, Func. Anal. Appl., 2 (1968), 1-19.

2. R. Arocena, M. Cotlar, and C. Sadosky, Weighted inequalities in L^2 and lifting properties, Math. Anal. & Appl., Part A, Adv. in Math. Suppl. Studies, 7A (1981), 95-128.

3. R. Arocena and M. Cotlar, Generalized Toeplitz kernels and moment problems of Adamjan-Arov-Krein, Integral Eq. and Operator Theory, 5 (1982), 37-55.

4. A. Benedek, A. P. Calderón, and R. Panzone, Convolution operators on Banach space valued functions, Proc. Nat. Acad. Sci. USA, 48 (1962), 356-365.

5. G. Bennet, Schur multipliers, Duke Math. J., 44 (1977), 603-639.

6. R. Blei, Uniformity property for $\Lambda(2)$ sets and Grothendieck's inequality, Symp. Math., 22 (1977), 321-337.

7. S. Bochner, Vorlesungen über Fouriersche Integrale, Akad. Verlagsgesellschaft, Leipszig, 1932. English transl., Ann. of Math. Studies, 42, Princeton University Press, Princeton, 1959.

8. S. Bochner, A theorem on Fourier-Stieltjes integrals, Bull. Amer. Math. Soc., 40 (1934), 271-276.

9. A. Córdoba and R. Fefferman, A weighted norm inequality for singular integrals, Studia Math., 57 (1976), 97-101.

10. M. Cotlar and C. Sadosky, On the Helson-Szegö theorem and a related class of modified Toeplitz kernels, Proc. Symp. Pure Math. AMS, 35: I (1979), 383-407.

11. M. Cotlar and C. Sadosky, On some L^p versions of the Helson-Szegö theorem, in Harmonic Analysis Conference in honor of Prof. A. Zygmund, Wadsworth Intl. Math. Series (1982), 306-317.

12. M. Cotlar and C. Sadosky, Majorized Toeplitz forms and weighted inequalities with general norms, in Harmonic Analysis (Ed.: F. Ricci & G. Weiss), Lecture Notes in Math. #908, Springer-Verlag (1982), 139-168.

13. M. Cotlar and C. Sadosky, Transformée de Hilbert, theorème de Bochner et le problème des moments, II, C.R. Acad. Sci. Paris, A, 285 (1977), 661-665.

14. N. Dincoleanu, Vector Measures, Pergamon Press, New York, 1967.

15. J. E. Gilbert, Harmonic analysis and the Grothendieck fundamental theorem, Symp. Math., 22 (1977), 393-420.

16. J. E. Gilbert, Nikisin-Stein theory and factorization with applications, Proc. Symp. Pure Math. AMS, 35:II(1979), 233-267.

17. A. Grothendieck, Resumé de la théorie métrique des produits tensoriels topologiques, Bol. da Soc. Mat. São Paulo, 8 (1956), 1-79.

18. R. A. Horn, Quadratic forms in Harmonic analysis and Bochner-Eberlein theorem. Proc. Amer. Math. Soc., 52 (1975), 263-270.

19. J. Marcinkiewicz and A. Zygmund, Quelques inégalités pour les operations linéaires, Fund. Math., 32 (1939), 115-121.

20. P. Masani, Propagators and dilatations, in Probability Theory in Vector Spaces (Ed.: A. Weron), Lecture Notes in Math. #656, Springer-Verlag (1977), 95-118.

21. Z. Nehari, On bounded bilinear forms, Ann. of Math., 65 (1957), 153-162.

22. J. L. Rubio de Francia, Weighted norm inequalities and vector valued inequalities, in Harmonic Analysis (Ed.: F. Ricci & G. Weiss), Lecture Notes in Math. #908, Springer-Verlag (1982), 86-101.

23. J. L. Rubio de Francia, Factorization and A_p weights, preprint.

24. C. Sadosky, Some applications of majorized Toeplitz kernels, Proc. Seminar Topics in Harmonic Anal. (Ed.: L. de Michele & F. Ricci), Milano, 1982. (In press.)

25. A. Weron, Prediction theory in Banach spaces, in Probability Winter School (Eds.: Z. Ciesielski, K. Urbanik, W. A. Woyczynski), Lecture Notes in Math. #472, Springer-Verlag (1975), 207-228.

26. A. Zygmund, Trigonometric Series, Second Edition, Cambridge University Press, Cambridge, 1959.

Functions of Bounded Variation and Fractional Dimension

Ron Blei*

This article casts some recent work ([2], [3], [4], [5], [7]) in a historical perspective and describes a development of ideas that led to 'fractional dimensions.' The paper is based on a talk given at the Cortona Conference (July 1982) and will follow closely the actual lecture -- hence the slightly informal style. I should like to thank the organizers, Professors G. Mauceri, F. Ricci, and G. Weiss for inviting me to partake in this very stimulating and enjoyable meeting.

The starting point is the Riesz Representation Theorem (1909). First, although we all know what is a function of bounded variation, let us give the definition, in a somewhat unconventional form: A function ϕ on $[0,1]$ is of bounded variation if

$$\sup_{\pi} \sup_{\substack{(\varepsilon_i) \\ \varepsilon_i = \pm 1}} \left|\sum_{\pi} (\Delta_i \phi) \varepsilon_i\right| < \infty ;$$

$\pi = \{x_o, \ldots, x_n\}$ denotes a partition of $[0,1]$ and

$$\Delta_i^{\pi}\phi = \Delta_i \phi = \phi(x_i) - \phi(x_{i-1}) , \quad i = 1, \ldots n .$$

The F. Riesz Theorem states that ν is a bounded linear functional on $C[0,1]$ the space of continuous functions on $[0,1]$, iff there is $\phi_\nu = \phi$ a function of bounded variation on $[0,1]$ so that

(*) Partially supported by NSF grant MCS 8002716

$$\lim_{\|\pi\|\to 0} \sum_{\pi} (\Delta_i \phi) f(t_i) = \nu(f) \quad \text{for all} \quad f \in C[0,1]$$

($\|\pi\| = \max\{x_i - x_{i-1} : i = 1,\dots,n\}$; the meaning of the statement above is clear from standard courses).

A year or so later, M. Fréchet quite naturally considered and solved the (two dimensional) problem of determining all bounded bilinear functionals on $C[0,1]$. First, Fréchet formalized a notion of variation for a function on $[0,1]^2$:

$$\text{(F)} \qquad \sup_{\pi_1,\pi_2} \sup_{\substack{(\varepsilon_i)_i,(\delta_j)_j \\ \varepsilon_i,\delta_j=\pm 1}} \Big| \sum_{\pi_1,\pi_2} (\Delta_{ij}\phi)\varepsilon_i\delta_j \Big| < \infty ;$$

$\pi_1 = \{x_o,\dots,x_n\}$, $\pi_2 = \{y_o,\dots,y_n\}$ denote partitions of $[0,1]^2$ and $\Delta_{ij}^{\pi_1\pi_2}\phi = \Delta_{ij}\phi = \phi(x_{i+1},y_{j+1}) - \phi(x_i,y_{j+1}) + \phi(x_i,y_j) - \phi(x_{i+1},y_j)$

He then proved: ν is a bounded bilinear functional on $C[0,1]$ iff there is $\phi_\nu = \phi$ a function on $[0,1]^2$ with bounded Fréchet variation ((F) above) so that

$$\lim_{\|\pi_1\|,\|\pi_2\|\to 0} \sum_{\pi_1,\pi_2} (\Delta_{ij}\phi) f(s_i) g(t_j) = \nu(f,g)$$

for all $f, g \in C[0,1]$ (this was announced in a note to the Comptes Rendus, 1910; details and applications appeared in [9]).

The next step was taken by J. Littlewood in a classical paper "On bounded bilinear forms in an infinite" number of variables" [13] which he began as follows: "Professor P. J. Daniell recently asked me if I could find an example of a function of two variables, of bounded variation according to a certain definition of Fréchet but not according to the usual definition.

The problem is equivalent to the following: to find a bi-linear form

$$Q(x,y) = \Sigma\,\Sigma\, a_{mn}\, x_m\, y_n$$

in an infinite number of variables x, y, such that $\Sigma\,\Sigma\, |a_{mn}|$ is divergent, but Q is bounded for all x, y belonging to the range ... defined by

$$|x_m| \leq 1 \quad (m = 1, \dots\ ), \quad |y_n| \leq 1 \quad (n = 1, \dots\ ) ."$$

Having so reduced the problem, Littlewood quickly answered it in the affirmative via "Hilbert's form,"

$$\sum_{m \neq n} \frac{u_m v_n}{m-n}, \quad \Sigma |u_m|^2, \quad \Sigma |v_n|^2 \leq 1 ,$$

and proceeded to consider a natural question: On one hand, every bounded bilinear form (a_{mn}) is square summable, $\sum_{m,n} |a_{mn}|^2 < \infty$ (trivial), and on the other hand, there are bounded bilinear forms (a_{mn}) so that $\sum_{m,n} |a_{mn}| = \infty$. What is the truth here? (Terminology: (a_{mn}) is a bounded bilinear form on ℓ^∞, abbrev. bounded bilinear form, if

$$\sup \{ | \sum_{m,n} a_{mn}\, x_m y_n | : (x_m),\ (y_n) \in \ell^\infty \} < \infty ;$$

equivalently,

$$\sup \{ | \sum_{m,n} a_{mn}\, \varepsilon_m \delta_n | : \varepsilon_m, \delta_n = \pm 1,\ m,n = 1, \dots \} < \infty) .$$

In this context, Littlewood gives a complete answer: All bounded bilinear forms (a_{mn}) satisfy

$(L)_1$ (Littlewood's inequality)

$$\sum_m (\sum_n |a_{mn}|^2)^{1/2} < \infty ,$$

which implies

$$(L)_2 \qquad \sum_{m,n} |a_{mn}|^{4/3} < \infty .$$

Moreover, there are bdd. bilinear forms (a_{mn}) so that

$$(L)_2^{\#} \qquad \sum_{m,n} |a_{mn}|^p = \infty \quad \text{for all} \quad p < 4/3 .$$

At this point, I confess that my overall view of developments is clearly biased, and with this disclaimer, in force throughout the article, let me make a few comments. During the 1930's and 1940's, some efforts were applied to the study of the Fréchet variation as well as some other notions of variation, but, as far as I can tell, Littlewood's paper [13] was not really paid much attention to at that time. For example, mathematicians like R. Adams, J. Clarkson ([1]), M. Morse and W. Transue ([14], [15]), tried to push classical ideas of measure theory to the framework of functions of bounded Fréchet variation--suggestively called bimeasures by Morse. And although success was limited in this particular direction, these works provided an important backdrop and inducement for the study of tensor products in functional and harmonic analysis. Indeed, by the mid 1950's we already have a well developed language and theory of topological tensor products much of which formulated by A. Grothendieck (e.g., [10] and [11]), and in this framework we see the next major step following Littlewood's inequality: In [11], Grothendieck proved what he called "théorème fondamental de la théorie métrique des produits tensoriels" and what has since then become known as

Grothendieck's inequality.

Let (a_{mn}) be a bounded bilinear form on ℓ^∞ and A be a bounded bilinear form on a Hilbert space H, i.e.

$$|A(x,y)| \le K \|x\| \|y\| \quad \text{for all} \quad x,y \in H,$$

and A bilinear. Let x_m, y_n, $m,n = 1,\ldots,$ be arbitrary vectors in the unit ball of H. Then,

$$\left| \sum_{m,n} a_{mn} A(x_m, y_n) \right| < \infty .$$

Of course Grothendieck's inequality easily implies Littlewood's: Let $H = \ell^2$, A be the usual inner product and $x_m = e_m$ the mth basic vector in ℓ^2.

The next step in the evolution of ideas in the present context is in the work of Varopoulos on tensor algebras and harmonic analysis. Continuing the Littlewood-Grothendieck line, Varopoulos showed in [17], en route to the failure of a multi-variable Von-Neumann inequality, that a Grothendieck type inequality could fail in dimensions higher than 2 -- more will be said of this later on.

And now I want to describe my own involvement here. My interest in all this began with a paper by Edwards and Ross, "p-Sidon sets" [8].

Definition.

$F \subset \Gamma$ is a p-Sidon set provided that

$$C_F(\hat{\Gamma})^{\wedge} \subset \ell^r$$

if and only if $r \geq p$ (Γ is a discrete abelian group and $C_F(\hat{\Gamma})$ denotes the space of continuous functions whose Fourier transform vanishes off F).

In what follows, we take for concreteness sake $\Gamma = \mathbb{Z}$, the additive group of integers and $\hat{\Gamma} = \mathbb{T}$, the circle group -- all that is said below holds equally well in any other commutative setting. To start, observe two extremal cases: (i) A lacunary set is 1-Sidon (a classical fact) and (ii) $\mathbb{Z}$ is 2-Sidon (exercise). In [8], Edwards and Ross transported Littlewood's results [13] to the harmonic analytic setting: Suppose $E_1 = \{\lambda_j\}$, $E_2 = \{\nu_j\}$ are infinite disjoint subsets of $\mathbb{Z}$ whose union is lacunary. Then $E_1 + E_2$ is 4/3-Sidon (proof: Observe that $f \in C_{E_1+E_2}(\mathbb{T})$ iff $(\hat{f}(\lambda_j + \nu_k))_{j,k}$ is a bdd. bilinear form. Now apply $(L)_2$ and $(L)_2^{\#}$). A little later on, this was generalized: Let $E_1, \ldots, E_J$ be infinite, mutually disjoint subsets of $\mathbb{Z}$ whose union is lacunary. Then $E_1 + \cdots + E_J$ is $2J/(J+1)$ - Sidon [12].(1) And so, a natural question was raised: Are there p-Sidon sets for $p \neq 2J/(J+1)$, $J = 1,2,\ldots$? I had worked on that question without success at that time (1974-75) and gave up -- the

(1) The requirement of disjointness is superfluous -- if E is an infinite lacunary subset of $\mathbb{Z}$ then $\underbrace{E + \cdots + E}_{J\text{-times}}$ is $2J/(J+1)$-Sidon.

question resembled the difficult and still unsolved $\Lambda(p)$ problem raised in Rudin's paper "Trigonometric series with gaps" [16].

And so, in the course of work, I developed an interest in Grothendieck's fundamental inequality and found another proof whose philosophy is roughly as follows: In one direction, Grothendieck's inequality implies Littlewood's which is equivalent to the classical Khintchin inequality --

Let $E = \{\lambda_j\}$ be a lacunary subset of $\mathbb{Z}$ and suppose $\phi \in \ell^2$. Then, there is $f \in L^\infty(\mathbb{T})$ so that $\hat{f}(\lambda_j) = \phi(j)$ for all j.

In the other direction, Grothendieck's inequality is implied by a 'uniformized' Khintchin inequality --

$E = \{\lambda_j\} \subset \mathbb{Z}$ lacunary and $\phi \in \ell^2$. Then, there is $f \in L^\infty(\mathbb{T})$ so that (i) $\hat{f}(\lambda_j) = \phi(j)$ for all j, (ii) $\|f\|_\infty \leq 3\|\phi\|_2$, and (iii) $\left(\sum_{n \notin E} |\hat{f}(n)|^2\right)^{\frac{1}{2}} \leq (1/2)\|\phi\|_2$.

Placed in an appropriate 'multidimensional' harmonic analytic framework, the proof quite naturally generalized and netted an assortment of multilinear inequalities -- I reported on all this at the 1976 Rome conference [2].

Going on, in the spring of 1977 I had a conversation about the results of [2] with N. Varopoulos (then visiting MIT). Pointing out the possible failure of a trilinear Grothendieck type inequality (shown in [17]), Varopoulos asked whether some sort of a characterization could be given. Let us now formalize matters:

Definition.

Let $N \geq 2$ and A be a bounded N-linear form on a Hilbert space H. A is said to be projectively bounded if for every bounded N-linear form on ℓ^∞, $(a_{i_1 \ldots i_N})_{i_1,\ldots,i_N}$, and arbitrary sequences of vectors in the unit ball of H, $(x_i^1)_i, \ldots, (x_i^N)_i \in H$,

$$|\Sigma\, a_{i_1 \ldots i_N} A(x_{i_1}^1, \ldots, x_{i_N}^N)| < \infty .$$

In this language, Grothendieck's inequality states that every bounded bilinear form on a Hilbert space is projectively bounded, while Varopoulos' theorem states that there are bounded trilinear forms on a Hilbert space which are not projectively bounded. In particular --

Theorem ([17])

Let $H = \ell^2(\mathbb{N}^2)$ (Hilbert Schmidt operators). There exists a choice of signs $\pm$ so that the trilinear form $A_\pm$ defined by

$$A_\pm(x,y,z) = \sum_{i,j,k} \pm\, x(i,j)y(j,k)z(i,k) , \quad x,y,z \in H ,$$

is not projectively bounded.

Before describing the general multilinear characterizations that resulted, to illustrate ideas I shall give an instance which implies and completes the theorem above. Let $E \subset \mathbb{Z}$ be (any) lacunary set and enumerate $E = \{\lambda_{ij}\}_{i,j\in\mathbb{N}}$. Define

$$E_{3,2} = \{(\lambda_{ij},\lambda_{jk},\lambda_{ik}) : i,j,k\in\mathbb{N}\}\subset\mathbb{Z}^3 .$$

Let $\phi\in\ell^\infty(\mathbb{N}^3)$ and say that $\phi\in B(E_{3,2})$ if there exists a measure $\mu\in M(\mathbb{T}^3)$ so that

$$\hat{\mu}\big((\lambda_{ij},\lambda_{jk},\lambda_{ik})\big) = \phi(i,j,k) \quad \text{for all} \quad (i,j,k)\in\mathbb{N}^3 .$$

Define A_ϕ a bounded trilinear form on $\ell^2(\mathbb{N}^2)$ by

$$A_\phi(x,y,z) = \sum_{i,j,k} \phi(i,j,k)\; x(i,j)y(j,k)z(i,k) .$$

Theorem ([3])

A_ϕ (defined above) is projectively bounded if and only if $\phi\in B(E_{3,2})$.

The multidimensional theory runs as follows: Let $J \geq K > 0$ be arbitrary fixed integers and suppose

$$S_1,\ldots,S_N\subset\{1,\ldots,J\} \quad \text{satisfy}$$

$$\text{(i)} \qquad |S_\alpha| = K , \qquad \alpha = 1,\ldots,N ,$$

and

$$\text{(ii)} \qquad |\{\alpha : i\in S_\alpha\}| \geq 2 \quad \text{for each} \quad i\in\{1,\ldots,J\}$$

($|\cdot|$ denotes cardinality). Furthermore, assume that each S_α is enumerated and ordered:

$$S_\alpha = (\alpha_1,\ldots,\alpha_K) .$$

Corresponding to the S_α's above, define projections $P_1,\ldots,P_N$ from $\mathbb{N}^J$ onto $\mathbb{N}^K$ by

$$P_\alpha(n_1,\ldots,n_J) = (n_{\alpha_1},\ldots,n_{\alpha_k}),$$

$(n_1,\ldots,n_J) \in \mathbb{N}^J$, $\alpha = 1,\ldots,N$. Now let $E \subset \mathbb{Z}$ be a lacunary set indexed by $\mathbb{N}^K$, $E = \{\lambda_k\}_{k\in\mathbb{N}^K}$.

Define

$$E_{J,K,N} = \{(\lambda_{P_1(n)},\ldots,\lambda_{P_N(n)}) : n\in \mathbb{N}^J\} \subset \mathbb{Z}^N .$$

Let $\phi \in \ell^\infty(\mathbb{N}^J)$ and say that $\phi \in B(E_{J,K,N})$ if there exists a measure $\mu \in M(\mathbb{T}^N)$ so that

$$\hat{\mu}\big((\lambda_{P_1(n)},\ldots,\lambda_{P_N(n)})\big) = \phi(n) \quad \text{for all} \quad n\in \mathbb{N}^J .$$

Define A_ϕ a bounded N-linear form on the Hilbert space $\ell^2(\mathbb{N}^K)$ by

$$A_\phi(x_1,\ldots,x_N) = \sum_{n\in\mathbb{N}^J} \phi(n)\, x_1\big(P_1(n)\big)\cdots x_N\big(P_N(n)\big) ,$$

$x_1,\ldots,x_N \in \ell^2(\mathbb{N}^K)$.

<u>Theorem ([3])</u>

A_ϕ (defined above) is projectively bounded if and only if $\phi \in B(E_{J,K,N})$.

The multilinear inequalities were first written down in an extremely cumbersome format ("Uniformizability in Harmonic Analysis and Applications," preprint). The formulation above (which appeared in [3]) was a result of an effort to make myself understood--I was lecturing about all this at Uppsala University in the Fall of 1977. And so, as the 'fractional product' features of $E_{J,K,N}$ came into focus, the spectral sets that had naturally surfaced in the characterizations of projectively bounded forms were naturally observed to fill the gaps in the p-Sidon problem: Write $\binom{J}{K} = N$ and let

$$S_1,\ldots,S_N \subset \{1,\ldots,J\}$$

be the collection of all K-subsets. Given a lacunary set $E = \{\lambda_k\}_{k\in\mathbb{N}^K}$ and following the notation of the previous paragraph, we define

$$E_{J,K} = \{(\lambda_{P_1(n)},\ldots,\lambda_{P_N(n)}) : n\in \mathbb{N}^J\}\subset \mathbb{Z}^N .$$

<u>Theorem ([4])</u>

$$E_{J,K} \text{ is } 2/(1+\tfrac{K}{J}) \text{ - Sidon (in } \mathbb{Z}^N) .$$

To obtain 'irrational'-Sidon sets we pass to limits in the appropriate sense, and to produce p-Sidon sets in $\mathbb{Z}$ (or any other Γ) we replace the Cartesian product operation by the group operation -- these matters are treated in [4].

At this point, the next natural task was to explain the underlying philosophy of the "fractional Cartesian products" described above. And so, we were led to the notion of 'combinatorial dimension:' Let $F\subset E^L$, where $L > 0$ is a fixed integer,

E is any set whatsoever, and E^L is the usual L-fold Cartesian product of E. Define the 'maximal distribution' of F by

$$\Psi_F(s) = \max\{|F \cap (A_1 \times \cdots \times A_L)| : A_i \subset E, |A_i| = s, i = 1,\ldots,L\},$$

where s is a positive integer. Define the 'combinatorial dimension' of F by

$$\dim F = \inf\{a : \overline{\lim_{s\to\infty}} \frac{\Psi_F(s)}{s^a} < \infty\};$$

if $\overline{\lim_{s\to\infty}} \frac{\Psi_F(s)}{s^{\dim F}} < \infty$, $\dim F$ is said to be exact, and otherwise $\dim F$ is said to be asymptotic. It turns out that $\dim E_{J,K} = J/K$ exactly; also, an abundance of α-dimensional subsets in E^L, $1 \leq \alpha \leq L$ arbitrary, was displayed in [7]. I spoke about all this at the "Seminars..." in Milano ([6]) and will give here only an illustration of the role that combinatorial dimension plays in a harmonic analytic context:

<u>Theorem ([5])</u>

Let $E \subset \mathbb{Z}$ be a lacunary set, $L > 0$ be an arbitrary integer, and F be an arbitrary subset of E^L.

(i) If $\dim F$ is exact then F is $2/(1+\frac{1}{\dim F})$ - Sidon (in $\mathbb{Z}^L$).

(ii) If $\dim F$ is asymptotic then

$$C_F(\mathbb{T}^L)^\wedge \subset \ell^r$$

if and only if

$$r > 2/(1+\frac{1}{\dim F}).$$

At this juncture, I ran out of lecture time and (recalling my promise to follow closely the talk's outline) let me conclude with a very brief remark: Returning to the classical Fréchet setting on $[0,1]^2$, we carry back with us the notion of combinatorial dimension, appropriately adapted to the 'continuous' setting and are led, in view of what was said above, to 'p-variation' of functions with bounded Fréchet variation ($p = 1$ corresponds to the usual total variation in the sense of Vitali). This will be considered in subsequent work.

References

1. C. R. Adams and J. A. Clarkson, On definitions of bounded variation of two variables, Trans. Amer. Math. Soc., Vol. 35 (1933), 824-854.

2. R. C. Blei, A uniformity property for $\Lambda(2)$ sets and Grothendieck's inequality, Symposia Mathematica, Vol. XXII (1977), 321-336.

3. ________, Multidimensional extensions of the Grothendieck inequality and applications, Arkiv för Matematik, Vol. 17 (1979) No. 1, 51-68.

4. ________, Fractional Cartesian products of sets, Ann. Institute Fourier, Grenoble 29, 2 (1979), 79-105.

5. ________, Combinatorial dimension and certain norms in harmonic analysis (to appear in Amer. J. of Math.).

6. ________, Combinatorial dimension: A continuous parameter (to appear in Alta Matematica).

7. ________ and T. W. Körner, Combinatorial dimension and random sets.

8. R. E. Edwards and K. A. Ross, p-Sidon sets, J. of Functional Analysis, 15 (1974), 404-427.

9. M. Fréchet, Sur les fonctionnelles bilinéaires, Trans. Amer. Math. Soc., Vol. 16 (1915), 215-234.

10. A. Grothendieck, Produits tensoriels topologiques et espaces nucléaires, Memoirs of the Amer. Math. Soc. 16 (1955).

11. ________, Résumé de la théorie métrique des produits tensoriels topologique, Bol. Soc. Matem. Sao Paulo, 8 (1956), 1-79.

12. G. W. Johnson and G. S. Woodward, On p-Sidon sets, Indiana Univ. Math. J., 24 (1974), 161-167.

13. J. E. Littlewood, On bounded bilinear forms in an infinite number of variables, Quart. J. Math. Oxford, 1 (1930), 164-174.

14. M. Morse and W. Transue, Functionals of bounded Fréchet variation, Canadian J. of Math, Vol. 1 (1949), 153-165.

15. M. Morse, Bimeasures and their integral extensions, Ann. Mat. Pura Appl., (4) 39 (1955), 345-356.

16. W. Rudin, Trigonometric series with gaps, J. Math. Mechanics, 9 (1960), 203-227.

17. N. Varopoulos, On an inequality of Von Neumann and an application of the metric theory of tensor products to operator theory, J. of Functional Analysis 16 (1974), 83-100.

R. C. Blei
Department of Mathematics
University of Connecticut
Storrs, CT 06268
USA

PATHOLOGICAL PROPERTIES AND S.I.P. MEASURES ON METRIZABLE GROUPS

C. Karanikas

Abstract

A construction is given for thin sets with group algebraic independence conditions on metrizable groups G . We list several pathological properties of measures concentrated on these sets and in particular we illustrate s.i.p. and i.p. measures in M(G) .

1. Introduction

Our aim here is to notify , that some properties of independent and dissociate sets in abelian groups (see for example in[2]) occur also for certain thin sets on metrizable groups . This work contains some new ideas in this direction , as well as a brief survey of technics and a simpler proof of the main result in[1] .

Let M(G) be the convolution measure algebra of a locally compact non-discrete metrizable group G . Let $\mu \in M(G)$, $y \in G$ and $\delta(y)$ be the point mass of y . If

$$\|\mu^n * \delta(y) - \mu^m\| = \|\delta(y) * \mu^n - \mu^m\| = \|\mu\|^n + \|\mu\|^m$$

either for $y \neq e$ (e is the unit of G) or $n \neq m$, we say that μ is an s.i.p. measure or μ has strongly independent powers.

in the case where this holds for $y = e$ and $n \neq m$, we say that μ is an i.p. measure or μ has independent powers . Notice that the powers of μ are convolution powers .

It is well known for many years that if G is a non-discrete abelian group , then M(G) contains s.i.p. measures ; e.g. continuous positive measures concentrated on independent sets (see.in [3]) . Recently we have seen in [1] that even in the non-abelian case there exist i.p. measures in M(G) . Note that the support of the measures we discuss are thin in a certain algebraic and topological sense .

In section 2 we shall illustrate the construction of the Cantor type thin sets P_X, P_Y and P_Z , having some algebraic independence conditions , namely the algebraic Properties X ,Y and Z respectively . We denote by $M_c(P)^+$ the set of all continuous positive measures concentrated on a Borel subset P of G ; we can verify the following :

(A) If $\mu \in M_c(P_Y)^+$ then μ is an s.i.p. measure (Theorem 2) .

(B) If $\mu \in M_c(P_X)^+$, then μ is an i.p. measure . This is a special case of (A) ; a different proof of (B) is contained in [1] . We should note here that i.p. measures on metrizable groups yield i.p. measures on any non-discrete group (section 6 in [1]) . Note also that the spectrum of an i.p. measure μ is exactly $\{z \in \mathbb{C} : |z| \leq \|\mu\|\}$. Thus provided that μ could be self-adjoint the asymmetry of M(G) follows ; a * algebra is symmetric iff any self-adjoint element has real spectrum .

(C) There exists an s.i.p. singular measure in $M_o(G)$. $M_o(G)$ for an abelian group G is thespace of all measures in M(G) whose Fourier-Stieltjes transforms tend to zero at infinity (of the dual group) . For the non abelian case let P(G) be the set of all continuous states on G and let $\mu \in M(G)$; we define the $\|\cdot\|'$ norm of M(G) : $\|\mu\|' = \sup(\langle f, \mu^* * \mu\rangle^{\frac{1}{2}} : f \in P(G))$. Now $M_o(G)$ is the closure in the $\|\cdot\|'$ norm of all absolutely continuous measures in M(G) . An equivalent definition given by Dunkl and Ramirez is the following : $M_o(G)$ is the space of all measures in M(G) such that the map $y \longrightarrow \mu * \delta(y) - \mu : G \longrightarrow (M(G), \|\cdot\|')$ is continuous .

(D) Given any $h > 0$, we can find a $\mu \in M(G)$ such that $\|\mu\| = 1$ and $\|\mu\|' < h$. This is an immediate consequence of (C) ; in fact choose any s.i.p. measure $\nu \in M_o(G)$ with $\|\nu\| = \frac{1}{2}$, then choose $y \in G$ such that $\|\nu * \delta(y) - \nu\|' < h$. Thus if $\mu = \nu * \delta(y) - \nu$, is as we claimed .

(E) If $\mu_1, \ldots, \mu_{n+1} \in M_c(G)$, then

$$\mu_1 * \ldots * \mu_{n+1}((P_Y^n)') = 0$$

where n=1,2,... and

$$(P_Y^n)' = \left\{ x_1 x_2 \ldots x_n : x_i \in P_Y,\ x_i \neq x_j^{\pm 1},\ i \neq j\ , 1 \leqslant i, j \leqslant n \right\}$$

2. Thin sets with algebraic independence conditions

We describe the sets P_X, P_Y and P_Z .

Definition: Template $\mathfrak{J}$ is a finite collection of open subsets of G with disjoint compact closures such that if $T \in \mathfrak{J}$ then $T^{-1} \in \mathfrak{J}$ and either $T \cap T^{-1} = \emptyset$ or $T = T^{-1}$. Note that the definition of templates in [1] is different .

Consider the templates $\mathfrak{J}$ and $\mathfrak{J}_o$, we say that $\mathfrak{J}$ is a refinement of $\mathfrak{J}_o$ if given any $I \in \mathfrak{J}$, there exists a $T \in \mathfrak{J}_o$ such that $I \subsetneq T$; we write $\mathfrak{J}_o < \mathfrak{J}$. We may suppose that if $\mathfrak{J}_o < \mathfrak{J}$ then each set in $\mathfrak{J}_o$ contains more than two elements of $\mathfrak{J}$.

Let G be a metrizable group and ($\mathfrak{J}_n$) , n = 1,2,... , be a sequence of templates such that :

(i) $\mathfrak{J}_1 < \mathfrak{J}_2 < \dots < \mathfrak{J}_n < \dots$

(ii) $\max(\,\mathrm{diam}(I) : I \in \mathfrak{J}_n\,) \longrightarrow 0$, as $n \longrightarrow \infty$

The set $P = \bigcap_{n=1}^{\infty}(\mathrm{cl}(I) : I \in \mathfrak{J}_n)$ is a Cantor type thin set ; P_X, P_Y and P_Z , will be as above , provided that :

(iii) each template $\mathfrak{J}_n$ (n= 1,2,...) has an algebraic independence condition (namely Property X,Y,Z respectively) .

We describe the independence conditions . Let $S = \{x_1, x_1^{-1}, \ldots, x_n, x_n^{-1}\}$ be a finite symmetric subset of G, containing at most 2n distinct elements ($x_i \neq x_j$, $i \neq j$, $1 \leqslant i,j \leqslant n$). A word w in the elements of S is called <u>restricted</u> if there is at most one entry in w from each set $\{x_i, x_i^{-1}\}$, $i = 1,2,\ldots,n$. For example the word $x_1x_2x_3$ is restricted , but the word $x_1x_2x_1$ is no-restricted . We say that S has <u>Property X</u> if there is no restricted word (in the elements of S) equal to a shorter word or to a non-restricted word of the same length .

For a template $\mathfrak{J}$, we say word in the elements of $\mathfrak{J}$ any finite product of sets from $\mathfrak{J}$. A restricted word , in the elements of $\mathfrak{J}$, has no entry from each set $\{T, T^{-1}\}$, $T \in \mathfrak{J}$. In a similar way as above we define templates having Property X .

Now, we say that S (resp. $\mathfrak{J}$) has <u>Property Z</u> if there is no pair of equal (resp. non-disjoint) restricted words . Finally S (resp. $\mathfrak{J}$) has <u>Property Y</u> if S (resp. $\mathfrak{J}$) has Properties X and Z . It is clear that Property Y does not occur in an abelian group .

<u>Definition</u> The independence Property W (say) <u>agrees with the structure of G</u> , if given any finite set $S = \{x_1, x_1^{-1}, \ldots, x_n, x_n^{-1}\}$ and any neighborhood N of e , there exist $y_i \in x_iN$ such that the set $\{y_1, y_1^{-1}, \ldots, y_n, y_n^{-1}\}$ has Property W .

It is obvious that if Property W (say) agrees with the structure of G , then we can construct a sequence ($\mathfrak{J}_n$) of templates as in (i) , (ii) and (iii) , satisfying Property W . In fact suppose that $\mathfrak{F}_n$ has Property W , choose on each set $I \in \mathfrak{F}_n$ at least two points . Let $\{x_1, x_1^{-1}, \dots, x_m, x_m^{-1}\}$ be the symmetric set generated by these points . Then substitute each point x_i by a point y_i in an arbitrarily small neighborhood of x_i , $i = 1, \dots, m$, providing that $\{ y_1, y_1^{-1}, \dots, y_m, y_m^{-1} \}$ has Property W . Finally with a suitable neighborhood around each point y_i, $i = 1, \dots, m$, we can construct a template $\mathfrak{F}_{n+1}$, having Property W .

For the Property Y we have the following :

Proposition 1. Let G be a locally compact non-discrete group . Then either the independence condition Y agrees with the structure of G or there exists in G an open abelian subgroup .

It is not difficult to see that the independence condition X agrees with any non-discrete group G (cf. [1] Proposition 3.3) . For any finite set S of G we have a finite collection w_1, w_2 of pairs (in the elements of S) where w_1 is restricted and w_2 is shorter or w_2 is a non-restricted word of the same length. Hence w_1 contains an element x S which does not appears in w_2 . Substitute x by an element in a neighborhood of x , providing that the words obtained from w_1 and w_2 (as above) are distinct . In this way we replace by a set S' which has Property X .

To see that the independence condition Z agrees with the structure of a group with no-open abelian subgroup , we need some trivial arguments similar with the following :

Lemma 1. Let $x,y \in G$ and N be a neighborhood of e , if

(i) $xnxn = x^2$, $n \in N$, or

(ii) $xnyu = yuxn$, $n,u \in N$,

then G contains an open abelian subgroup .

If G has an open abelian subgroup H , we can pick an independent set in H or we can find a set P_X .

3. Continuous measures on thin sets

We describe the proofs of (A),(B),(C) and (%) of section 1 .

Theorem 1 Let $\mu_1,\dots,\mu_{n+1}$, $n = 0,1,\dots$, be continuous measures on G . If $y,y_1 \in G$

(i) $\mu_1*\dots*\mu_n((P^n)'y \cap (P^n)'y_1) = 0$, whenever $y_1(y)^{-1} \neq e$

(ii) $\mu_1*\dots*\mu_{n+1}((P^n)') = 0$

Proof It is obvious that (i) implies (ii) . In fact ,

$$\mu_1* \dots*\mu_{n+1}((P^n)') = \int \mu_1*\dots*\mu_n((P^n)'y^{-1})d\mu_{n+1}(y)$$

since $\mu_1*\dots*\mu_n$ is finite , there exists a countable set of y's such that $\mu_1*\dots*\mu_n((P^n)'y^{-1}) \neq 0$; and so (ii) follows from the continuity of μ_{n+1} .

To establish (i) we use induction on n , so (ii) for $n-1$ will be available to us . We may assume that $y_1 = e$.

Let $t_1, t_2, \ldots, t_n$ and $z_1 z_2 \ldots z_n$ be restricted words in the elements of P . If there is no pair of restricted words (as above) such that

$$(1) \qquad y \in (z_1 \ldots z_n)(t_1 \ldots t_n)^{-1}$$

then (i) is trivial . We shall denote by M the set $\{ z_1, z_1^{-1}, \ldots, z_n, z_n^{-1}, t_1, t_1^{-1}, \ldots, t_n, t_n^{-1} \}$ which is considered fixed . Let $p_1 p_2 \ldots p_n$, $q_1 \ldots q_n$ be another pair of restricted words such that

$$(2) \qquad y \in (q_1 \ldots q_n)(p_1 \ldots p_n)^{-1}$$

From (1) and (2) we obtain

$$(3) \qquad (t_1 \ldots t_n)^{-1}(p_1 \ldots p_n) \cap (z_1 \ldots z_n)^{-1}(q_1 \ldots q_n) \neq \emptyset$$

We suppose that the word $p_1 \ldots p_n$ has no entry from the set M . then Property Y implies that $t_i = z_i$ ($i = 1, \ldots, n$) and so by (1), $y = e$ -contradiction . Thus the word $p_1 \ldots p_n$ has an entry from the set M and so

$$\mu_1 * \ldots * \mu_n ((P^n)' y^{-1} \cap (P^n)') \leqslant c\, \mu_1 * \ldots * \mu_n (P^{n-1}) ,$$

where $C = 4n^2$ and our inductive hypothesis implies (i) . Note that the cases $n = 0,1$ are trivial .

The following Lemma is an immediate consequence of Fubini's Theorem .

Lemma 2. Let $\mu_1,\dots,\mu_n$ be continuous measures concentrated on a Borel subset P of G . Then $\mu_1 * \dots * \mu_n$ is concentrated on the set $(P^n)'$ of all restricted words in the elements of P .

Theorem 2. If $\mu \in M_c(P_Y)^+$, then is an s.i.p. measure .

Proof By Lemma 2., $\mu^{n*}(y)$ is concentrated on $(P_y{}^n)'y$ and μ^m on $(P_Y{}^m)'$ and so for $n = m$ and $y \neq e$, Theorem 1. (i) implies Theorem 2. For the case where $n \neq m$ and $y \neq e$, we use arguments as in the proof of Theorem 2 (i) and Property X . Since P_Y has Property X , the last case ($n \neq m$, $y = e$) follows from Lemma 2 .

We have seen (a) ,(B) and (Ξ) of section 1. Finally to obtain a measure as in (C) , one could modify the construction of templates in section 5 of [1], using Property Y instead of Property X.

DEPARTMENT OF MATHEMATICS
UNIVERSITY OF CRETE
IRAKLION, GREECE

REFERENCES

[1] Brown, G., Karanikas, C., and Williamson, J. : " The asymmetry of $M_o(G)$". Math. Proc. Camb. Phill. Soc. 91 (1982), 407-433.

[2] Graham, C. and Mc Gehee, O. : "Essays in commutative Harmonic Analysis". Springer-Verlag No 238, New York.

[3] Williamson, J. : " Banach algebras elements with independent powers, and Theorems of Wiener-Pitt type ". Functional Algebra, ed. by Scott and Foresman (1966), 186-197.

UNIFORMLY BOUNDED REPRESENTATIONS AND L^p- CONVOLUTION OPERATORS ON A FREE GROUP

Anna Maria Mantero - Anna Zappa

1. INTRODUCTION.

Let G be a free group with r generators, $1 < r < \infty$. Every element x of G is a finite reduced word in the generators and their inverses. We denote by $|x|$ the length of x , i.e. the number of letters which compose x .

In a previous paper [MZ] we characterized the eigenfunctions of the operator L which plays, on the free group, the same role as the Laplace-Beltrami operator on a Riemannian symmetric space G/K. Moreover we consider an analytic family of representations π_z , $z \in \mathbb{C}$, introduced by A.Figà-Talamanca and M.Picardello [FP]. For some values of the parameter z , the representations π_z are naturally unitary and are called "principal series", for others they are unitarizable and are called "complementary series". We proved that, for all z , $0 < s = \mathrm{Re}\, z < 1$, π_z is uniformly bounded on an appropriate Hilbert space H_s.

In this paper we compute explicitly a constant c(s,t), such that $\| \pi_z(x) f \|_s \leq c(s,t) \|f\|_s$, $\forall f \in H_s$, $\forall x \in G$. From the properties of this constant, we deduce a stronger version of the Kaplanský density theorem, for a free group, i.e. we prove that if T is a right-invariant multiplier of $L^p(G)$, then T can be approximated by operators of convolutions by $C_c(G)$-functions, with controlled supports. The idea to prove such a result comes from papers of M.Cowling [C] and

U.Haagerup [Hg] .

In Sections 1 and 2 we briefly illustrate the results on the characterization of the operator L and the uniform boundedness of the representations π_z, $0 < s = \operatorname{Re} z < 1$, introducing the necessary definitions. In Section 3 we calculate explicitly the constant c(s,t) . Finally in Section 4 we prove the density theorem for multipliers of $L^p(G)$.

2. EIGENFUNCTIONS OF THE LAPLACE-BELTRAMI OPERATOR.

Let μ_1 denote the probability distribution of a simple random walk on G. We write (Ω,μ) for the Poisson boundary of G relative to μ_1 . Then Ω is the set of infinite reduced words in the generators of G and their inverses, compact in the product topology. To define μ we introduce for $|x| = n$ the sets

$$E(x) = \{\omega\in\Omega : \omega^{(n)} \ (= \text{first } n \text{ letters of } \omega) = x\}.$$

Let $\mathcal{M}$ be the σ-algebra generated by these sets for arbitrary n and $\mathcal{M}_n$ the sub-σ-algebra generated by $E(x)$, for $|x| \leq n$. We put $\mu(E(x)) = 1/2r(2r-1)^{n-1}$.

The free group G acts on Ω by left multiplication. We define $p(x,\omega)$ to be the Radon-Nikodym derivative:

$$p(x,\omega) = \frac{d\mu(x^{-1}\omega)}{d\mu(\omega)} = (2r-1)^{2N(x,\omega) - |x|},$$

where $N(x,\omega)$ is the largest integer n such that $x^{(n)} = \omega^{(n)}$, and we call $p(x,\omega)$ the Poisson kernel relative to the action of G on Ω. We now define the Poisson transform

$$P^z f(x) = \int_\Omega p^z(x,\omega)\, f(\omega)\, d\mu(\omega) , \quad \forall f \in L^2(\Omega).$$

To compute the "Laplace-Beltrami operator" on G , we define for every generator a_j the difference operator D_j: $D_j\varphi(x) = \varphi(xa_j) - \varphi(x)$.

We have $L\varphi = \sum_{j=1}^{r} D_j^* D_j \varphi = -2r\,(\varphi * \mu_1 - \varphi)$.

So, characterizing the eigenfunctions of L is the same as characterizing the eigenfunctions of the operator $*\mu_1$.

We denote by $\mathcal{F}(\Omega)$ the space of so-called cilindrical functions

$$\mathcal{F}(\Omega) = \{ f:\Omega \to \mathbb{C},\ \exists n \in \mathbb{N} \ /\ f \text{ is } \mathcal{M}_n\text{-measurable} \}.$$

The dual space $\mathcal{F}'(\Omega)$ of $\mathcal{F}(\Omega)$ is the space of all martingales $\underline{f} = (f_n)$ on Ω (f_n is $\mathcal{M}_n$-measurable) with respect to the duality

$$(\underline{f}, g) = \lim_{n\to\infty} \int_{\Omega} f_n(\omega)\, g(\omega)\, d\mu(\omega), \quad \forall\, \underline{f} \in \mathcal{F}'(\Omega),\ g \in \mathcal{F}(\Omega).$$

We can now state the analogue of Helgason theorem for free groups.

<u>THEOREM A. If</u> $z \neq ih\pi/\lg(2r-1)$, $h \in \mathbb{Z}$, <u>the function</u>

$$\varphi_z(x) = \int_{\Omega} P^z(x,\omega)\, \underline{f}(\omega)\, d\mu(\omega), \quad \forall\, \underline{f} \in \mathcal{F}'(\Omega)$$

<u>is an eigenfunction of the operator</u> $*\mu_1$ <u>with eigenvalue</u> $\gamma(z) = [(2r-1)^z + (2r-1)^{1-z}]/2r$. <u>Conversely, every eigenfunction arises in this way.</u>

A result of A.Figà-Talamanca and M.Picardello [FP] assures that the spherical functions are exactly all the functions obtained with $\underline{f}(\omega) \equiv 1$.

The Theorem A can be restated as follows: if $z \neq ih\pi/\lg(2r-1)$, $h \in \mathbb{Z}$, then the operator $P^z: \mathcal{F}'(\Omega) \to \mathcal{L}_{\gamma(z)}(G)$ is a bijection from $\mathcal{F}'(\Omega)$ onto the space of all eigenfunctions of the operator $*\mu_1$, associated to the eigenvalue $\gamma(z)$.

3. UNIFORM BOUNDEDNESS OF π_z.

We define, for $z \in \mathbb{C}$, a representation of G on $L^2(\Omega)$ by

$$[\pi_z(x) f](\omega) = P^z(x,\omega)\, f(x^{-1}\omega).$$

It is easy to see that for $\operatorname{Re} z = 1/2$, π_z is unitary (principal series) and for $0 < s = \operatorname{Re} z < 1$, $\operatorname{Im} z = t = h\pi\Psi^{-1}$, $h \in \mathbb{Z}$ and $\Psi = \lg(2r-1)$, π_z is unitarizable (complementary series). They are also irreducible (see [FP]). We call H_z the Hilbert space where the complementary series π_z is unitary. We have

THEOREM B. If $0 < s = \mathrm{Re}\, z < 1$, the representation π_z is uniformly bounded on $H_{-s}(\Omega)$.

The proof of this theorem relies upon the construction of appropriate intertwining operators. From Theorem A, we can deduce that, for $z, 1-z \neq ih\pi\gamma^{-1}$, $h \in \mathbb{Z}$, the operator $I_z = (P^z)^{-1} P^{1-z}$ is well defined on $\mathcal{F}'(\Omega)$. Moreover if we extend π_z as a representation on $\mathcal{F}'(\Omega)$, I_z is an intertwining operator, i.e.

$$\pi_{1-z}(x)\, I_z = I_z\, \pi_z(x).$$

We now define, for $f \in \mathcal{F}(\Omega)$

$$\|f\|_z^2 = \langle I_z f, f \rangle,$$

which is a norm if $0 < s = \mathrm{Re}\, z < 1$, $\mathrm{Im}\, z = h\pi\gamma^{-1}$, $h \in \mathbb{Z}$. Then the representation π_z is unitary on the completion $H_z(\Omega)$ of $\mathcal{F}(\Omega)$ with respect to this norm. Moreover we proved that, for $\mathrm{Re}\, z > 1/2$, I_z is an integral operator with kernel

$k_z(\omega,\omega') = \rho(z)\ (2r-1)^{2(1-\mathrm{Re}\, z)}\ N(\omega,\omega')$, $\omega \neq \omega'$, where

$\rho(z) = 2r\,[1-(2r-1)^{1-2z}]\,/\,(2r-1)\,[1-(2r-1)^{-2z}]$.

For every $\omega \in \Omega$, $k_z(\omega,\cdot)$ belongs to the Lorentz space $L^{1/2(1-s),\infty}(\Omega)$. This allows us to prove that the norm of the multiplier operator

$$p^{it}(x,\cdot)\ :\ H_s(\Omega) \longrightarrow H_s(\Omega),\ \forall t \in \mathbb{R},\ 0 < s < 1,$$

is bounded, uniformly in x, by the constant $c(s,t)$. The proof of Theorem B now reduces to the following estimate

$$\|\pi_{s+it}(x)\, f\|_s = \|p^{it}(x,\cdot)\pi_s(x)\, f\|_s \leq c(s,t)\,\|\pi_s(x)\, f\|_s$$

$$= c(s,t)\,\|f\|_s.$$

4. A CALCULATION.

We want to give an explicit evaluation of the constant $c(s,t)$ such that $\|\pi_z(x) f\|_s \le c(s,t) \|f\|_s$, $\forall f \in H_s(\Omega)$, $\forall x \in G$. We have the following :

THEOREM C. Let $z=s+it$, $0<s<1$, $s\neq 1/2$, $t \neq \pi h \Upsilon^{-1}$, $h \in \mathbb{Z}$. Then

$$\|\pi_z(x) f\|_s \le (2+\alpha(s,t)) \|f\|_s \quad , \quad \forall f \in H_s(\Omega), \forall x \in G,$$

where $\alpha(s,t)$ is such that

(i) for any fixed s, $0<s<1$, $s \neq 1/2$, $\alpha(s,\cdot)$ is a bounded, periodic function with period $\pi\Upsilon^{-1}$ and $\lim_{t\to 0} \alpha(s,t) = 0$;

(ii) for any fixed t, $t \neq \pi h \Upsilon^{-1}$, $h \in \mathbb{Z}$,

$$\lim_{s\to 0^+} \alpha(s,t) = \lim_{s\to 1^-} \alpha(s,t) = +\infty .$$

We first restrict ourselves to the case $0<s<1/2$ and evaluate the norm of the multiplier operators $p^{it}(x,\cdot)$ of $H_s(\Omega)$. We prove the following lemma.

LEMMA 1. Let $0<s<1/2$, $t \neq \pi h \Upsilon^{-1}$, $h \in \mathbb{Z}$. We define for $x \in G$, $\omega \in \Omega$,

$$F_x(\omega) = \rho_1(s) \int_\Omega |p^{it}(x,\omega) - p^{it}(x,\omega')|^2 (2r-1)^{2(1-s)N(\omega,\omega')} d\mu(\omega'),$$

where $\rho_1(s) = -\rho(s)$. Then $F_x \in L^{1/(1-2s),\infty}(\Omega)$, with $\|F_x\|^*_{1/(1-2s),\infty} \le \beta(s,t)$, where

(i) for any fixed s, $0<s<1$, $s \neq 1/2$, $\beta(s,\cdot)$ is a bounded, periodic function with period $\pi\Upsilon^{-1}$, and $\lim_{t\to 0} \beta(s,t) = 0$,

(ii) for any fixed t, $t \neq \pi h \Upsilon^{-1}$

$$\lim_{s\to 0^+} \beta(s,t) = +\infty .$$

Proof. The function F_x can be written in a more explicit way, as follows:

$$F_x(\omega)=\begin{cases}\rho_1(s)\int_{\Omega\setminus E(x^{(1)})}\left|(2r-1)^{2itN(\omega,\omega')}-1\right|^2 d\mu(\omega') & \text{, when } N(x,\omega)=0\\ \rho_1(s)\left[((2r-1)/2r)\left|(2r-1)^{2itk}-1\right|^2+((r-1)/r)\sum_{j=1}^{k-1}\left|(2r-1)^{2it(k-j)}-1\right|^2(2r-1)^{j-2sj}\right., & \\ & \text{when } 0<N(x,\omega)=k<|x|\ .\end{cases}$$

By the periodicity of the function $\left|(2r-1)^{2it(k-j)}-1\right|^2$, it is enough to consider $0<|t|\le \pi/2\Psi$. We get

$$\left|(2r-1)^{2it(k-j)}-1\right|^2=\begin{cases}4 & ,\ k-j\ge |t|^{-1}\Psi^{-1}\\ 4t^2(k-j)^2\Psi^2 & ,\ k-j<|t|^{-1}\Psi^{-1}.\end{cases}$$

So, for $0<k=N(x,\omega)\le|x|$, we have, for every $\omega\in\Omega$,

$$F_x(\omega)\le \rho_1(s)\left[4(2r-1)/2r+\left[4(r-1)/r\right]\sum_{j=1}^{k_t}(2r-1)^{(1-2s)j}\right.$$

$$\left.+\left[4t^2\Psi^2(r-1)/r\right]\sum_{j=k_t+1}^{k-1}(k-j)^2(2r-1)^{(1-2s)j}\right. ,$$

where $k_t=\mathrm{Int}\,(k-|t|^{-1}\Psi^{-1})_+$. Since

$$\sum_{j=1}^{k_t}(2r-1)^{(1-2s)j}\le\left[(2r-1)^{(1-2s)(k-|t|^{-1}\Psi^{-1})}-1\right]/\left[1-(2r-1)^{(2s-1)}\right]$$

(obviously this sum is zero if $k_t=0$) and

$$\sum_{j=k_t+1}^{k-1}(k-j)^2(2r-1)^{(1-2s)j}\le(k-1-k_t)\max_{k_t<j<k}\left[(k-j)^2(2r-1)^{(1-2s)j}\right]$$

$$\le 4e^{-2}\Psi^{-3}|t|^{-1}(1-2s)^{-2}(2r-1)^{(1-2s)k} ,$$

then

$$F_x(\omega)\le 4\rho_1(s)\left((2r-1)/2r\right)+\left[4\rho_1(s)(r-1)/r\right](2r-1)^{(1-2s)k}$$

$$\cdot\left[(2r-1)^{(2s-1)|t|^{-1}\Psi^{-1}}/(1-(2r-1)^{(2s-1)})+4|t|e^{-2}\Psi^{-1}(1-2s)^{-2}\right]$$

$$\equiv A(s)+B(s,t)(2r-1)^{(1-2s)k} ,$$

with obvious notation.

To prove that $F_x \in L^{1/(1-2s),\infty}(\Omega)$, we compute for $\lambda \in \mathbb{R}^+$,

$$\mu\{\omega\in\Omega,\ F_x(\omega)>\lambda\} \le \mu\{\omega\in\Omega,\ B(s,t)(2r-1)^{(1-2s)k} > \lambda - A(s)\}$$
$$\le ((2r-1)/2r)\ B(s,t)^{1/(1-2s)}(\lambda - A(s))^{1/(1-2s)} ,$$

and we obtain that

$$\|F_x\|^*_{1/(1-2s),\infty} \le ((2r-1)/2r)B(s,t)^{1/(1-2s)}$$
$$\equiv \beta(s,t) .$$

The properties (i) and (ii) follow now from the periodicity with respect to t and from the definitions of $B(s,t)$ and $\rho(s)$.

Proposition 6.2 of [MZ] can be now restated.

PROPOSITION 2. Let $0<s<1/2$, $t \ne \pi h \Psi^{-1}$, $h\in\mathbb{Z}$. Then for every $x\in G$, the $H_s(\Omega)$- multiplier operator $p^{it}(x,\cdot)$, satisfies

$$\|p^{it}(x,\cdot)f\|_s \le (2+\beta(s,t))\|f\|_s,\quad \forall f\in H_s(\Omega) ,$$

where $\beta(s,t)$ is defined in Lemma 1.

Proof of Theorem C. If $0<s<1/2$, then

$$\|\pi_z(x)f\|_s = \|p^{it}(x,\cdot)\pi_s(x)f\|_s \le (2+\beta(s,t))\|f\|_s .$$

If $1/2<s<1$, since $p^{it}(x,\cdot)$ as $H_s(\Omega)$-multiplier operator is the adjoint of $p^{-it}(x,\cdot)$ as multiplier operator on $H_{1-s}(\Omega) = H^*_s(\Omega)$, then

$$|||p^{it}(x,\ )|||_{op} \le 2+\beta(1-s,-t) = 2+\beta(1-s,t) ,$$

and

$$\|\pi_z(x)f\|_s \le (2+\beta(1-s,t))\|f\|_s .$$

The theorem is now proved if we set

$$\alpha(s,t) = \begin{cases}\beta(s,t) , & 0<s<1/2\\ \beta(1-s,t) , & 1/2<s<1 .\end{cases}$$

5. A DENSITY THEOREM FOR MULTIPLIERS OF $L^p(G)$.

The following theorem applies the results of Theorem C.

THEOREM D. *Let* $1<p<\infty$, *and* $T : L^p(G) \to L^p(G)$ *an operator commuting with (let us say) right translations. Then there exists a net* $(f_\alpha)_{\alpha \in A}$ *of functions in* $C_c(G)$ *such that*

(i) $\lim_\alpha f_\alpha * f = Tf$ *uniformly on compacta,* $\forall f \in L^p(G)$,

(ii) $|||f_\alpha *|||_p \leq 4\,|||T|||_p$.

Moreover $\operatorname{supp} f_\alpha \subseteq \operatorname{supp} T$, $\forall \alpha \in A$.

To prove this theorem we give some definitions and results. We remark that the results of the first part of the paper [C] are true for a free group G. In fact such a results, proved for connected semisimple Lie groups, with finite center, pass to closed subgroups, and G is a closed subgroup of $SL(2,\mathbb{R})$.

DEFINITION 1. (See [C]). *A function* $f: G \to \mathbb{C}$ *is a Herz-Schur multiplier, written* $f \in HS(G)$, *if* Mf *defined by* $Mf(x,y) = f(xy^{-1})$, $\forall x,y \in G$, *multiplies pointwise the projective tensor product space* $L^2(G) \otimes_\gamma L^2(G)$ *considered as a subspace of* $L^2(G\times G)$.

We denote by $END_p(G)$ the Banach algebra of bounded linear operators on $L^p(G)$. Then any element $k \in END_p(G)$ can be represented as a map $k : G\times G \to \mathbb{C}$ such that $|||k|||_p < \infty$ where $|||k|||_p$ is the best constant C such that

$$\Big|\sum_{x,y \in G} k(x,y)\, u(y)\, v(x)\Big| \leq C \left(\sum |u(y)|^p\right)^{1/p} \left(\sum |v(y)|^{p'}\right)^{1/p'}$$

$\forall\, u \in L^p(G)$, $v \in L^{p'}(G)$.

DEFINITION 2. *For* $1<p<\infty$, *we define* $B_p(G)$ *to be the set of (continuous) maps*

$f : G \to \mathbb{C}$ such that, if $Mf(x,y) = f(xy^{-1})$, then $Mf \in V_p(G)$, i.e. $Mf: G \times G \to \mathbb{C}$ and there exists a constant $\|Mf\|_{V_p}$ such that

$$|||(Mf)k|||_p \leq \|Mf\|_{V_p} |||k|||_p$$

for all $k \in END_p(G)$ with finite support in $G \times G$. Here $(Mf)k$ denotes the pointwise multiplication.

While it is easy to check that $B_2(G)$ and $HS(G)$ are isometrically isomorphic, it has been proved by Gilbert [G] that $HS(G) \subsetneq B_p(G)$, with $\|\cdot\|_{B_p} \leq \|\cdot\|_{HS}$. We also know from Theorem 2 of [H], that the Banach algebra $B_p(G)$ operates on the space of $L^p(G)$-convolütor operators in such a way that if T is a convolution operator and $\varphi \in B_p(G)$, then

$$|||\varphi T|||_p \leq \|\varphi\|_{B_p} |||T|||_p .$$

To understand how the study of uniformly bounded representations is tied to the proof of Theorem D, we recall that, by Proposition 1.2 of [C], if f is a matrix coefficient of a uniformly bounded represehtation π of G on some Hilbert space H, i.e. $f(x) = \langle \pi(x)\xi, \eta\rangle$ for some $\xi, \eta \in H$, then $f \in HS(G)$ and

$$\|f\|_{HS} \leq \sup_{x \in G} \|\pi(x)\|^2 \|\xi\| \|\eta\| .$$

PROPOSITION 1. Let $0 < s < 1$, $s \neq 1/2$, $t \neq \pi h \gamma^{-1}$, $h \in \mathbb{Z}$. Then the spherical functions φ_{s+it} satisfy

$$\|\varphi_{s+it}\|_{HS} \leq (2 + \alpha(s,t))^2$$

where $\alpha(s,t)$ has been defined in Theorem C.

Proof. It follows by Proposition 1.2 of [C] and Theorem C.

Proof of Theorem D. On discrete groups every operator $T : L^p(G) \to L^p(G)$ which commutes with (let us say) right translation, i.e. such that $T(f*g) = Tf*g$ for all $f,g \in L^p(G)$, is a convolution operator on $L^p(G)$. So, there exists a function $h \in L^p(G)$ such that $Tf = h*f$, $\forall f \in L^p(G)$. By Proposition 1, the spherical functions φ_{s+it} satisfy the hypothesis of Theorem 1.3 of [C]. So there exists a net $(\Psi_\alpha)_{\alpha \in A}$, $\Psi_\alpha \in C_c(G)$, such that $\Psi_\alpha \to 1$ uniformly on compacta and $\|\Psi_\alpha\|_{HS} \leq 4$. Let define $f_\alpha = \Psi_\alpha h$. Then $f_\alpha \to h$ uniformly on compacta and $\operatorname{supp} f \subseteq \operatorname{supp} T$: Moreover $f_\alpha * f \to h*f = Tf$ uniformly on compacta, and

$$
\begin{aligned}
|||f_\alpha *|||_p &\leq \|\Psi_\alpha\|_{B_p} \, |||h*|||_p \leq \|\Psi_\alpha\|_{HS} \, |||h*|||_p \\
&\leq 4 \, |||T|||_p .
\end{aligned}
$$

ADDENDUM

We observe that ideally the value of the constant in (ii) of Theorem D should be 1 (instead of 4). In fact this can be easily obtained if we use the following suggestion of G.Fendler (personal communication). For $a, b \in \mathbb{R}$ and $\lambda > 0$, then $(a+b)^2 \leq (1+\lambda) a^2 + (1+1/\lambda) b^2$. Moreover $\inf_{\lambda > 0} \{(1+\lambda) a^2 + (1+1/\lambda) b^2\} = (a+b)^2$, whenever $a \geq 0$, $b \geq 0$. So we can rewrite the proof of Proposition 6.2 of [MZ] as follows

$$
\begin{aligned}
\|p^{it}(x,\cdot)\, f\|_s^2 &= \|p^{it}(x,\cdot)\, f\|_2^2 - 2^{-1} \int_\Omega \int_\Omega |p^{it}(x,\omega)\, f(\omega) - p^{it}(x,\omega')\, f(\omega')|^2 \\
&\qquad \cdot k_s(\omega,\omega')\, d\mu(\omega)\, d\mu(\omega') \\
&\leq \|f\|_2^2 - 2^{-1} \int_\Omega \int_\Omega \Big[(1+\lambda)\, |f(\omega)\, p^{it}(x,\omega') - f(\omega')\, p^{it}(x,\omega')|^2 \\
&\qquad + (1+1/\lambda)\, |f(\omega)|^2 |p^{it}(x,\omega) - p^{it}(x,\omega')|^2\Big] k_s(\omega,\omega')\, d\mu(\omega)\, d\mu(\omega') \\
&= \|f\|_2^2 - (1+\lambda)/2 \int_\Omega \int_\Omega |f(\omega) - f(\omega')|^2 k_s(\omega,\omega')\, d\mu(\omega)\, d\mu(\omega') \\
&\qquad - (1+1/\lambda)/2 \int_\Omega \int_\Omega |f(\omega)|^2\, |p^{it}(x,\omega) - p^{it}(x,\omega')|^2\, k_s(\omega,\omega')\, d\mu(\omega)\, d\mu(\omega') \\
&= -\lambda \|f\|_2^2 + (1+\lambda) \|f\|_s^2 + \langle |f|^2, F_x \rangle\, (1+1/\lambda)/2 .
\end{aligned}
$$

Since $\|F_x\|^*_{1/(1-2s),\infty} \le \beta(s,t)$, then

$$\left\|P^{it}(x,\cdot)\, f\right\|_s^2 \le (1+\lambda)\|f\|_s^2 + (1 + 1/\lambda)/2 \;\; \|f\|^*_{1/s,2}\beta(s,t)$$
$$\le \left[(1+\lambda)+(1 + 1/\lambda)\beta(s,t)/2\right]\|f\|_s^2.$$

Finally, taking the infimum over all $\lambda > 0$, we get

$$\left\|P^{it}(x,\cdot)\, f\right\|_s^2 \le (1 + (\beta(s,t)/2)^{1/2})^2 \|f\|_s^2 \quad .$$

REFERENCES

[C] - M.G. COWLING, Harmonic Analysis on some nilpotent groups, Atti del Seminario Intensivo "Topics in Modern Harmonic Analysis" I.N.D.A.M. (to appear).

[FP] - A. FIGA'-TALAMANCA and M. PICARDELLO, Spherical functions and harmonic analysis on free groups, J.Funct.Anal. 47 n.3 (1982).

[G] - J.E. GILBERT, L^p-convolution operators and tensor products of Banach spaces III, (to appear).

[Hg] - U. HAAGERUP, An example of a non nuclear C^*-algebra which has the metric approximation property, Invent.Math. 50 (1979).

[H] - C. HERZ, Une généralisation de la notion de transformée de Fourier-Stieltjes, Ann.Inst.Fourier (Grenoble) 24 (1974).

[MZ] - A.M. MANTERO and A. ZAPPA, The Poisson transform and representations of a free group, J.Funct.Anal. (to appear).

AUTHORS' ADDRESS: Istituto di Matematica,

Via L.B. Alberti 4

16132 GENOVA - ITALY.

"El universo... se compone de un numero indefinido, y tal vez infinido, de galerias hexagonales..."
J.L. Borges, La Biblioteca de Babele.

SPHERICAL FUNCTIONS ON SYMMETRIC GRAPHS

by

Alessandra Iozzi and Massimo A. Picardello

1. Introduction. The theory of representations of free groups with finitely many generators has been recently considered in analogy with the semisimple theory [12, 16, 13]. This analogy arises from the realization of a free group as a homogeneous tree, playing the role of the symmetric space G/K for a semisimple Lie group G with maximal compact subgroup K, and relies upon the use of the Poisson boundary and spherical functions. The results of [12] have been lately extended to every group acting isometrically and simply transitively on a homogeneous tree. If the tree is homogeneous of degree d, every such group is isomorphic with a free product of the type $(\ast_{i=1}^{s} \mathbb{Z}) * (\ast_{j=1}^{m} \mathbb{Z}_2)$, with 2s+m = d [3]. The aim of the present paper is to establish a similar theory for groups acting on suitable graphs.

To define and use spherical functions, we restrict attention to "symmetric graphs". A graph Γ is symmetric of type $k \geqslant 2$ and order $r \geqslant 2$ if every vertex v belongs exactly to r polygons, with k sides each, contained in the graph, with no sides and no vertex in common except v, and if every nontrivial loop in Γ runs through all the edges of at least one polygon. In other words, a symmetric graph of type k and order r can be thought of as a "homogeneous tree of order r built up with polygons with k sides". Notice that, if $k = 2$, Γ is a homogeneous tree of degree 2r.

Different notions of length on a graph were introduced in [15]. Here we define the length $|v|$ of a vertex v, with respect to a reference vertex e, as the minimal number of polygons crossed by paths in Γ connecting e with v. If $k > 2$, it is easy to see, as in [3, thm. 1], that every group acting simply transitively on Γ and isometrically with respect to the metric induced by this length is isomorphic with the free product $G = \mathop{*}\limits_{i=1}^{r} \mathbb{Z}_k$ (the results of this paper also hold in the case $k = 2$, considered in [3]). Every vertex of Γ can be identified with an element of G, and every polygon of the graph corresponds, under this identification, to an orbit under right translations by one of the factors $\mathbb{Z}_k$. In this setting, it is natural to define the length $|g|$ of an element

$g \in G$ as follows. Denote by a_i, $1 \leq i \leq r$, the generators of G. Every element g of G has a unique representation as a *reduced word* $g = a_{i_1}^{n_1} \ldots a_{i_m}^{n_m}$, with $0 < n_j < k$, for $j = 1,\ldots,m$: then $|g| = m$. The m factors in the reduced expression for g are called the *blocks* of g.

Let μ_1 be the probability measure equidistributed on words of length one. In section 2 we introduce the convolution algebra $\ell^1_{\#}$ of all summable functions which are *radial*, that is, which depend only on the length of a word, and we define *spherical functions* as the radial eigenfunctions φ of the operator of convolution by μ_1, normalized in such a way that $\varphi(e) = 1$. We prove that the algebra $\ell^1_{\#}$ is commutative, is generated by μ_1, and the multiplicative functionals on $\ell^1_{\#}$ are given by the bounded spherical functions. Spherical functions can be parameterized by a complex number z, so that, if φ_z is spherical, then $\mu_1 * \varphi_z = \gamma(z)\ \varphi_z$, with $\gamma(z) = \frac{1}{(k-1)r}\ (q^z + q^{1-z} + k - 2)$, where $q = (k-1)(r-1)$. Furthermore, we show that every spherical function φ_z decomposes as a linear combination of exponentials, that φ_z is bounded if and only if $0 \leq \operatorname{Re} z \leq 1$, and is positive definite if and only if it is bounded and $\gamma(z)$ is real. It follows that the Gelfand spectrum of $\ell^1_{\#}$, that is, the spectrum of μ_1 in $\ell^1_{\#}$, is the ellipse $E = \{\gamma(z): 0 \leq \operatorname{Re} z \leq 1\}$. In particular, $\ell^1_{\#}$ is not symmetric.

By different means, special cases of this result have been obtained by R.A. Bonic [2] for the groups $\mathbb{Z}_2 * \mathbb{Z}_2$ and $\mathbb{Z}_2 * \mathbb{Z}_2 * \mathbb{Z}_2$. The same problem for the group $\overset{r}{\underset{i=1}{*}} \mathbb{Z}_2$ has been considered by W. Betori and M. Pagliacci [3] in the framework of spherical functions on trees: by the same approach, they also treat the free group $\mathbb{F}_r = \overset{r}{\underset{i=1}{*}} \mathbb{Z}$, and, more generally, the group $\mathbb{F}_r * (\overset{m}{\underset{j=1}{*}} \mathbb{Z}_2)$. Previous results for free groups had been obtained in [4,5,6,9,1], and, by a different approach, in [17,11].

In section 3 we compute the spectrum of μ_1 in ℓ^p, $1 \leq p < \infty$, in the C^*-algebra $C^*_\lambda(G)$ of left convolution operators on ℓ^2, and in the full C^*-algebra $C^*(G)$. For the special case of two generators, the spectral radius of μ_1 in C^*_λ had been computed by P. Gerl [14], by different means.

In section 4 we define, for every complex number z, a representation π_z of G on the space K of *cylindrical functions* on the Poisson boundary Ω of G (that is, functions which depend only on finitely many letters of the infinite reduced words $\omega \in \Omega$), in such a way that $\varphi_z(x) = (\pi_z(x)\,\mathbb{1}, \mathbb{1})$, where $\mathbb{1}$ is the constant function $\mathbb{1}$ on Ω, and (,) denotes the inner product in $L^2(\Omega)$ with respect to the Poisson measure. The representation π_z extends to a unitary representation on $L^2(\Omega)$ if and only if Re $z = \frac{1}{2}$ (*principal series* of representations). On the other hand, if φ_z vanishes at infinity and is positive definite but Re $z \neq \frac{1}{2}$, we consider the infinite dimensional Hilbert space H_z defined as the completion, in the norm defined by φ_z, of the subspace M of K generated by $\{\pi_z(x)\,\mathbb{1},\ x \in G\}$. Then the representation $\pi_z|_M$ extends to a unitary representation on H_z, again denoted by π_z (*complementary series*). The unitary representation π_z is an extension of the representation originally defined on K if and only if $M = K$: this happens exactly when $\gamma(z) = \frac{1}{1-k}$.

We show that all the representations of the principal and complementary series are irreducible, except for the case when $k \geqslant r$ and $\gamma(z) = \frac{1}{1-k}$. As in the case of free groups, the proof of this result is considerably more difficult than in the semisimple theory (because G does not act transitively on its Poisson boundary), and relies upon precise estimates on the distribution of values of two-sided translates of spherical functions (Lemma 4).

Finally, we point out that, although this theory bears a close similarity with the free group theory [12,13], it is different in some respects. First of all, some of the unitary representations of the principal or complementary series are reducible: indeed, the representations π_z with $\gamma(z) = \frac{1}{1-k}$ are subrepresentations of the regular representations for $k > r$ (in this case, φ_z belongs to ℓ^2). Secondly, the ellipse E, i.e., the spectrum of μ_1 in ℓ^1, is not centered in the origin: the real point +1 is a vertex, but the point -1 is not. Observe that every character χ of $G = \mathop{*}\limits_{i=1}^{r} \mathbb{Z}_k$ is an eigenfunction of μ_1 with eigenvalue $<\chi, \mu_1>$. If χ is a *radial* character, the eigenvalue is equal to ± 1. The trivial character corresponds to the eigenvalue 1. If $k = 2$ (or in the case of a free group) there is another radial character, namely, the alternating character $\chi(x) = (-1)^{|x|}$, which corresponds to the eigenvalue -1. The shift of the ellipse for $k > 2$ can be related to the fact that, because of torsion, no nontrivial radial character exists in this case. Essentially for the same reasons, for certain values of p and k, the spectrum of μ_1 in ℓ^p contains isolated points.

2. <u>Spherical functions</u>. Let W_n be the set of words of length n in $G = \mathop{*}\limits_{i=1}^{r} \mathbb{Z}_k$ and observe that $\# W_n = (k-1)^n r(r-1)^{n-1}$. A function defined on G is called radial if it depends only on the length of $x \in G$.

Denote by $\mathcal{R}$ the convolution algebra of finitely supported radial function, and by μ_n the probability measure equidistributed over W_n; we shall often write δ_e instead than μ_o.

LEMMA 1. *For* $n \geqslant 1$,

$$\mu_1 * \mu_n = \frac{1}{(k-1)r}\mu_{n-1} + \frac{k-2}{(k-1)r}\mu_n + \frac{r-1}{r}\mu_{n+1} \tag{1}$$

Proof. Let χ_n be the characteristic function of W_n and let $x \in W_m$, $m > 0$. Then $|xy| = m+1$ for $(k-1)(r-1)$ choices of $y \in W_1$, and $|xy| = m-1$ for only one choice of $y \in W_1$; the remaining $k-2$ words y of length one satisfy the identity $|xy| = m$.

Therefore

$$\chi_1 * \chi_n(x) = \sum_y \chi_1(y)\ \chi_n(xy) =$$

$$= \chi_{n-1}(x) + (k-2)\chi_n(x) + (k-1)(r-1)\chi_{n+1}(x). \qquad \square$$

COROLLARY 1. *$\mathcal{R}$ is a commutative algebra with identity, generated by* μ_1.

DEFINITION. A function φ on $G = \mathop{*}\limits_{i=1}^{r} \mathbb{Z}_k$ is *spherical* if

i) φ is radial

ii) $\varphi(e) = 1$

iii) for every $f \in R$, there exists $c \in \mathbb{C}$ such that $\varphi * f = c\varphi$.

We adopt the following notation: if h, k are functions on G, then $\langle h, k\rangle = \sum_{x \in G} h(x)\, k(x)$. Observe that the constant c in part iii) of the above definition is given by $c = \langle f, \varphi\rangle$. We shall make frequent use of the projection $\mathcal{E}$ onto radial functions defined by

$$\mathcal{E}f(x) = \frac{1}{\#W_n} \sum_{y \in W_n} f(y)$$

where $n = |x|$. Observe that $\mathcal{E}$ is an expectation, i.e., $\langle \mathcal{E}f, g\rangle = \langle f, \mathcal{E}g\rangle = \langle \mathcal{E}f, \mathcal{E}g\rangle$ for all functions f and g. The following result is useful in section 3.

LEMMA 2. *The expectation $\mathcal{E}$ maps positive definite functions into positive definite functions.*

Proof. By [7, thm. 8] there exists a positive projection T, mapping VN(G) onto the subalgebra generated by R, which maps positive definite functions into positive definite functions. This projection satisfies the rule $\langle f, Tg\rangle = \langle f, g\rangle$ whenever $f \in R$, hence it coincides with $\mathcal{E}$.

An explicit construction of this map can be obtained

as in the proof of [7, theorem 8]. □

Let us denote by λ the left regular representation $\lambda(x)\, f(y) = f(x^{-1}y)$. The following proposition can be proved as in [12, Lemma 2].

PROPOSITION 1. *For a nonzero function* φ *the following are equivalent:*

i) φ *is spherical*

ii) $\mathcal{E}(\lambda(x)\varphi)(y) = \varphi(x)\ \varphi(y)$

iii) φ *is radial and the functional* $Lf = \langle f, \varphi \rangle$ *is multiplicative on* R.

We wish to point out two consequences of this result. First of all, by part iii above and identity (1), it follows that a spherical function is uniquely determined by the value $\langle \varphi, \mu_1 \rangle$ which it attains on words of length one. Secondly, let us denote by $\ell^1_\#$ the completion of R in the norm of ℓ^1. Then the Gelfand spectrum of the commutative Banach algebra $\ell^1_\#$ is in one-to-one correspondence with the set of bounded spherical functions.

Let now Ω be the set of all infinite reduced words in the generators of G and, for $x \in G$, let us denote by $E(x)$

the subset of Ω of all words beginning with the finite word x. The family $\{E(x), x \in G\}$ is a basis for a topology which makes Ω compact. Let ν be the measure on Ω defined by $\nu(E(x)) = 1/\#W_n$ if $n = |x|$. It is easy to see, as in [10], that, with the above topology, the space (Ω,ν) is the Poisson boundary of G relative to μ_1. In fact, it is immediately seen that ν is μ_1-stationary. For $x \in G$, let ν_x be the translate of ν by x: $\nu_x(A) = \nu(x^{-1}A)$ for every Borel subset A of Ω. It is easy to see that ν_x is absolutely continuous with respect to ν. Indeed, the Radon-Nikodym derivative $d\nu_x/d\nu(\omega) = P(x,\omega)$, called the *Poisson kernel*, can be computed as follows.

For $m \geq 0$, denote by ω_m the word of length m consisting of the first m blocks of ω. Then, if $m > |x|$, the expression $\delta(x,\omega) = m - |x^{-1}\omega_m|$ is independent of m, and $P(x,\omega) =$
$= \nu_x(E(\omega_m))/\nu(E(\omega_m)) = \nu(E(x^{-1}\omega_m))/\nu(E(\omega_m)) = ((k-1)(r-1))^{\delta(x,\omega)}$.

The next proposition yields a handier expression for the Poisson kernel. We introduce the following notations. Let x be a word of length n, and, for every $j=1,\ldots n$, denote by x_j the word of length j consisting of the first j blocks of x; set $x_o = e$. Denote by a_i, $i=1,\ldots r$, the generators of G, and let $a_{i_j}^{m_j}$ be the j-th block of x. For each $j=1,\ldots n$, we write $B(x_j) = \cup\{E(x_{j-1}\, a_{i_j}^{p}): 0 < p < k,\ p \neq m_j\}$, and

$C(x_{j-1}) = E(x_{j-1}) - E(x_j) - B(x_j)$. The characteristic functions of $E(x_j)$, $B(x_j)$, $C(x_j)$ will be denoted by $\chi_{E(x_j)}, \chi_{B(x_j)}, \chi_{C(x_j)}$, respectively. Then the expression of $P(x,\omega)$ computed above yields:

PROPOSITION 2. *Let* $|x| = n$ *and* $q = (k-1)(r-1)$. *Then, with notations as above, the Poisson kernel is given by the following disjoint expansion:*

$$P(x,\omega) = q^n \chi_{E(x)} + \sum_{j=1}^{n} q^{2j-n-1} \chi_{B(x_j)} + \sum_{j=0}^{n-1} q^{2j-n} \chi_{C(x_j)}.$$

We recall that the Poisson kernel, being a Radon-Nikodym derivative, satisfies the following cocycle identities:

$$P(e,\omega) = 1$$

$$P(xy,\omega) = P(y,x^{-1}\omega)\, P(x,\omega).$$

We can now describe the connection between the Poisson kernel and spherical functions.

THEOREM 1. *Let* $z \in \mathbb{C}$ *and* $\varphi_z(x) = \int_\Omega P^z(x,\omega)\, d\nu(\omega)$. *Then* φ_z *is spherical and* $\varphi_z * \mu_1 = \gamma(z)\, \varphi_z$, *with*

$\gamma(z) = \frac{1}{(k-1)r}(q^z + q^{1-z} + k - 2) = \varphi_z(x)$ *where* $q = (k-1)(r-1)$ *and* $|x| = 1$. *Conversely, every spherical function arises in this way. Furthermore,* $\mathcal{E}(P^z(x,\omega))$ *is independent of* ω, *and* $\varphi_z(x) = \mathcal{E}(P^z(x,\omega))$.

Proof. The argument bears a close similarity with the case of a free group [12,13] : it is outlined here for the sake of completeness. Let φ_z be as in the statement. Then, by the second cocycle identity, we have:

$$\varphi_z * \mu_1(x) = (k-1)^{-1} r^{-1} \sum_{|y|=1} \varphi_z(xy) =$$

$$= (k-1)^{-1} r^{-1} \int_\Omega \sum_{|y|=1} P^z(y,x^{-1}\omega)\, P^z(x,\omega)\, d\nu(\omega).$$

On the other hand, the explicit expression for the Poisson kernel, $P(x,\omega) = q^{\delta(x,\omega)}$, allows us to write

$$\sum_{|y|=1} P^z(y,\omega) = q^z + q^{1-z} + k - 2.$$

Since the right hand side is independent of ω, we obtain:

$$\varphi_z * \mu_1(x) = (k-1)^{-1} r^{-1} \sum_{|y|=1} P^z(y,x^{-1}\omega)\, \varphi_z(x) =$$

$$= \gamma(z)\, \varphi_z(x).$$

Furthermore, φ_z is radial, because the probability distribution of $\delta(x,\omega)$ depends only on the length of x, and $\varphi_z(e) = 1$, by the first cocycle identity. Thus, by Corollary 1, φ_z is spherical.

Conversely, let φ be a spherical function and choose $z \in \mathbb{C}$ such that, if $|x| = 1$, then $\varphi(x) = \gamma(z)$ (observe that γ is surjective). Then φ and φ_z coincide on words x with $|x| \leq 1$: hence, by Corollary 1, they coincide everywhere. Finally, notice that we have already proved that $\mathcal{E}(P^z(x,\omega)) = \gamma(z)$ if $|x| = 1$. Then it is easy to check that $\mathcal{E}(P^z(\cdot,\omega))$ is the spherical function φ_z whose convolution eigenvalue under μ_1 is $\gamma(z)$. □

The next theorem gives an explicit expansion of the spherical function φ_z as a linear combination of exponentials. We adopt the following notations: let $q = (k-1)(r-1)$, and, for $z \neq \frac{1}{2} + m\pi i/\ln q$, $m \in \mathbb{Z}$, let c_z, c_z' be such that

$$c_z + c_z' = 1, \qquad c_z\, q^{-z} + c_z'\, q^{z-1} = \gamma(z) \tag{2}$$

Thus $\quad c_z = \dfrac{q^{1-z} - (r-1)^{-1}\, q^{z} + k-2}{(k-1)r\,(q^{-z}-q^{z-1})}$, and $c_z' = c_{1-z}$.

For future purposes, observe that $c_z = 0$ if $q^{1-z} = 1-k$.

THEOREM 2. *With notations as above, let* $h_z(x) = q^{-z|x|}$.
Then

i) if $z \neq \frac{1}{2} + m\pi i/\ln q$, $\varphi_z = c_z h_z + c_{1-z} h_{1-z}$

ii) if $z = \frac{1}{2} + m\pi i/\ln q$, $m \in \mathbb{Z}$, *then* $\varphi_z(x) =$

$$= \left(1 + \frac{2q + (k-2)\sqrt{q} - (k-1)r}{(k-1)r} |x|\right)(-1)^{m|x|} q^{-\frac{1}{2}|x|}$$

Proof. Suppose $z \neq \frac{1}{2} + m\pi i/\ln q$. Then the linear system (2) is nonsingular. Let $f = c_z h_z + c_{1-z} h_{1-z}$: then $f(e) = 1$ and $f * \mu_1(e) = \langle f, \mu_1 \rangle = \gamma(z)$.
To complete the proof of i), it is enough to show that $f * \mu_1(x) = \gamma(z) f(x)$ for $|x| > 1$. Since $\gamma(z) = \gamma(1-z)$, it suffices to show that $h_z * \mu_1(x) = \gamma(z) h_z(x)$. We have $h_z * \mu_1(x) = (k-1)^{-1} r^{-1} \sum_{|y|=1} h_z(xy)$. The argument of Lemma 1 now yields:

$$h_z * \mu_1(x) = \frac{1}{(k-1)r} (q^z + k-2 + q^{1-z}) h_z(x) = \gamma(z) h_z(x)$$

This proves i). Let now $z = \frac{1}{2} + m\pi i/\ln q$, $k_z(x) = |x| h_z(x)$ and $g = h_z + \frac{2q + (k-2)\sqrt{q} - (k-1)r}{(k-1)r} k_z$. We now show that $g = \varphi_z$.
It is obvious that $g(e) = 1$. Furthermore, by the choice of z, $q^z = q^{1-z}$: it follows easily that $g(x) = \gamma(z)$ for $|x| = 1$.
It remains to show that $g * \mu_1 = \gamma(z) \mu_1$. For this, it is

enough to prove the identity $k_z * \mu_1 = \gamma(z)\, k_z$, which can be easily verified. □

The estimates in the next Corollary are immediate consequences of Theorem 2.

COROLLARY 2. *The spherical function* φ_z *is bounded if and only if* $0 \leq \mathrm{Re}\, z \leq 1$. *For every* $p > 2$, $\varphi_z \in \ell^p(G)$ *if and only if* $\frac{1}{p} < \mathrm{Re}\, z < 1 - \frac{1}{p}$, *or* $\gamma(z) = (1-k)^{-1}$.

3. <u>The spectrum of radial functions</u>. We denote by $\ell^1_{\#}$ the completion of R in the norm of $\ell^1(G)$. Then, by Corollary 1, $\ell^1_{\#}$ is a commutative Banach algebra with identity, generated by μ_1; its Gelfand space E is the spectrum of μ_1 in $\ell^1_{\#}$. By Proposition 1, the multiplicative functionals on $\ell^1_{\#}$ are given by the bounded spherical functions; it follows from Corollary 2 that $E = \{<\mu_1, \varphi_z> : 0 \leq \mathrm{Re}\, z \leq 1\} = \{\gamma(z) : 0 \leq \mathrm{Re}\, z \leq 1\}$. It is easy to check that E is the ellipse whose foci are the real points $\frac{k-2\pm 2\sqrt{q}}{(k-1)r}$. The substrip $1/p \leq \mathrm{Re}\, z \leq 1-1/p$ maps under γ onto a subellipse of E: the line $\mathrm{Re}\, z = \frac{1}{2}$ maps onto the segment I connecting the two foci of E. For future reference, we now describe the inverse images under γ of the real point of E which do not belong to I. Recall that γ is

periodic along the imaginary axis, with period $2\pi/\ln q$, where $q = (k-1)(r-1)$. The segment connecting the right focus with the right vertex of E is the image under γ of the segments $0 \leq \mathrm{Re}\, z \leq 1$, $\mathrm{Im}\, z = 2m\pi i/\ln q$, $m \in \mathbb{Z}$. On the other hand, the segment connecting the left vertex with the left focus is the image of $0 \leq \mathrm{Re}\, z \leq 1$, $\mathrm{Im}\, z = (2m+1)\pi i/\ln q$. The remarks concerning the Gelfand space of $\ell^1_\#$ are summarized in the next statement.

PROPOSITION 3. *The spectrum of* μ_1 *in* ℓ^1 *is the ellipse* $E = \{\gamma(z) : 0 \leq \mathrm{Re}\, z \leq 1\}$.

In what follows, we denote by C^* the full C^*-algebra of the group G, and by C^*_λ the C^*-algebra generated by the left regular representation λ of G (that is, the C^*-algebra generated by finitely supported left convolution operators on ℓ^2).

PROPOSITION 4. *The spectrum of* μ_1 *in* $C^*(G)$ *is the real axis of the ellipse E, i.e., the interval* $D = [\frac{2(k-2)}{(k-1)r} - 1, 1]$. *Moreover, a spherical function* φ_z *is positive definite if and only if* $\gamma(z) \in D$.

Proof. Observe that, by Lemma 1, $\gamma(z) \in D$ if and only if φ_z is real valued. Let $f \in \ell^1_{\#}$ and write $f^*(x) = \overline{f(x^{-1})}$. Then, if $\gamma(z) \in D$, $\langle f^* * f, \varphi_z \rangle = \langle f^*, \varphi_z \rangle \langle f, \varphi_z \rangle = |\langle f, \varphi_z \rangle|^2 \geq 0$. Thus the spherical function φ_z gives rise to a positive functional on the enveloping C^*-algebra of $\ell^1_{\#}$, which is still multiplicative. Hence the segment D is contained in the spectrum of μ_1 in $C^*(G)$. On the other hand, if $\psi \in \ell^\infty(G)$ gives rise to a multiplicative functional on $C^*(G)$, then ψ is also multiplicative on $\ell^1_{\#}$. Therefore $\&\psi$ is a bounded spherical function, and the corresponding eigenvalue belongs to the ellipse E. As μ_1 is self-adjoint, its spectrum in $C^*(G)$ is real, hence it coincides with D. To complete the proof, let $\gamma(z) \in D$, and suppose f is a positive element in $\ell^1(G)$: we want to show that $\langle f, \varphi_z \rangle \geq 0$. By Lemma 2, $\&f$ is a positive element in $\ell^1_{\#}$. Thus $\langle f, \varphi_z \rangle = \langle \&f, \varphi_z \rangle \geq 0$. □

Our next goal is to describe the spectrum of μ_1 as a convolution operator on ℓ^p, $1 < p < \infty$. We denote by cv_p the algebra of left convolution operators on ℓ^p, that is, the algebra generated by finitely supported functions in the norm $\|f\|_{cv_p} = \sup \{\|f * g\|_p : \|g\|_p \leq 1\}$. Similarly, the algebra of right convolution operators on ℓ^p is denoted by cv'_p. The subalgebras of cv_p and cv'_p generated by μ_1 are denoted by

$cv_p^{\#}$, $(cv'_p)^{\#}$, respectively.

LEMMA 3. *If* $\frac{1}{p}+\frac{1}{q}=1$, *then* $cv_p^{\#}=cv_q^{\#}$.

Proof. We first observe that $cv_p = cv'_q$. Indeed, let $f \in cv_p$, $h \in \ell^p$, and $g \in \ell^q$. Then:

$$|\langle g * f, h\rangle| = |g * f * h(e)| \leq \|f * h\|_p \|g\|_q$$

$$\leq \|f\|_{cv_p} \|h\|_p \|g\|_q .$$

Thus it suffices to show that $cv_p^{\#} = (cv'_p)^{\#}$.

Let f be a radial function, and, for every function g, let us write $\check{g}(x) = g(x^{-1})$. Then one readily sees that $\check{f} * \check{g} = (g * f)^{\vee}$. Since the map $g \to \check{g}$ is an isometry of ℓ^p, the lemma is proved. □

THEOREM 3. *Let* $1 < p < 2$, *and let* p' *be the conjugate exponent,* $\frac{1}{p'} = 1 - \frac{1}{p}$. *Then the spectrum of* μ_1 *as a convolution operator on* ℓ^p, *or on* $\ell^{p'}$, *is the set* $D_p \cup E_p$, *where* E_p *is the ellipse* $\{\gamma(z): \frac{1}{p'} \leq \mathrm{Re}\, z \leq \frac{1}{p}\}$, *which degenerates into a segment for* $p = 2$, $D_p = \{(1-k)^{-1}\}$ *if* $k > 2$ *and* $p > 1 + \frac{\ln(r-1)}{\ln(k-1)}$, *and* D_p *is empty otherwise.*

If $\mathrm{Re}\, z > \frac{1}{p}$, $\gamma(z) \neq (1-k)^{-1}$ *and* $q = (k-1)(r-1)$, *the function* $f_z(x) = \frac{(k-1)r}{(k-1)q^{-z} - q^{z} - k + 2}\, q^{-z|x|}$ *defines a bounded con-*

volution operator on ℓ^p *and on* $\ell^{p'}$ *such that*

$$(\mu_1 - \gamma(z)\,\delta_e) * f_z = \delta_e \;. \tag{3}$$

Proof. We consider first the case $p = 2$. In order to show that the interval $E_2 = \{\gamma(z): \operatorname{Re} z = \frac{1}{2}\}$ is contained in the spectrum of μ_1, we shall prove that the multiplicative functionals on R determined by the spherical functions φ_z with $\operatorname{Re} z = \frac{1}{2}$ extend to bounded linear functionals on the C^*-algebra C^*_λ of left convolution operators on ℓ^2. In other words, we shall show that $\varphi_z \in B_\lambda$ if $\operatorname{Re} z = \frac{1}{2}$. By Theorem 2 and Proposition 4, each such φ_z is positive definite and belongs to ℓ^p for every $p > 2$; furthermore if $\frac{1}{2} < s < 1$, then φ_s is positive definite and $\varphi_s\,\varphi_z \in \ell^2 \subset A(G)$.

Since $\|\varphi_s\varphi_z\|_A = \varphi_s(e)\,\varphi_z(e) = 1$, and $\lim_{s\to 1} \varphi_s(z)\,\varphi_z(x) = \varphi_z(x)$, it follows that $\varphi_z \in B_\lambda$ (see [8]). We now show that, except for $\gamma(z)=\frac{1}{1-k}$, the complement of E_2 belongs to the resolvant of μ_1. Suppose $\gamma(z) \notin E_2$, that is, $\operatorname{Re} z \neq \frac{1}{2}$: since $\gamma(z) = \gamma(1-z)$, we can assume $\operatorname{Re} z > \frac{1}{2}$. Recall than χ_n denotes the characteristic function of the set $W_n = \{x: |x| = n\}$, and $h_z(x) = q^{-z|x|}$, where $q = (k-1)(r-1)$. Then:

$$\|\chi_n h_z\|_2 = q^{-n \operatorname{Re} z} (\# W_n)^{\frac{1}{2}} =$$

$$= \left(\frac{r}{r-1}\right)^{\frac{1}{2}} q^{n(\frac{1}{2}-\operatorname{Re} z)} .$$

Hence $\sum_{n=0}^{\infty} (1+n) \|\chi_n h_z\|_2 < \infty$. By [15, thm. 1], $h_z \in C_\lambda^*$. If $\gamma(z) \neq (1-k)^{-1}$, to complete the proof for the case $p = 2$, it suffices to prove the identity (3). For $x \neq e$, this identity was obtained in the proof of Theorem 2. On the other hand,

$$[(\mu_1 - \gamma(z)\delta_e) * h_z](e) = q^{-z} - \gamma(z) =$$

$$= \frac{1}{(k-1)r} [(k-1)^{1-z}(r-1)^{-z} - (k-1)^z (r-1)^z - k+2] .$$

Thus (3) holds. If $\gamma(z) = (1-k)^{-1}$, the right hand side vanishes, and $h_{1-z} \in C_\lambda^*$ if and only if $k>r$ (and $k>2$) [15, thm.1]. This yields the exceptional point $(1-k)^{-1}$. Let now $1<p<2$, $p' = 1-\frac{1}{p}$ and $\gamma(z) \neq \frac{1}{1-k}$. By Corollary 2, $\varphi_z \in \ell^{p'}$ if $\frac{1}{p'} < \operatorname{Re} z < \frac{1}{p}$. Thus the identity $\mu_1 * \varphi_z = \gamma(z)\varphi_z$ implies that $\mu_1 - \gamma(z)\delta_e$ is not invertible as a convolution operator on $\ell^{p'}$, and hence on ℓ^p, by Lemma 3. Therefore the set $\{\gamma(z): \frac{1}{p'} < \operatorname{Re} z < \frac{1}{p}\}$ is contained in the spectrum of μ_1 in cv_p and in cv_q and the same is true for its closure E_p. It remains to show that the complement of E_p belongs to the resolvant of μ_1. This will be accomplished if we prove identity (3) and show that $f_z \in cv_p$. The proof of (3) is the same as in the case Re $z = \frac{1}{2}$. It is enough to show that $h_z \in cv_p$, because f_z is a multiple of h_z. One has

$$\|h_z\|_{cv_p} \leq 1 + \frac{r}{r-1} \sum_{n=1}^{\infty} q^{n(1-\mathrm{Re}\, z)} \|\mu_n\|_{cv_p} \tag{4}$$

The norm of μ_n as a convolution operator on ℓ^p can be estimated by the Riesz convexity theorem:

$$\|\mu_n\|_{cv_p} \leq \|\mu_n\|_1^{1/p} \|\mu_n\|_{C_\lambda^*}^{2/p'} = \|\mu_n\|_{C_\lambda^*}^{2/p'}$$

On the other hand, $\|\mu_n\|_{C_\lambda^*} = \sup \{|<\mu_n, \varphi_z>|: z \in sp_{C_\lambda^*}(\mu_1)\} = \sup \{|\varphi_z(n)|: \mathrm{Re}\, z = \frac{1}{2}\}$ by the first part of the proof (by abuse of notation, we denote by $\varphi_z(n)$ the value taken by the radial function φ_z on words of length n). It follows from Theorem 1 that $|\varphi_z(n)| \leq \varphi_{\frac{1}{2}}(n)$ if Re $z = \frac{1}{2}$. Therefore Theorem 2.ii yields:

$$\|\mu_n\|_{cv_p} \leq (\varphi_{\frac{1}{2}}(n))^{2/p'} \leq (1 + C\, n)^{2/p'} q^{-n/p'},$$

where C is a suitable constant. Combining this inequality with (4), we have that $h_z \in cv_p$. This completes the proof. □

As a consequence of the theorem, we can evaluate the

spectral radius of μ_1 as a convolution operator on ℓ^p. For the special case p=2 and r=2, this result was obtained by P. Gerl [G], by a different approach.

COROLLARY 3. *The spectral radius of* μ_1 *as a convolution operator on* ℓ^p, $1 \leq p \leq \infty$, *is* $\gamma(\frac{1}{p}) = \frac{q^{1/p} + q^{1/p'} + k - 2}{(k-1)r}$ *where* $\frac{1}{p} + \frac{1}{p'} = 1$ *and* $q = (k-1)(r-1)$. *In particular, the spectral radius of* μ_1 *in* C^*_λ *is* $\gamma(\frac{1}{2}) = \frac{2\sqrt{q} + k-2}{(k-1)r}$. *If* $k > r \geq 2$, C^*_λ *contains nontrivial radial idempotents.*

4. Principal and complementary series of representations. We shall now relate spherical functions to representations of G. A function ξ on the boundary Ω of G is called *cylindrical* if it belongs to the linear span K of the characteristic functions of the sets E(x), $x \in G$. For every $z \in C$ and $x \in G$, define an operator on K by the rule

$$\pi_z(x)\,\xi(\omega) = P^z(x,\omega)\,\xi(x^{-1}\omega).$$

It is easy to see, by the cocycle identities, that each π_z is a representation of G on K. The spherical function φ_z is a coefficient of π_z: namely, $\varphi_z(x) = (\pi_z(x)\,\mathbb{1}, \mathbb{1})$, where $\mathbb{1}$ denotes the function identically one on Ω, and (,) is the inner product in $L^2(\Omega,\nu)$.

It follows immediately, from the definition of the Poisson kernel as a Radon-Nikodym derivative, that π_z extends to a *unitary* representation on the Hilbert space $L^2(\Omega,\nu)$ (again denoted by π_z, by abuse of notation) if and only if $\mathrm{Re}\, z = \frac{1}{2}$. If $\mathrm{Re}\, z = \frac{1}{p}$, then π_z extends to an isometric representation on $L^p(\Omega)$; however, we shall see that some of these representations admit a unitary extension on suitable Hilbert spaces. In the sequel, we investigate the irreducibility of those representations π_z which admit a unitary extension. A relevant step into this direction is provided by the following result (recall that K denotes the space of cylindrical functions).

PROPOSITION 5. *The linear span of* $\{\pi_z(x)\,\mathbb{1}: x \in G\}$ *is all of* K *if and only if* $z \neq 2m\pi i/\ln q$ *and* $z \neq [\ln(k-1)+(2m+1)\pi i]/\ln q$, $m \in \mathbb{Z}$.

Proof. Assume that $z \neq 2m\pi i/\ln q$ and $z \neq [\ln(k-1)+(2m+1)\pi i]/\ln$ We want to show that the characteristic function $\chi_{E(x)}$ of every set $E(x)$, $x \in G$, is a linear combination of the form $\chi_{E(x)}(\omega) = \sum_i c_i P^z(x_i,\omega)$, for suitable $x_i \in G$. We proceed by induction on the length of x. The claim is obviously true for $x = e$. Suppose $|x| = n > 0$. For $1 \leq j \leq n$, let x_j be the word consisting of the first j blocks of x ($x_0 = e$), and let $B(x_j)$ and $C(x_j)$ be as in Proposition 2. Observe that $\chi_{B(x_j)}$ and $\chi_{C(x_{j-1})}$ belong to the linear span of $\{\chi_{E(x)} : |x| \leq j\}$. It follows from Proposition 2 and the definition of the sets $B(x_j)$, $C(x_j)$ that

$$(1-q^{-2z})\ \chi_{E(x)} + (q^{-z} - q^{-2z})\ \chi_{B(x_n)} \tag{5}$$

is a linear combination of $P^z(x,\omega)$ and of characteristic functions $\chi_{E(t)}$ with $|t| < n$. The expression in (5) can be written as

$$(1-q^{-2z})\ \chi_{E(x)} + (q^{-z} - q^{-2z}) \sum \chi_{E(y)} \tag{6}$$

where the summation ranges over all words y of length n, $y \neq x$, such that $y_{n-1} = x_{n-1}$. In other terms, the expression (6) is a linear combination over all words of the same length n as x which coincide with x at least for the first n-1 blocks. Interchanging the role of x and each of the y's, we obtain k-1 similar expressions, which can all be expanded as linear combinations of $P^z(x,\omega)$ and the functions $\chi_{E(t)}$ with $|t| < n$. Let B_z be the (k-1)-dimensional matrix of the coefficients of the k-1 linear expressions analogous to (6). That is, the coefficients b_{ij} of the matrix B_z are given by $b_{ii} = 1-q^{-2z}$ and $b_{ij} = q^{-z} - q^{-2z}$, for $i \neq j$.

Suppose that B_z is nonsingular. Then, by solving the system of k-1 linear equations corresponding to B_z, one obtains an expansion of each function $\chi_{E(x)}$, with $|x| = n$, as a linear combination of function $\chi_{E(t)}$ with $|t| < n$ and of $P^z(x,\omega)$. Thus the induction step follows, provided we show that $\det B_z \neq 0$, or, equivalently, that every eigenvalue of B_z is nonzero.

Let $v = (v_1,\ldots,v_{k-1})$ be an eigenvector of B_z, with eigenvalue λ. Then, for $j=1,\ldots k-1$,

$$\lambda v_j = (1-q^{-2z})v_j + (q^{-z} - q^{-2z}) \sum_{\substack{i=1 \\ i\neq j}}^{k-1} v_i$$

Thus, for every j,

$$(\lambda + q^{-z} - 1)v_j = (q^{-z} - q^{-2z}) \sum_{i=1}^{k-1} v_i.$$

Notice that the right hand side of the above equality is independent of j. Thus either $\lambda = 1 - q^{-z}$ or $v_1 = v_2 = \ldots = v_{k-1}$ and $\lambda + q^{-z} - 1 = (k-1)(q^{-z} - q^{-2z})$. Thus B_z admits the eigenvalue $\lambda = 0$ if and only if $q^{-z} = 1$ or $(k-1)q^{-2z} - (k-2)q^{-z} =$

In the first case, $z = 2m\pi i/\ln q$, $m \in \mathbb{Z}$ and $\pi_z(x)\,\mathbb{1} = \mathbb{1}$ for every $x \in G$. In the latter case, $z = [\ln(k-1) + (2m+1\pi i]/\ln q$, and $q^{-2z} - 1 = (k-2)(q^{-z} - q^{-2z})$. By this equality and Proposition 2 it follows that $\pi_z(x)\,\mathbb{1}$ is a linear combination of $\mathbb{1}$ and of functions of the form

$$(k-2)\,\chi_{E(y)} - \chi_{B(y)} \tag{7}$$

with $0 < |y| \leq |x|$. To show that, in this case, the linear span of $\{\pi_z(x)\,\mathbb{1},\ x \in G\}$ is properly contained in K, it is enough to exhibit a nonzero function $\xi \in L^2(\Omega)$ which is orthogonal to $\mathbb{1}$ and to every function of the form (7). Vectors with these properties are easily found. For instance, let x, t be two words of the same length $n > 0$ with the property that at least one block of x and the corresponding block of t correspond to different generators; that is, suppose that $\chi_{E(x)} + \chi_{B(x)}$ and $\chi_{E(t)} + \chi_{B(t)}$ have disjoint supports. Then it is clear that the function $\xi = \chi_{E(x)} + \chi_{B(x)} - \chi_{E(x)} - \chi_{B(t)}$ has the required properties. □

In the sequel, the set of complex numbers $2m\pi i/\ln q$ and $[\ln(k-1) + (2m+1)\pi i]/\ln q$, $m \in \mathbb{Z}$ will be denoted by Υ. Observe that $\gamma(\Upsilon) = \{1, (1-k)^{-1}\}$.

We have already observed that the representations π_z with Re $z = \frac{1}{2}$ extend to unitary representations on $L^2(\Omega)$. On the other hand, we know, by Proposition 4, that the spherical function φ_z is positive definite for every z which belongs to a set which is larger than $\{\mathrm{Re}\, z = \frac{1}{2}\}$: namely, the set $\{z: \frac{2(k-2)}{r} - 1 \leq \gamma(z) \leq 1\} = \{z: \gamma(z) \in D\}$ (notations as in Proposition 4). As in section 3, we denote by I the segment connecting the foci of the ellipse E, and recall that I is the image under γ of the set $\{\mathrm{Re}\, z = \frac{1}{2}\}$. Let us consider the set of all complex numbers z such that $\gamma(z) \in D - I$, that is, the set $C = \{z: 0 < |\mathrm{Re}\, z - \frac{1}{2}| \leq \frac{1}{2},\ \mathrm{Im}\, z = m\pi i/\ln q,\ m \in \mathbb{Z}\}$. For every such z, the representation π_z gives rise to a unitary representation on a Hilbert space as follows. Let $\xi = \sum c_i \pi_z(x_i) \mathbb{1}$ be an element of the linear span of $\{\pi_z(x) \mathbb{1}: x \in G\}$, and define

$$\|\xi\|_z^2 = \sum_{i,j} c_i \bar{c}_j \varphi_z(x_i x_j^{-1}) \tag{8}$$

Then $\| \ \|_z$ is a norm, because φ_z is positive definite, and π_z is isometric with respect to this norm. Thus π_z gives rise to a unitary representation on the completion H_z of the linear span of $\{\pi_z(x)\mathbb{1}, x \in G\}$. As a consequence of Proposition 5, if $z \notin \Upsilon$ the Hilbert space H_z is the completion of K

in the norm $\| \|_z$, and the representation thus obtained on H_z is an extension of π_z. Of course, this extension could also be considered for Re $z = \frac{1}{2}$, i.e., $\gamma(z) \in I$. In this case, however, Proposition 5 implies that H_z is isomorphic with $L^2(\Omega)$ and the unitary representation π_z coincides with the representation on $L^2(\Omega)$ described above, except for $k=r\geq 2$ and $z = \frac{1}{2} +(2m+1)\pi i/\ln q, m \in \mathbb{Z}$ (in the latter case, H_z is a proper closed subspace of $L^2(\Omega)$).

In analogy with the current terminology for Lie groups [18] and free groups [13], the family of unitary representations π_z on $L^2(\Omega)$ with Re $z = \frac{1}{2}$ (i.e., $\gamma(z) \in I$), is called the *principal series*, whereas the family of unitary representations π_z on H_z such that $\gamma(z)$ lies in the interior of D-I is called the *complementary series*.

In this convenient to realize the representations π_z on function spaces defined on the group G, rather than on its boundary. For this purpose, we define the *Poisson transform* P_z as follows: for $\xi \in K$, $P_z\, \xi(x) = (\pi_z(x)\, \mathbb{1}, \bar{\xi}) =$
$= \int_\Omega P^z(x,\omega)\; \xi(\omega)\; d\nu(\omega)$.
It is readily seen, by means of the cocycle identities, that the range of P_z is the linear span $\tilde{K}_z$ of the translates $\{\lambda(x)\; \varphi_z,\; x \in G\}$. Moreover, the Poisson transform P_{1-z}

intertwines the representation π_z on K with the representation by left translations on $\widetilde{K}_{1-z} = \widetilde{K}_z$, that is,

$\lambda(x)\ P_{1-z} = P_{1-z}\ \pi_z(x)$ for every x in G.

Let us now restrict attention to the complex numbers z such that $\gamma(z) \in D$. Then φ_z is positive definite and (8) defines a norm on K. If P_z is injective, then it extends to an isometry between the completion H_z of K in the norm (8) and the completion $\widetilde{H}_z$ of $\widetilde{K}_z$ in the norm

$$\|\sum_j c_j\ \lambda(x_j)\ \varphi_z\|_z^2 = \sum_{i,j} c_i\ \bar{c}_j\ \varphi_z(x_i\ x_j^{-1}).$$

This shows that π_z is unitarily equivalent to π_w if P_z and P_w are injective and the positive definite spherical functions φ_z and φ_w coincide, i.e., $\gamma(z) = \gamma(w)$. Indeed, the representations π_z and π_w can be realized as translation representations on the same Hilbert space $\widetilde{H}_z = \widetilde{H}_w$.

The injectivity of the Poisson transform can be easily obtained as a consequence of Proposition 5.

COROLLARY 4. *The Poisson transform* P_z *is injective on* K *if and only if* $z \notin \Upsilon$.

Proof. Choose a function $\xi \in K$ such that $P_z\ \xi(x) = (\pi_z(x)\mathbb{1},\bar{\xi})=0$

for every $x \in G$. This amounts to say that $\bar{\xi}$ belongs to the annihilator of K in $L^2(\Omega)$. By Proposition 5, this annihilator is nontrivial if and only if $z \in \Upsilon$. □

The remainder of this section is devoted to proof that, except for some special values of z, the representations π_z of the principal and complementary series are irreducible. Indeed, we shall show that the representations π_z are all irreducible, except when $k \geqslant r$ and $\gamma(z) = (1-k)^{-1}$ (that is, when $z, 1-z \in \Upsilon$). The main result towards the proof of the irreducibility is the following. Recall that $\varphi_z(n)$ denotes the value attained by the spherical function φ_z on the words of length n, and $q = (k-1)(r-1)$.

THEOREM 4. *Let z be a complex number, with* $\gamma(z) \neq (1-k)^{-1}$ *if* $k>r$.

i) If π_z *is a representation of the complementary series, let* $T_n = \varphi_z(n)^{-1}\,\pi_z(\mu_n)$. *Then, for every* ξ *in* H_z, $T_n\,\xi$ *converges, in the norm of* H_z, *to the constant* $\int_\Omega \xi\, d\nu$.

ii) If π_z *is a representation of the principal series, choose a sequence* n_j *such that* $\varphi(n_j)\, q^{-\frac{1}{2}n_j}$ *converges to a finite limit. Then, for every* ξ *in* $L^2(\Omega)$, $T_{n_j}\,\xi$ *converges to the constant* $\int_\Omega \xi\, d\nu$ *in the weak topology of* $L^2(\Omega)$.

The argument of the proof of Theorem 4, which requires some rather long and involved computations, is similar to the argument used for free groups (see [13, Chapter 5], [12, § 3]). We omit the proof, and limit ourselves to provide, in the next lemma, the estimates which are significantly different from the free group theory.

The lemma describes the probability distribution of the length of two-sided translates of words of length n. For x, y, $w \in G$, $|w| = n$, we denote by $r_n(j;x,y)$ the probability that $|x\,w\,y| = |x| + |y| + n - j$.

LEMMA 4. *For every* x, $y \in G$, *the limit* $\lim_n r_n(j;x,y) = p_j(x,y)$ *exists and depends only on* j,x,y. *Moreover*

$$r_n(j;x,y) - p_j(x,y) = O((r-1)^{-n}).$$

Proof. Let x, $y \in G$, with $|x| = m$, $|y| = p$, and write $A_n(j;x,y) = \{w \in G: |w| = n, |x\,w\,y| = n + m + p - j\}$. Then $r_n(j;x,y) = \#A_n(j;x,y) / (k-1)^n r(r-1)^{n-1}$. We want to estimate the cardinality $\#A_n(j;x,y)$ for n large. Without loss of generality, we may assume $p \leqslant m < n$. Let s,t be two words of length one, and let h be an integer, $h > 2$. Define $C_n(s,t) = \{w \in G: |w| = n, |swt| = n-2\}$, and $c_n(s,t) = \#C_n(s,t)$.

Denote by a_i, $1 \leqslant i \leqslant r$, the generators of G, and by b_d, $1 \leqslant d \leqslant (k-1)r$ the words of length one, ordered in such a way that $b_{(k-1)(i-1)+\ell} = a_i^{\ell}$, $1 \leqslant i \leqslant r$, $1 \leqslant \ell \leqslant k-1$. Let S be the $(k-1)r$-dimensional matrix whose entries are $S_{uv} = 0$ if $u = (k-1)(i-1)+h$, $v = (k-1)(i-1)+h'$, $1 \leqslant h,h' \leqslant k-1$, and $S_{uv} = 1$ otherwise. Let $w = b_{i_1} \dots b_{i_n}$ be a word of length n. Then $w \in C_n(s,t)$ if and only if $b_{i_1} = s^{-1}$, $b_{i_n} = t^{-1}$ and $S_{i_j i_{j+1}} \neq 0$ for $1 \leqslant j \leqslant n-1$. The latter condition is equivalent to $\prod_{j=1}^{n-1} S_{i_j i_{j+1}} = 1$. Therefore $c_n(s,t) = \sum_{i_2,\dots,i_{n-1}=1}^{(k-1)r} \prod_{j=1}^{n-1} S_{i_j i_{j+1}} = (S^{n-1})_{i_1 i_n}$. We now estimate $c_n(s,t)$ by computing the eigenvalue of S. We claim that the only rigenvalues of S are $(k-1)(r-1) = q$, $1-k$ and 0, and that the largest eigenvalue is simple. Indeed, let λ be an eigenvalue of the (self-adjoint) matrix S, and let $\underset{\sim}{v} \in \mathbb{R}^{(k-1)r}$ be the corresponding eigenvector. Then, for $1 \leqslant i \leqslant r$ and $1 \leqslant h \leqslant k-1$,

$$\lambda v_{(k-1)(i-1)+h} = Sv_{(k-1)(i-1)+h} = \sum_{j=1}^{(k-1)r} v_j - \sum_{d=1}^{k-1} v_{(k-1)(i-1)+d} \tag{9}$$

The right hand side of this equality does not depend on h.

Therefore, either there exist $c_i \in D$, $1 \leqslant i \leqslant r$, such that $v_{(k-1)(i-1)+h} = c_i$ for $1 \leqslant h \leqslant k-1$, or $\lambda = 0$ (and the corresponding eigenspace has dimension $(k-2)r$). If $\lambda \neq 0$, (9) becomes

$$(\lambda + k-1)\ c_i = (k-1) \sum_{j=1}^{r} c_i \tag{10}$$

Again the right hand side does not depend on i. Therefore either $\lambda = 1-k$ (and the corresponding eigenspace has dimension $r-1$) or all the c_i are equal, so that λ is a simple eigenvalue. In the latter case (10) yields $\lambda = (k-1)(r-1)=q$. This proves our claim. Let now $\underset{\sim}{v}_h$, $1 \leqslant h \leqslant (k-1)r$, be an orthonormal basis in $\mathbb{R}^{(k-1)r}$ consisting of eigenvectors of S, chosen so that $\underset{\sim}{v}_1$ corresponds to the eigenvalue q. The canonical basis of $\mathbb{R}^{(k-1)r}$ is denoted by $\underset{\sim}{e}_h$, $1 \leqslant h \leqslant (k-1)r$. Then, as $c_n(s,t) = (S^{n-1})_{i_1 i_n}$, we have

$$|c_n(s,t) - q^{n-1}\ (\underset{\sim}{e}_{i_1}, \underset{\sim}{v}_1)(\underset{\sim}{v}_1\ \underset{\sim}{e}_{i_n})| \leqslant$$

$$\leqslant (k-1)^{n-1} \sum_{h=2}^{(k-1)r} |(\underset{\sim}{e}_{i_1}, \underset{\sim}{v}_h)(\underset{\sim}{v}_h, \underset{\sim}{e}_{i_n})|$$

We have already observed that $\underset{\sim}{v}_1 = \dfrac{1}{\sqrt{(k-1)r}} \sum_{h=1}^{(k-1)r} \underset{\sim}{e}_h$

Therefore $(\underset{\sim}{e}_1, \underset{\sim}{v}_1)(\underset{\sim}{v}_1, \underset{\sim}{e}_{i_n}) = ((k-1)r)^{-1}$, and

$$\frac{c_n(s,t)}{(k-1)^n\ r(r-1)^{n-1}} - \frac{1}{r^2(k-1)^2} = 0((r-1)^{-n})$$

We now estimate $\#A_n(j;x,y)$ in terms of the numbers $c_n(s,t)$. Recall that $|x| = m$, $|y| = p$, and $p \leqslant m \leqslant n$. Moreover, we can assume $0 \leqslant j \leqslant 2(m+p)$, otherwise $A_n(j;x,y)$ is empty. In the remainded of the proof, we restrict attention to the case $j < 2p \leqslant 2m$ (so that x and y do not cancel out completely when the word x w y is written in reduced form). Only minor modifications are necessary to treat to other cases, which are left to the reader.

Write $x = x_1 \ldots x_m$ and $y = y_1 \ldots y_p$, with $|x_i| = |y_i| = 1$. We first treat the case when j is odd, $j = 2h+1$. Denote by B_t the set of words of length n which give rise to h-t cancellations on the left with x and t cancellations and one reduction on the right with y. On the other hand, denote by B'_t the set of words of length n which give rise to t cancellations and one reduction on the left with x and h-t cancellations on the right with y. In other words,

$$B_t = \{w \in G: |w|=n,\ |xw| = m+n-2(h-t),$$
$$|wy| = n+p-2t-1\}$$

$$B'_t = \{w: |w|=n;\quad |xw| = m+n-2t-1,\quad |wy| = n+p-2(h-t)\}.$$

Then we can write $A_n(j;x,y)$ as a disjoint union as follows:

$$A_n(j;x,y) = \bigcup_{t=0}^{h} (B_t \cup B'_t)$$

Let $0 < t < h$. Then B_t is in one-to-one correspondence with the set

$$\cup\{C_{n-h}(u,v): |u| = |v| = 1,\ |x_{m-(h-t)}\, u^{-1}| =$$

$$= |x^{-1}_{m-(h-t)+1}\, u^{-1}| = |v^{-1}\, y^{-1}_t| = 2,\ |v^{-1}\, y_{t+1}| = 1\}.$$

Similarly, there is a bijection between B'_t, $0 < t < h$, and the set

$$\cup\{C_{n-h}(u,v): |u| = |v| = 1,\ |x_{m-(h-t)}\, u^{-1}| = 1,$$

$$|x^{-1}_{m-(h-t)+1}\, u^{-1}| = |v^{-1}\, y^{-1}_t| = |v^{-1}\, y_{t+1}| = 2\}.$$

In the case $t = 0$ and $t = h > 0$, the sets B_t and B'_t can be written as disjoint union of sets of the type $C_{n-h}(u,v)$ as follows:

$$B_o \approx \cup \{C_{n-h}(u,v) : |u| = |v| = 1, |x_{m-h}u^{-1}| = |x^{-1}_{m-h+1}\ u^{-1}| =$$

$$= 2, |v^{-1}\ y_1| = 1\},$$

$$B_o \approx \cup \{C_{n-h}(u,v) : |u| = |v| = 1,$$

$$|x_{m-h}\ u^{-1}| = 1, |x^{-1}_{m-h+1}\ u^{-1}| = |v^{-1}\ y_1| = 2\},$$

$$B_h \approx \cup \{C_{n-h}(u,v) : |u| = |v| = 1,$$

$$|x_m\ u^{-1}| = |v^{-1}\ y_h^{-1}| = 2, \quad |v^{-1}\ y_{h+1}| = 1\},$$

$$B'_h \approx \cup \{C_{n-h}(u,v) : |u| = |v| = 1,$$

$$|x_m\ u^{-1}| = 1, |v^{-1}\ y_h^{-1}| = |v^{-1}\ y_{h+1}| = 2\}.$$

(If $h = 0$, the conditions involving x_{m-h+1} and y_h should be dropped from all these expressions).

Thus, for every t, $0 \leqslant t \leqslant h$, we conclude, by (11), that there exists the limit $p_j(x,y)$ of the sequence $r_n(j:x,y) =$

$= \#A_n(j;x,y) \,/\, (k-1)^n\, r(r-1)^{n-1}$, and $r_n(j;x,y) - p_j(x,y) =$
$= O((r-1)^{-n})$. This completes the proof for odd values of j.
Let now j be even, j = 2h. If h = 0, then

$$A_n(j;x,y) = \cup \{C_n(u,v): |u| = |v| = 1,$$

$$|x_m\, u^{-1}| = 2 = |v^{-1}\, y_1|\},$$

and the statement follows again by means of (11).
If h > 0, let us consider the disjoint decomposition

$$A_n(j;x,y) = (\bigcup_{t=0}^{h} D_t) \cup (\bigcup_{t=0}^{h-1} D_t')$$

where

$$D_t = \{w: |w| = n,\ |xw| = m+n-2(h-t),\ |wy| =$$

$$= n + p - 2t\},$$

$$D_t' = \{w: |w| = n,\ |xw| = m+n-2(h-t) + 1,\ |wy| =$$

$$= n + p - 2t - 1\}.$$

In other words, D_t is the set of words w of length n which give rise to h-t cancellations under left multiplication by x and t cancellations under right multiplication by y, whereas D'_t is the set of words w which give rise to h-t-1 cancellations and one reduction on the left with x and t cancellations and one reduction on the right with y.

As before, if $0 < t < h$ one has

$$D_t \approx \cup \{C_{n-h}(u,v) : |u| = |v| = 1, |x_{m-(h-t)} u^{-1}| =$$

$$= |x^{-1}_{m-(h-t)+1} u^{-1}| = |v^{-1} y_{t+1}| = |v^{-1} y_t^{-1}| = 2\};$$

moreover, for $0 < t < h-1$

$$D'_t \approx \cup \{C_{n-h}(u,v) : |u| = |v| = 1, |x^{-1}_{m-h+t+2} u^{-1}| =$$

$$= |v^{-1} y_t^{-1}| = 2, |x_{m-h+t+1} u^{-1}| = |v^{-1} y_{t+1}| = 1\}.$$

Similar decompositions hold for $D_\circ$ and $D'_\circ$, provided we delete the conditions involving y_t (and also $x_{m-h+t+2}$, if h=1), for D_h by deleting the conditions involving $x_{m-h+t+1}$ and for D'_{h-1} by deleting the condition involving $x_{m-h+t+2}$. Thus the statement follows again by (11). □

We have already remarked that the unitary representation π_z is unitarily equivalent to π_w if $\gamma(z) = \gamma(w)$ and the Poisson transforms P_z and P_w are injective, i.e., if $z,w \notin \Upsilon$. In particular, $\pi_z \sim \pi_{1-z}$ if $z, 1-z \notin \Upsilon$. If z or $1-z \in \Upsilon$, then π_z and π_{1-z} are two inequivalent representations of the principal or complementary series, corresponding to the same eigenvalue $\gamma(z) = 1$ or $\gamma(z) = (1-k)^{-1}$. As we are dealing with representations of the principal and complementary series, we restrict attention to the eigenvalue $(1-k)^{-1}$, which lies in the interior of D: it belongs to the spectrum of the left complementary series if $k \neq r$, and is the left boundary point of the principal series (i.e., the left focus of E) if $k = r$.

We are now ready to discuss the irreducibility of the unitary representations of the principal and complementary series. The argument requires some modifications with respect to the case of a free group (cfr. [13, chapter 4]).

THEOREM 5. *If* $k < r$, *every representation* π_z *of the principal or complementary series of* G *is irreducible. If* $k \geqslant r$, *then* π_z *is irreducible if and only if* $\gamma(z) \neq (1-k)^{-1}$.

Proof. Let Q be a projection, on the Hilbert space H of π_z, which commutes with π_z, and let $\xi = Q\,\mathbb{1}$. Then

$$\pi_z(\mu_n)\xi = Q\,\pi_z(\mu_n)\,\mathbb{1} = \varphi_z(n)\,Q\,\mathbb{1} = \varphi_z(n)\xi \tag{12}$$

Let us first consider representations π_z of the complementary series. Then $H = H_z$, and (12) amounts to $T_n\xi = \xi$, where $T_n = \varphi_z(n)^{-1}\,\pi_z(\mu_n)$. If $k<r$ or $\gamma(z) \neq \frac{1}{1-k}$, $T_n\xi$ converges to $(\xi, \mathbb{1})\mathbb{1}$ in the norm of H_z. Thus $\xi = (\xi, \mathbb{1})\mathbb{1}$. Since Q is a projection, either $Q\,\mathbb{1} = \mathbb{1}$ or $Q\,\mathbb{1} = 0$. On the other hand, H_z is the completion of the space generated by $\mathbb{1}$ under the action of π_z. Therefore $\mathbb{1}$ is a cyclic vector for π_z, and either Q is the identity or $Q = 0$. By Schur's lemma, π_z is irreducible. If $k>r$ and $\gamma(z) = \frac{1}{1-k}$, $\varphi_z \in \ell^2$, by Thm. 2 : thus π_z is reducible [8,13.10.5].

Let us now restrict attention to representations π_z of the principal series. If $k \neq r$ and $z \in \Upsilon$, then π_z and π_{1-z} belong to the complementary series: thus we can suppose that $z,\ 1-z \notin \Upsilon$.

Let n_j be the sequence of Theorem 4.ii: then (12) yields $T_{n_j}\xi = \xi$. We compute now the Poisson transform of ξ:

$$\mathcal{P}_z\, \xi(x) = (\xi,\, \pi_z(x)\, \mathbb{1}) = (T_{n_j}\, \xi,\, \pi_z(x)\, \mathbb{1}) =$$

$$= (\xi,\, T_{n_j}\, \pi_z(x)\, \mathbb{1}),$$

because T_{n_j} is a self-adjoint operator on $H = L^2(\Omega)$. By Theorem 4, $T_{n_j}\, \pi_z(x)\, \mathbb{1} \to \varphi_z(x)\, \mathbb{1}$ weakly in $L^2(\Omega)$. Therefore $\mathcal{P}_z\, \xi(x) = (\xi, \mathbb{1})\, \varphi_z(x)$.

By Corollary 4, $\mathcal{P}_z$ is injective, because $1-z \notin \Upsilon$. Therefore $\xi = (\xi, \mathbb{1})\, \mathbb{1}$, and again $Q\, \mathbb{1} = \mathbb{1}$ or $Q\, \mathbb{1} = 0$. Now the assumption $z \notin \Upsilon$ and Proposition 5 yield that $\mathbb{1}$ is a cyclic vector under π_z, and, as before, π_z is irreducible.

Finally, if $k = r$ and π_z belongs to the principal series, then $z \in \Upsilon$ if and only if $1-z \in \Upsilon$, and $\pi_z = \pi_{1-z}$. For this choice of z, Proposition 5 yields that $\mathbb{1}$ is not a cyclic vector under the action of π_z in $H = L^2(\Omega)$: therefore π_z is reducible. For the other values of z, we have $z,\, 1-z \notin \Upsilon$, and the previous argument applies. □

References

1. J.P. Arnaud, *Fonctions sphériques et fonctions définies positive sur l'arbre homogène*, C.R. Acad. Sc. Paris 290 (1980), 99-191.

2. R.A. Bonic, *Symmetry in group algebras of discrete groups*, Pacif. J. Math. 11 (1961), 73-94.

3. W. Betori, M. Pagliacci, *Harmonic analysis for groups acting on trees*, to appear in Boll. Un. Mat. It.

4. P. Cartier, *Géomètrie et analyse sur les arbres*, Lecture Notes in Math. 317, Springer-Verlag, Berlin, 1973, 123-140.

5. P. Cartier, *Fonctions harmoniques sur un arbre*, Symp. Math. 9 (1972), 203-270.

6. P. Cartier, *Harmonic analysis on trees*, Proc. Symp. Pure Math. Amer. Math. Soc. 26 (1972), 419-424.

7. J. Dixmier, *Formes linéaires sur un anneau d'operateurs*, Bull. Soc. Math. France 81 (1953), 9-39.

8. J. Dixmier, "Les C^*-algèbres et leurs représentations", Gauthier-Villars, Paris, 1969.

9. J.L. Dunau, *Etude d'une classe de marches aléatoires sur l'arbre homogène*, Public. Laboratoire de Statistique et Probabilités, Université Paul Sabatier, Toulouse, 1976.

10. E.B. Dynkin, M.B. Maliutov, *Random walks on groups with a finite number of generators*, Soviet Math. (Doklady) 2 (1961), 399-402.

11. J. Duncan, J.H. Williamson, *Spectra of elements in the measure algebra of a free group*, Proc. Royal Irish Acad. 82 (1982), 109-120.

12. A. Figà-Talamanca, M.A. Picardello, *Spherical functions and harmonic analysis on free groups*, J. Functional Anal. 47 (1982), 281-304.

13. A. Figà-Talamanca, M.A. Picardello, "Harmonic analysis on free groups", Lecture Notes in Pure and Applied Mathematics, Marcel Dekker, New York (in print).

14. P. Gerl, *A local central limit theorem on some groups*, Lecture Notes in Stat., Springer-Verlag, Berlin, 1981, 73-82.

15. A. Iozzi, M.A. Picardello, *Graphs and convolution operators*, to appear in "Topics in Modern Harmonic Analysis", Indam, Roma.

16. A.M. Mantero, A. Zappa, *The Poisson transform and uniformly bounded representations of free groups*, preprint.

17. T. Pytlik, *Radial functions on free groups and a decomposition of the regular representation into irreducible components*, J. Reine Angew. Math. 326 (1981), 124-135.

18. G. Warner, "Harmonic analysis on semisimple Lie groups", I and II, Springer-Verlag, Berlin, 1972.

A REMARK ON MAPPINGS OF BOUNDED SYMMETRIC DOMAINS INTO BALLS

by

Marco Rigoli e Giancarlo Travaglini

ABSTRACT: Let F be a biholomorphic mapping of a n-dimensional bounded symmetric domain D (with rank ℓ) into the unit ball of $\mathbf{C}^n$. Then F(D) cannot contain a ball of radius greater than $\ell^{-1/2}$ and this estimate is sharp. The above result provides a "quantitative" version of a well known result of H. Poincaré.

1.

H. Poincaré in 1907 first showed the failure of Riemann mapping theorem in higher dimension: he proved that the bidisc and the ball in $\mathbf{C}^2$ are not biholomorphically equivalent. This suggests to study geometric properties of the image of one of the above domains into the other under a biholomorphic mapping.

Recently the following significative result was proved by J.E. Foarness and E.L. Stout [2]: let M be a paracompact connected n-dimensional complex manifold, then there is an open set in M that is biholomorphically equivalent to the polidisc in $\mathbf{C}^n$ and that contains almost all of M. As a corollary of their proof, there exists a positive number $r<1$ with the following property: for any biholomorphic mapping F from the polidisc U^n into the unit ball B_n of $\mathbf{C}^n$ the image $F(U^n)$ cannot contain a ball of radius greater than r (observe that this is enough to get Poincaré's result). Foarness and Stout asked for the best value of r. The answer was independently found by H. Alexander [1] and L. Lempert [5], who showed that the best value, let say $r(U^n)$, is $n^{-1/2}$. The technique of Alexander and Lempert was recently applied to the classical

Cartan domains by Y. Kubota [4]. Actually the above technique works in general and we can prove that $r(D)=\ell^{-1/2}$ whenever D is a n-dimensional bounded symmetric domain with rank ℓ. Our result is independent of classification theory and is nothing but an application of Harish-Chandra's theory of bounded symmetric domains.

We state the result.

THEOREM. *Let D be a n-dimensional bounded symmetric domain with rank ℓ, and let $F:D\to B_n$ be a biholomorphic mapping into the unit ball of C^n. Then F(D) cannot contain a ball of radius greater than $\ell^{-1/2}$ and there is mapping $\tilde{F}$ attaining this value.*

2.

The results quoted from the theory of bounded symmetric domains can be found in [3].

Proof. Let D=G/K be realized according to the standard Harish-Chandra's imbedding. Since G=KAK we have D=KA·O. The orbit A·O is a unit (ℓ-dimensional) cube around O, and we can choose a toral subgroup $\mathbf{T}^\ell$ in K such that $\mathbf{T}^\ell A\cdot O$ is a polidisc U^ℓ and the coordinates so that U^ℓ is exactly the projection of D on the space of the first ℓ coordinates.

Now let $F=(F_1,\ldots,F_n)$ be as in the statement of the theorem. We can suppose F(O)=O, otherwise we precede F by an automorphism of D. For $j=1,\ldots,n$ we write $G_j(z_1,\ldots,z_\ell)=F_j(z_1,\ldots,z_\ell,0,\ldots,0)$. Every G_j can be expanded as $G_j(z_1,\ldots,z_\ell)=\sum_{\nu_1,\ldots,\nu_\ell} a^{(j)}_{\nu_1,\ldots,\nu_\ell} z_1^{\nu_1}\ldots z_n^{\nu_n}$. Now we denote by F and G also the almost everywhere defined boundary values of F and G. Then we have

$$(1)\qquad 1 \geq \frac{1}{(2\pi)^\ell}\int_0^{2\pi}\cdots\int_0^{2\pi}\sum_{j=1}^{n}|F_j(e^{i\theta_1},\ldots,e^{i\theta_\ell},0,\ldots,0)|^2\, d\theta_1\ldots d\theta_\ell$$

$$= \int_{\mathbb{T}^\ell} \sum_{j=1}^{n} |G_j|^2 = \sum_{j=1}^{n} \sum_{\nu_1,\dots,\nu_\ell} |a^{(j)}_{\nu_1,\dots,\nu_\ell}|^2 .$$

Now we observe that for any $j=1,\dots,\ell$ the point $Q_j=(\underbrace{0,\dots,e^{i\theta_j},\dots,0,0,\dots,0}_{n})$ belongs to ∂D (since $(\underbrace{0,\dots,e^{i\theta_j},\dots,0}_{\ell})$ belongs to ∂U^ℓ). Hence $F(Q_j)$ belongs to $\partial F(D)$ (since F is biholomorphic, hence proper).

Now, if $F(D)$ contains a ball of radius r, we have, for $j=1,\dots,\ell$,

$$r^2 \leq \frac{1}{2\pi}\int_0^{2\pi} \sum_{j=1}^{n} |F_j(Q_j)|^2 = \frac{1}{2\pi}\int_0^{2\pi} \sum_{j=1}^{n} |G_j(0,\dots,e^{i\theta_j},\dots,0)|^2 \, d\theta_j$$

$$\leq \sum_{j=1}^{n} \sum_{\nu_j} |a^{(j)}_{0,\dots,\nu_j,\dots,0}|^2 .$$

If we add the above ℓ inequalities and we look at (1) as well as $F(0)=0$ we get $\ell r^2 \leq 1$, i.e. $r \leq \ell^{-1/2}$.

We end by writing a biholomorphic mapping $\tilde{F}:D\to B_n$ such that

$$\ell^{-1/2} B_n \subseteq \tilde{F}(D) . \tag{2}$$

To get (2) we simply recall that $D=KA\cdot 0$, where K is a group of unitary transformations. Hence D contains a unit ball and is contained in a ball of radius $\ell^{1/2}$. Clearly the mapping

$$\tilde{F}(z_1,\dots,z_n) = (\ell^{-1/2} z_1,\dots,\ell^{-1/2} z_n)$$

satisfies (2).

REFERENCES

[1] H. Alexander, *Extremal holomorphic imbeddings between the ball and the polydisc*, Proc. Amer. Math. Soc., 68 (1978), 200-202.

[2] J.E. Foarness and E.L. Stout, *Polydiscs in complex manifolds*, Math. Ann., 227 (1977), 145-154.

[3] A. Korányi, *Holomorphic and harmonic functions on bounded symmetric domains*, CIME Summer Course on Geometry of Bounded Homogeneous Domains, Cremonese, Roma (1968), 125-197.

[4] Y. Kubota, *A note on holomorphic imbeddings of the classical Cartan domains into the unit ball*, Proc. Amer. Math. Soc., 85 (1982), 65-68.

[5] L. Lempert, *A note on mapping polydiscs into balls and vice versa*, Acta Math. Acad. Scient. Hung., 34 (1979), 117-119.

Marco Rigoli
Dept. of Mathematics
Washington University
Box 1146
St. Louis
MO 63130

Giancarlo Travaglini
Istituto Matematico
Università degli Studi
Via Saldini 50
20133 Milano
ITALY

and

Dept. of Mathematics
Washington University
Box 1146
St. Louis
MO 63130

The Cauchy-Ahlfors operator,

an invariant differential operator for vector fields.

H.M. Reimann, Bern

1 Introducing the S-operator

In [1] Ahlfors introduced the differential operator

$$(Su)_{ij} = \frac{1}{2}\left(\frac{\partial u_i}{\partial x_j} + \frac{\partial u_j}{\partial x_i}\right) - \frac{\delta_{ij}}{n}\sum_{k=1}^{n}\frac{\partial u_k}{\partial x_k}$$

which maps vector fields in $\mathbb{R}^n$ into tensor fields of symmetric tensors with vanishing trace. The definition can be motivated by conformality considerations (cf.[3]). A bijective (sense preserving) mapping $h:\mathbb{R}^n \to \mathbb{R}^n$ is conformal, if the Jacobian matrix

$H(x) = \left(\frac{\partial h_i}{\partial x_j}(x)\right)$ is a multiple of a rotation. This can be

expressed by the equality

$$\frac{H^*H}{(\det H)^{2/n}} = 1 \qquad (H^* \text{ is the transpose of } H)$$

More generally, the expression $H^*H\,(\det H)^{-2/n}$ can be used as a measure of conformality.

If u is a vector field in $\mathbf{R}^n$ with sufficient regularity properties, then the differential equation $x^{\cdot} = u(x)$ determines a flow h_t. The curves $h_t(z)$ with $z \in \mathbf{R}^n$ fixed are the solutions of this differential equation with initial condition $x(0) = z$. The variational equation associated with this flow is the equation

$$H_t^{\cdot} = U\left(h_t(z)\right) H_t$$

with initial condition $H_0 = I$. Here $H_t = \left(\frac{\partial (h_t)_i}{\partial z_j}\right)$ and $U(x) = \left(\frac{\partial u_i}{\partial x_j}(x)\right)$ are the Jacobians of the mapping h_t and of the vector field u respectively. The first variation of the measure of conformality can be calculated to be

$$\frac{d}{dt} \frac{H_t^* H_t}{(\det H_t)^{2/n}} = H_t^*[U^* \circ h_t + U \circ h_t - \frac{2}{n} I \operatorname{tr} (U \circ h_t)] H_t$$

In this setting, the differential operator S appears in a very natural way.

It is worthwhile to note the special case $n = 2$. The differential operator S applied to the vector field (u,v) gives

$$\frac{1}{2} \begin{pmatrix} u_x - v_y & u_y + v_x \\ v_x + u_y & v_y - u_x \end{pmatrix}$$

and the system $S(u,v) = 0$ coincides with the Cauchy-Riemann equations. In particular, the solutions of $S(u,v) = 0$ are the analytic functions.

Considerations of invariance provide us with a further motivation for the introduction of the S-operator.

The action of the group of Euclidean motions on the vector fields is defined by

$$u_g(x) = gu(g^{-1}x) \qquad g \in SO(n)$$

$$u_t(x) = u(x-t) \qquad t \in \mathbf{R}^n$$

The motion group then also acts on the Jacobian matrices (the gradients) $U = \left(\frac{\partial u_i}{\partial x_j} \right)$

$$U_g(x) = g\, U(g^{-1}x) g^{-1} \qquad g \in SO(n)$$

$$U_t(x) = U(x-t) \qquad t \in \mathbf{R}^n$$

This action on matrix valued functions is however not irreducible. The irreducible subspaces are the space of symmetric matrices with vanishing trace, the space of anti-symmetric matrices and the space of traces. By projecting U into these subspaces one arrives at the following decomposition:

$$U(x) = Su(x) + Au(x) + \frac{1}{n} (\text{div } u(x))\, I$$

with

$$Su = \frac{1}{2} (U+U^*) - \frac{1}{n} (\text{div } u)\, I$$

$$Au = \frac{1}{2} (U-U^*)$$

The general version of this decomposition procedure, where $u : \mathbf{R}^n \to U$ is a function with values in a representation space for the group $SO(n)$, was described by Stein and Weiss [14]. The S-operator appears in their paper, however their interest concentrated on the Riesz system

$$Au = 0$$

$$\text{div } u = 0$$

and its generalizations.

2 The equation $Su = \phi$

We already have seen that the homogeneous equation $Su = 0$ is connected with conformal mappings. In higher dimensions ($n \geq 3$) the only conformal mappings are the Möbius transformations. Since the vector fields u satisfying $Su = 0$ give rise to (local) one-parameter families of conformal mappings, the equation $Su = 0$ will describe the Lie algebra of the Möbius group. (The Möbius group $M(n)$ is the identity component of the transformation group of $\overline{\mathbb{R}}^n = \mathbf{R}^n \cup \{\infty\}$, which is generated by reflections in spheres and hyperplanes). This fact seems to have been known to Cauchy. The explicit form of the solutions of the equation $Su = 0$, namely

$$u(\mathbf{x}) = a + B\mathbf{x} + \lambda\mathbf{x} + 2\mathbf{x}(c,\mathbf{x}) - c|\mathbf{x}|^2$$

with $a,c \in \mathbf{R}^n$, $\lambda \in \mathbf{R}$ and B a constant matrix with $B^* = -B$, is contained in his paper [6] - at least for the case $n = 3$.

In this context let us mention that the operator S has its significance in classical field theory (see e.g. Truesdell and Muncaster [15]).

For the inhomogeneous equation $Su = \phi$ there is again an essential difference between the cases $n = 2$ and $n = 3$. For $n = 2$ the equation is equivalent to

$$\frac{\partial}{\partial \bar{z}} f = \phi$$

for complex valued functions f, ϕ defined on $\mathbf{C}$. As is well known, for $\phi \in C_c^\infty$ this equation can always be solved and the solution is determined up to an analytic function. In $\mathbf{R}^n$, $n \geq 3$, the equation $Su = \phi$ with ϕ a matrix valued function can only be solved if ϕ satisfies an integrability condition. According to Ahlfors [2] this integrability condition can be stated as a singular integral equation with a Calderon-Zygmund type kernel. As in the 2-dimensional case, the solution of $Su = \phi$ then has the

form of a convolution integral.
The integrability condition can be interpreted as an orthogonality statement with respect to the space of square integrable matrix valued functions (Reimann [10]).

3 Quasiconformal deformations

A vector field $u : \mathbf{R}^n \to \mathbf{R}^n$ with

$$\| Su \|_\infty = \sup_{x \in \mathbf{R}^n} \| Su(x) \| < \infty$$

is called a quasiconformal deformation. The norm of the matrix $Su(x)$ is taken to be

$$\| Su(x) \| = (\operatorname{tr} Su(x)\ Su(x))^{\frac{1}{2}}$$

but any equivalent norm will do. As for the regularity hypothesis, the continuous vector field is assumed to have generalized derivatives which are locally integrable functions.

It turns out that quasiconformal deformations in $\mathbf{R}^n$, $n \geqslant 3$ have very specific growth properties (Sarvas [11]):
Modulo a solution of the equation $Su = 0$ they satisfy

$$u(x) = O(|x| \log |x|) \qquad \text{for} \quad |x| \to \infty$$

Quasiconformal deformations generate flows of quasiconformal mappings. If $\| Su \|_\infty < \infty$ and $u(x) = O(|x|\log|x|)$ for $|x| \to \infty$, then the differential equation $\dot{x} = u(x)$ has unique solutions $h_t(z)$ with initial condition $h_0(z) = z$. For fixed t, the mapping h_t is quasiconformal with maximal dilatation $K_t \leq e^{|t|\sqrt{n}\,\|Su\|_\infty}$

Equivalent versions of this result were obtained by Ahlfors [4] Semenov [12] and the author [7]. The 2-dimensional version of this theorem is due to Schwartz [13].

Quasiconformal deformations u with $u(x) = O(|x|\log|x|)$ can be characterized by a "discrete" condition, which is halfway between a Lipschitz- and a Zygmund-condition. The continuous vector field u is a quasiconformal deformation if and only if

$$\left| \frac{(a,\ v(x+a) - v(x))}{|a|^2} - \frac{(b,\ v(x+b) - v(x))}{|b|^2} \right| \leq c$$

for all $x \in \mathbf{R}^n$ and for all $a,b \in \mathbf{R}^n$ with $|a| = |b| \neq 0$. (Reimann [7]).

4 Invariance and duality

The exceptional features of the S-operator are the invariance properties with respect to the Möbius group $M(n)$. The action of the Möbius transformations $a : \mathbf{R}^n \to \mathbf{R}^n$ on vector fields is defined by

$$u_a(x) = A(a^{-1}x)\, u(a^{-1}x)$$

(A is the Jacobian matrix of a). The operator S then satisfies

$$S u_a(x) = A(a^{-1}x)\, Su(a^{-1}x)\, A^{-1}(a^{-1}x)$$

(Ahlfors [4]). The transformation formulas for Euclidean motions are seen to be a special case of this formula.

By considering the subgroup in $M(n)$ of transformations which map the unit ball B onto itself (this subgroup is isomorphic to $M(n-1)$) the operator S becomes an invariant operator on hyperbolic space B.

On $\mathbf{R}^n$ there is an operator S^* dual to S. It is a differential operator which maps functions with values in the space of symmetric tensors with vanishing trace into vector fields:

$$(S^*\phi)_i = \sum_{j=1}^{n} \frac{\partial \phi_{ij}}{\partial x_j}$$

S^* is dual to S in the sense that

$$(Su,\phi) = -(u,S^*\phi)$$

for smooth vector- and tensor-valued functions with compact support. The scalar products are the customary L^2-products:

$$(u,v) = \int_{\mathbf{R}^n} \sum_{i=1}^{n} u_i v_i \, dx \qquad (\phi,\psi) = \int_{\mathbf{R}^n} \operatorname{tr} \phi\psi \, dx$$

The second order operator S^*S then is invariant with respect to Euclidean motions. It is in formal analogy to the Laplace operator $\Delta = \operatorname{div} \operatorname{grad}$.

The corresponding operator on hyperbolic space involves the invariant metric

$$|ds| = \frac{|dx|}{1-|x|^2} = \rho(x)|dx|$$

The operator takes the form $\rho^{-n-2} S^* \rho^n S$. It is invariant under Möbius transformations mapping B onto itself. (Ahlfors [4]).

Invariance under Möbius transformations is the property required for the investigation of the quotient spaces B/Γ, where Γ is a discrete subgroup of Möbius transformations. In this direction Ahlfors [5] proved, what he called a weak finiteness theorem:
If Γ is finitely generated, then the space of quasiconformal deformations of B, which give rise to densities ϕ on B/Γ satisfying special growth and boundary conditions and such that $S^*\phi = 0$, is finite dimensional.

5 The Dirichlet Problem

The invariant operators $\Delta_E = S^*S$ on Euclidean space and $\Delta_H = \rho^{-n-2} S^* \rho^n S$ on hyperbolic space can be treated much in the same way as the usual Laplace operator.

The Dirichlet problem for Δ_H on B was completely solved by Ahlfors [3]:

To any continuous vector field v on ∂B there exists a unique solution u of the equation $\Delta_H u = 0$ with boundary values $u|_{\partial B} = v$. The solution is given as a Poisson type integral

$$u(y) = c_n \int_{\partial B} \frac{(1 - |y|^2)^{n+1}}{|x - y|^{2n}} (I - 2Q(y-x))(1 - nQ(x))\, v(x)d\sigma($$

$$\text{with} \quad (Q(x))_{ij} = \frac{x_i x_j}{|x|^2}$$

The deduction of this beautiful explicit formula is essentially based on the fact that the Möbius group acts transitively on B.

In three-dimensional Euclidean space, the elasticity operator is given by $S^*S + c \text{ grad div}$ with $c > 0$. The Dirichlet problem for this operator was solved in great generality by Weyl [16]. An explicit solution (based on Fourier analysis) for the Dirichlet problem

$$S^*Su = 0, \qquad u|_{\partial B} = v,$$

where the domain B is a ball in $\mathbb{R}^n$, is given in [8].

6 Variants of the S operator

The method of Stein and Weiss referred to in section 1 can be applied to functions on $\mathbb{R}^n$ with values in an arbitrary finite dimensional representation space U of the group $SO(n)$. The generalized S operators S_k are obtained by taking for U the spaces H_k of spherical harmonics of degree k. The operator S_k then maps functions with values in H_k into functions with values in H_{k+1}.

In hyperbolic space a unified approach is obtained by considering the cosphere bundle X of B. The hyperbolic space itself is represented as the quotient space

$$SO_0(1,n) / SO(n)$$

The Möbius group $M(n-1)$ is isomorphic to the identity component $SO(1,n)$ in the generalized Lorentz group $SO(1,n)$ and $SO(n)$ is the stabilizer of $0 \in B$. The cosphere bundle X is then isomorphic to

$$SO_0(1,n) / SO(n-1)$$

The non-commutative algebra of invariant differential operators on the homogeneous space X is generated by a first order differential operator D and a second order operator D_{Σ}. Locally, this second order operator is essentially the Laplace operator on the sphere $\Sigma = SO(n) / SO(n-1)$. A function on X which is an eigenfunction of D_{Σ} can be viewed as a function on B with values in some eigenspace H_k of the Laplace operator on the sphere. The generalized S operators are then differential operators from one eigenspace of D_{Σ} into another. They can be expressed in terms of D and D_{Σ} (see [9]).

References

[1] Ahlfors, L.V. Kleinsche Gruppen in der Ebene und im Raum. Festband zum 70. Geburtstag von Rolf Nevanlinna, Springer, Berlin, 1966, 7 - 15.

[2] Ahlfors, L.V. Conditions for quasiconformal deformations in several variables. Contributions to Analysis. Academic Press, New York, 1974, 19 - 25.

[3] Ahlfors, L.V. Invariant operators and integral representations in hyperbolic space. Math. Scand. 36 (1975) 27 - 43.

[4] Ahlfors, L.V. Quasiconformal deformations and mappings in $\mathbb{R}^n$. J. d'Analyse Math. 30 (1976) 74 - 97.

[5] Ahlfors, L.V. Möbius transformations in several dimensions. University of Minnesota, 1981.

[6] Cauchy, A.L. Sur les corps solides ou fluides dans lesquels la condensation ou dilatation linéaire est la même en tous sens autour de chaque point. Ex. de math. 4 (1829) 214 - 216 = Oeuvres (2) 9, 254 - 258.

[7] Reimann, H.M. Ordinary differential equations and quasiconformal mappings. Inventiones math. 33 (1976) 247 - 270

[8] Reimann, H.M. A rotation invariant differential equation for vector fields. Ann. Sc. Norm. Sup. Pisa, IV 9 (1982) 159 - 174

[9] Reimann, H.M. Invariant differential operators in hyperbolic space. Comment. Math. Helv. (to appear)

[10] Reimann, H.M. Invariant systems of differential operators (to appear)

[11] Sarvas, J. Singularities of quasiconformal deformations in $\mathbf{R}^n$. Indiana Univ. Math. J. 31 (1982) 121 - 134

[12] Semenov, V.I. One parameter groups of quasiconformal mappings in euclidean space (Russian). Sibir. Mat. Zurn. 17 (1976) 177 - 193

[13] Schwartz G.P. Parametric representations of plane quasiconformal mappings. Thesis, University of Minnesota, 1970.

[14] Stein, E.M. and Weiss, G. Generalization of the Cauchy-Riemann equations and representations of the rotation group. Am. J. of Math. 90 (1968) 163 - 196

[15] Truesdell, C. and Muncaster, R.G. Fundamentals of Maxwell's kinetic theory of a simple monatonic gas. Academic Press, New York, 1980.

[16] Weyl, H. Eigenschwingungen eines beliebig gestalteten elastischen Körpers. Rend. Circ. Mat. Palermo 39 (1915) 1 - 50 or Selecta Hermann Weyl, Birkhäuser, Basel 1956.

A KERNEL FOR GENERALIZED CAUCHY-RIEMANN SYSTEMS

by

J.E. GILBERT
University of Texas

R.A. KUNZE
University of California, Irvine

R.J. STANTON
Ohio State University

P.A. Tomas
University of Texas

0. <u>INTRODUCTION</u>. In this paper we construct an integral transform which produces global solutions for certain generalized Cauchy-Riemann systems associated with representations of "motion" groups. The systems we consider are natural generalizations of those considered by Stein and Weiss in [1]. The integral transform is constructed in analogy with the Knapp-Wallach Szego transform, which was used to construct discrete series representations of semi-simple groups [2].

1. <u>PRINCIPAL SERIES FOR MOTION GROUPS</u>. In this section, we describe a family of infinite dimensional Hilbert space representations for motion groups. These representations need not be unitary and may be regarded as the analogues of the (non-unitary) principal series representations for semi-simple groups.

To begin with we assume only that the group G is the semi-direct product of a closed normal abelian subgroup V and a compact subgroup K. Thus, $G = VK = KV$ and $V \cap K = \{1\}$. It follows that for each $x \in G$ there is a unique $\nu(x) \in V$ and

$\kappa(x) \in K$ such that

$$(1.1) \qquad x = \nu(x)\kappa(x) .$$

Let λ be a continuous character of V, not necessarily unitary, and M a closed subgroup of K such that

$$(1.2) \qquad \lambda(m v m^{-1}) = \lambda(v)$$

for all $m \in M$ and $v \in V$. Let σ be a continuous irreducible unitary representation of M with representation space $\mathscr{H}_\sigma$. Then K acts unitarily by right translation on the space $L^2(K,\sigma)$ of square-integrable λ covariants on K, i.e., on the space of measurable $f: K \to \mathscr{H}_\sigma$ such that

$$f(mk) = \sigma(m)f(k)$$

for all $(m,k) \in M \times K$, and

$$\int_K |f(k)|^2 dk < \infty$$

dk denoting normalized Haar measure on K. We extend each $f \in L^2(K,\sigma)$ to G by setting

$$(1.3) \qquad f(x) = \lambda(\nu(x))f(\kappa(x)) , x \in G .$$

Let $\mathscr{H}_{\lambda\sigma}$ denote the set of all such extensions. Then $\mathscr{H}_{\lambda\sigma}$ is a Hilbert space with inner product

$$(1.4) \qquad (f|g) = \int_K (f(k)|g(k))dk .$$

The functions f in $\mathscr{H}_{\lambda\sigma}$ have the property that

$$(1.5) \qquad f(v\,m\,x) = \lambda(v)\sigma(m)f(x)$$

for all $v \in V$, $m \in M$, and $x \in G$. To see this note that

$$(1.6) \qquad \nu(kx) = k\nu(x)k^{-1}\ ,\ \kappa(kx) = k\kappa(x)$$

for all $k \in K$ and $x \in G$. Now suppose $f \in \mathscr{H}_{\lambda\sigma}$. Then for $m \in M$ and $x \in G$

$$\begin{aligned} f(mx) &= \lambda(\nu(mx))f(\kappa(mx)) \\ &= \lambda(m\nu(x)m^{-1})f(m\kappa(x)) \\ &= \lambda(\nu(x))\sigma(m)f(\kappa(x)) \\ &= \sigma(m)f(x) \end{aligned}$$

by (1.2) and (1.6). Similarly, for $v \in V$ and $x \in G$, we have

$$(1.7) \qquad \nu(vx) = v\nu(x)\ ,\ \kappa(vx) = \kappa(x)$$

and this implies

$$f(vx) = \lambda(v\nu(x))f(\kappa(x)) = \lambda(v)f(x)\ .$$

It follows that f satisfies (1.5). Moreover, it is easy to see that if $f:G \to \mathscr{H}_\sigma$ is square-integrable over K and transforms according to (1.5), then $f \in \mathscr{H}_{\lambda\sigma}$.

Now it is obvious that the right G translates of any function in $\mathscr{H}_{\lambda\sigma}$ satisfy (1.5). Thus to show that G acts on $\mathscr{H}_{\lambda\sigma}$ by right translation, it suffices to show that any such translate is square-integrable over K. For this let $v \in V$, $u \in K$, and $f \in \mathscr{H}_{\lambda\sigma}$. Then because K normalizes V and f transforms to the left by (1.5), it follows that

$$(R(vu)f)(k) = \lambda(k\,v\,k^{-1})f(ku) \tag{1.8}$$

for all $k \in K$, $R(vu)$ denoting right translation by vu. Because λ is continuous and K is compact, it follows that

$$\int_K |\lambda(k\,v\,k^{-1})|^2 |f(ku)|^2 dk$$

$$\leq (\sup_k |\lambda(k\,v\,k^{-1})|^2) \int_K |f(k)|^2 dk < \infty .$$

Let $R = R(\cdot,\lambda,\sigma)$ denote the representation of G on $\mathscr{H}_{\lambda\sigma}$ defined by right translation. The restriction of R to K is always unitary, but R is a unitary representation of G iff λ is a unitary character of V.

A few final comments are in order. First, suppose λ is a unitary character and that M is the full isotropy group of λ. Then $R(\cdot,\lambda,\sigma)$ is one of the irreducible unitary representations of G given by the standard Mackey theory [3]. These are precisely the representations that occur in the Plancherel formula for G [4]. Second, suppose σ is the trivial 1-dimensional representation of M. Then as λ varies over the unitary dual of V, the given representations are analogous to spherical principal series representations in the semi-simple case and provide a natural direct integral decomposition of $L^2(V)$. For this case, the map S to be introduced next already appears in [5, Thm. 4.3].

2. THE τ-QUOTIENT MAP S . In this section we define a general class of linear maps S taking $\mathcal{H}_{\lambda\sigma}$ to continuous K covariants on G . Upon appropriate specialization, the image of S will consist of solutions of an associated Cauchy-Riemann system.

For this purpose, let τ be an irreducible unitary representation of K and $\mathcal{H}_\tau$ the representation space for τ . Assume that the restriction of τ to M contains σ . Let $C:\mathcal{H}_\sigma \to \mathcal{H}_\tau$ be a linear map intertwining σ and the restriction of τ to M , i.e., a linear map such that

$$C\,\sigma(m) = \tau(m)C\ ,\ m \in M\ . \tag{2.1}$$

Then, by the Frobenius reciprocity theorem, there is an operator

$$B:\mathcal{H}_\tau \to \mathcal{H}_{\lambda\sigma}$$

that is canonically associated with C and intertwines τ with the restriction of $R(\cdot,\lambda,\sigma)$ to K . In fact, in the present context, $B = B_C$ may be defined quite explicitely.

2.2. FROBENIUS RECIPROCITY. *Let* $n_{\sigma\tau} = (\dim \mathcal{H}_\tau/\dim \mathcal{H}_\sigma)^{\frac{1}{2}}$ *and for* C *satisfying (2.1) define* $B = B_C$ *for* $\varphi \in \mathcal{H}_\tau$ *and* $x \in G$ *by*

$$(B\varphi)(x) = n_{\sigma\tau}\lambda(\nu(x))C^*\tau(\kappa(x))\varphi\ .$$

Then the map

$$C \to B_C$$

is a conjugate linear isometry of $\mathrm{Hom}_M(\mathcal{H}_\sigma,\mathcal{H}_\tau)$ *onto* $\mathrm{Hom}_K(\mathcal{H}_\tau,\mathcal{H}_{\lambda\sigma})$.

To prove this suppose $v \in V$, $x \in G$, and $\varphi \in \mathscr{H}_\tau$. Then by (1.7)

$$(B\varphi)(vx) = n_{\sigma\tau}\lambda(v\nu(x))C^*\tau(\kappa(x))\varphi = \lambda(v)(B\varphi)(x) .$$

Suppose $m \in M$. Then by (1.6) and (1.2)

$$(B\varphi)(mx) = n_{\sigma\tau}\lambda(m\nu(x)m^{-1})C^*\tau(m\kappa(x))\varphi$$

$$= n_{\sigma\tau}\lambda(\nu(x))C^*\tau(m)\tau(\kappa(x))\varphi .$$

Because σ and τ are unitary representations, it follows from (2.1) that $C^*\tau(m) = \sigma(m)C^*(m \in M)$. From this and the above, we see that

$$(B\varphi)(mx) = \sigma(m)(B\varphi)(x)$$

for all $m \in M$ and $x \in G$. Since λ, ν, τ, and κ are continuous, it follows that $B\varphi$ is a continuous map of G to $\mathscr{H}_\sigma$. Thus (2.2) defines a linear map B of $\mathscr{H}_\tau$ into $\mathscr{H}_{\lambda\sigma}$. From the formula for B and the relations

(2.3) $$\nu(xk) = \nu(x) , \quad \kappa(xk) = \kappa(x)k$$

which are valid for $x \in G$ and $k \in K$, it follows that B intertwines τ and the restriction of $R(\cdot,\lambda,\sigma)$ to K. To show that B is an isometry, we first compute its adjoint $B^*:\mathscr{H}_{\lambda\sigma} \to \mathscr{H}_\tau$. For $f \in \mathscr{H}_{\lambda\sigma}$ and $\varphi \in \mathscr{H}_\tau$, we have

$$(B^*f|\varphi) = (f|B\varphi) = n_{\sigma\tau}\int_K (f(k)|C^*\tau(k)\varphi)dk$$
$$= n_{\sigma\tau}\int_K (\tau(k^{-1})Cf(k)|\varphi)dk .$$

It follows that

$$B^*f = n_{\sigma\tau}\int_K \tau(k^{-1})Cf(k)dk \tag{2.4}$$

for all f in $\mathcal{H}_{\lambda\sigma}$. Now take $f = B\varphi$ with φ in $\mathcal{H}_\tau$. Then by (2.4)

$$B^*B = n_{\sigma\tau}^2\int_K \tau(k^{-1})CC^*\tau(k)dk \ .$$

This implies

$$\mathrm{tr}(B^*B) = n_{\sigma\tau}^2\,\mathrm{tr}(C^*C) \ . \tag{2.5}$$

Since σ and τ are irreducible, Schur's lemma implies that B and C are scalar multiples of isometries. Specifically

$$B^*B = \|B\|_\infty^2 I_\tau \ , \ C^*C = \|C\|_\infty^2 I_\sigma$$

where I_τ is the identity operator on $\mathcal{H}_\tau$, $\|B\|_\infty$ is the operator norm of B , and I_σ, $\|C\|_\infty$ are defined analogously. Thus, by (2.5)

$$\|B\|_\infty^2 \dim \mathcal{H}_\tau = n_{\sigma\tau}^2 \|C\|_\infty^2 \dim \mathcal{H}_\sigma \ .$$

It follows that $\|B\|_\infty = \|C\|_\infty$. Thus, since the map $C \to B_C$ is evidently conjugate linear, it is necessarily injective. Finally, if $B \in \mathrm{Hom}_K(\mathcal{H}_\tau,\mathcal{H}_{\lambda\sigma})$, then $B = B_C$ where C is the adjoint of the map

$$\varphi \to (B\varphi)(1) \ , \ \varphi \in \mathcal{H}_\tau \ .$$

For our purpose it is convenient to assume that C is an isometry. Then $B = B_C$ is an isometry that intertwines τ and the restriction of $R(\cdot,\lambda,\sigma)$ to K, $B^*B = I_\tau$, and BB^* is the orthogonal projection of $\mathscr{H}_{\lambda\sigma}$ onto $B(\mathscr{H}_\tau)$.

Now we define $S = S_B$ on $\mathscr{H}_{\lambda\sigma}$ by

$$(2.6) \qquad (Sf)(x) = B^*R(x)f\ , x \in G\ .$$

Since $R = R(\cdot,\lambda,\sigma)$ is, in fact, a continuous representation, it is immediate that Sf is a continuous function on G. Because R is unitary on K

$$(2.7) \qquad B^*R(k) = \tau(k)B^*\ , k \in K\ .$$

Thus, for $k \in K$ and $x \in G$

$$(Sf)(kx) = B^*R(k)R(x)f = \tau(k)(Sf)(x)\ .$$

Hence, (2.6) defines a linear map

$$S:\mathscr{H}_{\lambda\sigma} \to C(G,\tau)$$

$C(G,\tau)$ denoting the space of continuous τ-covariants on G. Now G acts on $C(G,\tau)$ by right translation. In this context right translation will be denoted by T. Then it follows from (2.6) that

$$(2.8) \qquad T(y)S = SR(y)\ , y \in G\ .$$

Thus, S maps $\mathcal{H}_{\lambda\sigma}$ onto a G-invariant subspace of $C(G,\lambda)$ and intertwines the two G actions. We call S a τ-quotient map. The general theory of τ-quotients will not be presented here.

To complete this section we derive an integral formula for Sf. First note that since $G = KV$ and $Sf \in C(G,\lambda)$, it suffices to compute $(Sf)(v)$ for $v \in V$. By (2.4) and (2.6)

$$(Sf)(v) = \int_K \tau(k^{-1})C(R(v)f)(k)dk .$$

Hence, by (1.8)

$$(Sf)(v) = \int_K \lambda(k\,v\,k^{-1})\tau(k^{-1})Cf(k)dk \tag{2.9}$$

for arbitrary $f \in \mathcal{H}_{\lambda\sigma}$ and $v \in V$. From (1.2), (1.5), and (2.1), it follows that the integral in (2.9) is a function on $M\backslash K$, the space of all cosets Mk with $k \in K$. Hence, we may also write

$$(Sf)(v) = \int_{M\backslash K} \lambda(k\,v\,k^{-1})\tau(k^{-1})Cf(k)d(Mk) . \tag{2.10}$$

3. GENERALIZED CAUCHY-RIEMANN SYSTEMS. Now we specialize the foregoing, assuming that V is a real inner product space of dimension n and that K is a closed subgroup of $O(v)$, the orthogonal group on V. Multiplication in G is specified by the requirement that for $k \in K$ and $v \in V$

$$k\,v\,k^{-1} = kv$$

kv denoting the image of v under the linear transformation k. Let $x,w \to z \cdot w$ denote the complex bilinear form on $V_{\mathbf{C}}$ that extends the inner product on V. Then the characters of V are all of the form

$$v \to e^{iv \cdot z} \;, \; v \in V \;.$$

If z is the point in $V_{\mathbf{C}}$ that defines the character λ, then (2.9) may be written in the form

$$(3.1) \qquad (Sf)(v) = \int_K e^{ikv \cdot z}\tau(k^{-1})Cf(k)dk \;.$$

It follows that the τ-covariant Sf is a C^∞, in fact real analytic, function on G.

Let $C^\infty(G,\tau)$ denote the space of all C^∞ τ-covariants on G and ρ the standard representation of K on $V_{\mathbf{C}}$. Then the gradient operator ∇ on $\mathcal{H}_\tau$-valued functions on V extends uniquely to a right invariant linear map, again denoted ∇, of $C^\infty(G,\mathcal{H}_\tau)$ to $C^\infty(G,\mathcal{H}_\tau \otimes V_{\mathbf{C}})$. Moreover, it is easy to check that

$$\nabla : C^\infty(G,\tau) \to C^\infty(G,\tau \otimes \rho) \;.$$

Thus, ∇ intertwines the right translation action of G on $C^\infty(G,\tau)$ with the corresponding action of G on $C^\infty(G,\tau\otimes\rho)$.

Now suppose W is a subspace of $\mathcal{H}_\tau \otimes V_C$ that is invariant under $\tau\otimes\rho$. Let

(3.2) $$C^\infty_W(G,\tau) = \{F \in C^\infty(G,\tau) : (\nabla F)(x) \in W, \forall x \in G\} .$$

This space can also be characterized as the kernel of a differential operator $\mathcal{T}_W$. For this purpose, let $P:\mathcal{H}_\tau \otimes V_C \to \mathcal{H}_\tau \otimes V_C$ be the orthogonal projection of $\mathcal{H}_\tau \otimes V_C$ on $W^\perp$ and for $F \in C^\infty(G,\tau)$ set

(3.3) $$(\mathcal{T}_W F)(x) = P(\nabla F)(x) , \quad x \in G .$$

Because P commutes with $\tau\otimes\rho$, it follows that $C^\infty_W(G,\tau)$ is a G-invariant subspace of $C^\infty(G,\tau)$. One may refer to the functions F in $C^\infty_W(G,\tau)$ as the <u>solutions</u> of the <u>Cauchy-Riemann system</u>

(3.4) $$\mathcal{T}_W F = 0$$

<u>that corresponds to</u> $\tau\otimes\rho$ <u>and</u> W.

Next we wish to choose σ, τ, C, W, and z so that each of the integral transforms Sf in (3.1) is a solution of (3.4). The first step is to compute ∇Sf. For this let $e_1,\cdots,e_n$ be an orthonormal base for V. Write

$$v = \sum_{j=1}^n v_j e_j , \quad k^{-1}z = \sum_{j=1}^n (k^{-1}z)_j e_j .$$

Then $kv\cdot z = \sum_j v_j (k^{-1}z)_j$, and

$$\nabla_v(e^{ikv\cdot z}\tau(k^{-1})Cf(k))$$

$$= \sum_j ie^{iv\cdot k^{-1}z}(k^{-1}z)_j\tau(k^{-1})Cf(k) \otimes e_j$$

$$= ie^{ikv\cdot z}\tau(k^{-1})Cf(k) \otimes k^{-1}z .$$

It follows from this and (3.1) that

(3.5) $$(\nabla Sf)(v) = i\int_K e^{ikv\cdot z}(\tau \otimes \rho)(k^{-1})Cf(k) \otimes z\, dk .$$

The form of this integral suggests the following procedure. Let γ be any non-zero vector in $\mathscr{H}_\sigma$, $\omega = C\gamma$, and W the $\tau \otimes \rho$ stable subspace of $\mathscr{H}_\tau \otimes V_{\mathbf{C}}$ that is generated by $\omega \otimes z$. Then for any $m \in M$

$$(\tau \otimes \rho)(m)(\omega \otimes z) = \tau(m)C\gamma \otimes z = (C\sigma(m)\gamma) \otimes z .$$

Because σ is irreducible, it follows that

(3.6) $$C(\mathscr{H}_\sigma) \otimes z \subset W .$$

Thus, in (3.5), $Cf(k) \otimes z \in W$ for every k. Since W is closed and $\tau \otimes \rho$ invariant, it results that

(3.7) $$\mathscr{T}_W Sf = 0 .$$

At this point σ, τ, C, and z are subject only to the initial assumptions, and W is specified by the choice of a vector γ in $\mathscr{H}_\sigma$. But (3.6) shows that W is independent of γ, i.e., that W is simply the $\tau \otimes \rho$ stable subspace of $\mathscr{H}_\tau \otimes V_{\mathbf{C}}$ that is generated by the subspace $C(\mathscr{H}_\sigma) \otimes z$.

If z is an isotropic vector in $V_{\mathbb{C}}$ $(z^2 = z \cdot z = 0)$ then the integral transforms Sf in (3.1) are automatically harmonic as functions on V. In fact, if Δ is the Laplace operator on V, then for any z

$$\Delta_v e^{ikv \cdot z} = -(k^{-1}z)^2 e^{ikv \cdot z} = -z^2 e^{ikv \cdot z}$$

so that

$$\Delta Sf = -z^2 Sf \tag{3.8}$$

for all f in $\mathcal{H}_{\lambda\sigma}$. Thus, in general, a τ-quotient map S will produce solutions Sf of a generalized Cauchy-Riemann system (as here defined) and these solutions Sf are necessarily eigen functions of the Laplacian when viewed as functions on V.

In our context, the Cauchy-Riemann systems considered by Stein and Weiss [1] are those in which $K = SO(V)$ and W is given apriori as a subspace in which $\tau \otimes \rho$ acts by the Cartan composition of τ with ρ.

To briefly illustrate these ideas in the simplest case, let $K = SO(V)$ and $\tau = \tau_m$ the representation of K on the space $\mathcal{H}_m = \mathcal{H}_m(V)$ of homogeneous harmonic polynomials of degree m on V. Let z be any non-zero isotropic vector in $V_{\mathbb{C}}$ and M the subgroup of $V_{\mathbb{C}}$ that fixes z (M is isomorphic to $SO(n-2)$). Then the homogeneous polynomial

$$\omega(v) = (v \cdot z)^m, \quad v \in V \tag{3.9}$$

is harmonic and $\tau(M)$ invariant $(\omega(m^{-1}v) = \omega(v) \ \forall\, m \in M)$. Let

σ be the trivial 1-dimensional representation of M on $\mathbb{C}$ and C the obvious isomorphism of $\mathbb{C}$ onto $\mathbb{C}\omega$. Then $\mathcal{H}_{\lambda\sigma}$ is a space of complex-valued functions on G, which is isomorphic to $L^2(K/M)$, and (3.1) takes the form

$$(3.10) \qquad (Sf)(v) = \left(\int_K e^{ikv\cdot z} f(k)\tau(k^{-1})dk\right)\omega .$$

We remark that when f is identically 1 on K, then in (3.10) the operator

$$\int_K e^{ikv\cdot z} f(k)\tau(k^{-1})dk$$

is precisely $J_\tau(v,z)$ where J_τ is the generalized Bessel function studied in [5] and [6].

REFERENCES

[1] E.M. STEIN and G. WEISS, Generalization of the Cauchy-Riemann equations and representations of the rotation group, Amer. J. Math. 90 (1968) 163-196.

[2] A.W. KNAPP and N.R. WALLACH, Szegö kernels associated with discrete series, Invent. Math. 34 (1976) 163-200.

[3] G. MACKEY, Infinite-dimensional group representations, Bull. Amer. Math. Soc. 69 (1963) 628-686.

[4] K.I. GROSS and R.A. KUNZE, Fourier decompositions of certain representations, Symmetric Spaces, 119-139, Marcel Dekker, New York, 1972.

[5] K.I. GROSS and R.A. KUNZE, Bessel functions and representation theory I, J. Funct. Anal. 22 (1976) 73-105.

[6] K.I. GROSS and R.A. KUNZE, Bessel functions and representation theory II, J. Funct. Anal. 25 (1977) 1-49.

HARMONIC ANALYSIS ON GROUPS OF HEISENBERG TYPE

by

A. Kaplan and F. Ricci

Groups of Heisenberg (or simply H-) type are 2-step stratified nilpotent Lie groups whose Lie algebras carry a suitably compatible inner product ([Ka 1]). They include the nilpotent Iwasawa subgroups of semisimple Lie groups of split rank one [Ko2] as well as many more examples supplied by the theory of Clifford algebras.

Previous work on these groups has pointed out their similarities with Heisenberg groups (existence and properties of elementary homogeneous norms [C], [Ko], and fundamental solutions for their sublaplacians [Ka 1]) as well as their differences (mainly concerning the geometry of the associated left-invariant metrics [Ka 2], [R], [Ś], [TV]).

In this paper we show on the one hand how a unified representation theory can be built up for all H-type groups in an intrinsic way, modeled after the Bargmann representations of the Heisenberg group. On the other hand we point out some of the differences among H-type groups that appear in the harmonic analysis of their symmetry groups. More specifically, we concentrate on the question of whether the algebra of L^1-functions on an H-type group N which are invariant under a given group K of "rotations" (= isometric automorphisms) is commutative - in other words, when $(K \ltimes N, N)$ is a Gelfand pair. Via the Bargmann-like representations we reduce the problem to one involving symmetric powers of irreducible representations of certain subgroups of K.

In many cases this analysis can be carried out completely and we meet both positive and negative answers to the commutativity question; moreover, we expect no further positive cases to arise. It is known that commutativity occurs if for every $n \in N$ there is a $k \in K$ such that $k(n) = n^{-1}$; this happens with the Iwasawa groups and in some other instances. On the other hand, a 13-dimensional H-type group with a 5-dimensional center and with $K = Spin(5)$ shows that the condition above is not necessary for the commutativity of $L^1_K(N)$.

The authors wish to thank the Consiglio Nazionale delle Ricerche and the National Science Foundation for their support during the preparation of this work.

§1. Algebras and groups of Heisenberg type. Let $\mathfrak{n}$ be a 2-step nilpotent Lie algebra over $\mathbb{R}$ endowed with a positive definite inner product $\langle\ ,\ \rangle$. Let $\mathfrak{z}$ be the center of $\mathfrak{n}$, $\mathfrak{v}$ its orthogonal complement,

$$\mathfrak{n} = \mathfrak{v} \oplus \mathfrak{z}$$

and set $|x| = \langle x,x\rangle^{1/2}$. Then $\mathfrak{n}$ - as well as the corresponding simply connected Lie group - is said to be of *Heisenberg type* if for every $v \in \mathfrak{v}$, $|v| = 1$, the map $\mathrm{ad}\,v\colon \mathfrak{v} \to \mathfrak{z}$ is an isometric submersion: that is, it becomes a surjective isometry when restricted to the orthogonal complement of its kernel [Ka 1].

The equation

$$(1.1) \qquad \langle J_z v, v' \rangle = \langle z, [v,v'] \rangle \qquad (z \in \mathfrak{z}\,; v,v' \in \mathfrak{v})$$

defines a linear map $J: \mathfrak{z} \to \mathrm{End}(\mathfrak{v})$ satisfying

(1.2)
(i) J_z is skew symmetric $\qquad (z,z' \in \mathfrak{z})$

(ii) $J_z J_{z'} + J_{z'} J_z = -2\langle z,z' \rangle 1$.

In particular, $|z| = 1$, $z \perp z'$ imply $J_z^2 = -1$, $J_z J_{z'} = -J_{z'} J_z$.

Set $m = \dim \mathfrak{z}$, $n = \dim \mathfrak{v}$ and let $C(m)$ denote the Clifford algebra over the quadratic space $(\mathfrak{z}, -|\ |^2)$. The properties above show that J extends to a (unitary) representation of $C(m)$ on $\mathfrak{v}$. Moreover, every module $\mathfrak{v}$ over $C(m)$ arises from an H-type algebra in this fashion.

We will say that an H-type algebra $\mathfrak{n} = \mathfrak{v} \oplus \mathfrak{z}$ is *reducible* if there exist a proper decomposition $\mathfrak{v} = \mathfrak{v}_1 \oplus \mathfrak{v}_2$ such that $\mathfrak{v}_1 \oplus \mathfrak{z}$, $\mathfrak{v}_2 \oplus \mathfrak{z}$ are both algebras of H-type relative to the induced structures. Otherwise, $\mathfrak{n}$ will be called *irreducible*. It is clear that these notions correspond to the standard ones for the associated Clifford module.

Two H-type algebras $\mathfrak{n}_1$, $\mathfrak{n}_2$ will be said to be *isomorphic* if they are isometrically so. Writing $\mathfrak{n}_j = \mathfrak{v}_j \oplus \mathfrak{z}_j$, this amounts to the existence of isometries $\phi: \mathfrak{z}_1 \to \mathfrak{z}_2$, $\psi: \mathfrak{v}_1 \to \mathfrak{v}_2$ such that

(1.3) $$\psi(J_z v) = J_{\phi(z)}\psi(v) \qquad (z \in \mathfrak{z},\ v \in \mathfrak{v}).$$

Since we do not insist that ϕ be the identity, inequivalent Clifford modules may yield isomorphic H-type algebras. For example, from the standard theory of Clifford algebras (see e.g. [H]) one concludes that for each $m \geq 1$ there exist just one *irreducible* H-type algebra with $\dim \mathfrak{z} = m$, modulo equivalence. The corresponding dimensions for $\mathfrak{v}$ are as follows:

$m = \dim \mathfrak{z}$	$8k$	$8k+1$	$8k+2$	$8k+3$	$8k+4$	$8k+5$	$8k+6$	$8k+7$
$n = \dim \mathfrak{v}$	2^{4k}	2^{4k+1}	2^{4k+2}	2^{4k+2}	2^{4k+3}	2^{4k+3}	2^{4k+3}	2^{4k+3}

The group $A(\mathfrak{n})$ of orthogonal automorphisms of $\mathfrak{n}$ has been studied in great detail by C. Riehm [R]. Let U be the subgroup of those automorphisms acting trivially on the center; by restriction to one can identify U with the orthogonal transformations on $\mathfrak{v}$ which intertwine the representation J of $C(m)$. For the irreducible H-type algebras,the corresponding groups U are as follows:

m(mod 8)	0	1	2	3	4	5	6	7
U	±1	U(1)	SU(2)	SU(2)	SU(2)	U(1)	±1	±1

To complete the description of $A(\mathfrak{n})$, one first observes that for each unit vector $z \in \mathfrak{z}$, the linear map $J_z : \mathfrak{v} \longrightarrow \mathfrak{v}$ can be extended to an isometric automorphism of $\mathfrak{n}$ by defining it on $\mathfrak{z}$ as minus the reflection with respect to the hyperplane $z^{\perp}$. Since $\mathrm{Pin}(m) \overset{\text{def.}}{=}$ multiplicative group generated in $C(m)$ by the unit vectors of $\mathfrak{z}$, the map $z \longrightarrow J_z$ extends to a group homomorphism $\mathrm{Pin}(m) \longrightarrow A(\mathfrak{n})$, which has finite kernel. Also, the image of $\mathrm{Pin}(m)$ in $A(\mathfrak{n})$ commutes with U and the product of these subgroups has finite index in $A(\mathfrak{n})$. Let $A_o(\mathfrak{n})$, U_o, $\mathrm{Spin}(m)$, denote the identity connected component of $A(\mathfrak{n})$, U, $\mathrm{Pin}(m)$ respectively. Then

$$(1.4) \qquad A_o(\mathfrak{n}) \cong (U_o \times \mathrm{Spin}(m))/(\text{finite subgroup}).$$

§2. <u>Harmonic analysis on N</u>. Since, for example, groups of Heisenberg type are among those 2-step nilpotent groups considered in [M], all the unitary irreducible representations that are not one-dimensional are parametrized by the elements $a \in \mathfrak{z} \setminus (o)$. It is convenient in any case to have the following explicit realization of these representations, a variant of the Bargmann-Fock model of the representations of the Heisenberg group.

Given $a \in \mathfrak{z} - (0)$, let $b = a/|a|$; then $J_b : \mathfrak{v} \longrightarrow \mathfrak{v}$ is a complex structure on $\mathfrak{v}$. Consider now the Hilbert space (Fock space) $\mathcal{F}_a(\mathfrak{v})$ of all entire holomorphic functions F on $(\mathfrak{v}, J_b)$ such that

$$\|F\|^2 = \int |F(w)|^2 e^{-\frac{|a||w|^2}{2}} dw < \infty .$$

The corresponding irreducible unitary representation π_a of N is realized on $\mathcal{F}_a(\mathfrak{v})$ by

$$(\pi_a(\exp(v+z))F)(w) = F(w+v)e^{i<a,z> - \frac{1}{4}|a|(|v|^2 + 2(<w,v> - i<b,[w,v]>))}$$

and it is uniquely determined up to equivalence by the condition

$$\pi_a(\exp z) = e^{i<a,z>}\mathrm{Id}, \quad z \in \mathfrak{z} .$$

As before, let $A(\mathfrak{n})$ = group of isometric automorphisms of $\mathfrak{n}$, acting on N by exponentiation: $k(\exp(v+z)) = \exp(k(v)+k(z))$ (note that $A(\mathfrak{n})$ preserves the decomposition $\mathfrak{n} = \mathfrak{v} \oplus \mathfrak{z}$). Each $k \in A(\mathfrak{n})$ defines a unitary $T_k : \mathcal{F}_a \longrightarrow \mathcal{F}_{k(a)}$ for each $a \in \mathfrak{z} - (o)$, by $(T_k F)(w) = F(k^{-1}w)$. We shall also consider the one-parameter group of (non-isometric) automorphisms $\{D_\lambda, \lambda \in \mathbb{R}_+\}$, acting by $D_\lambda(\exp(v+z)) = \exp(\lambda v + \lambda^2 z)$. Then D_λ commutes with $A(\mathfrak{n})$. Define $U_\lambda : \mathcal{F}_a \longrightarrow \mathcal{F}_{\lambda^2 a}$ by $U_\lambda F(w) = F(\lambda w)$.

(2.1) Lemma. Let $k \in A(\mathfrak{n})$, $a \in \mathfrak{z}^*$, $n \in N$ and $\lambda \in \mathbb{R}_+$. Then

(i) $\pi_a(k^{-1}(n)) = T_k^{-1} \circ \pi_{k(a)}(n) \circ T_k$

(ii) $\pi_a(D_\lambda^{-1}(n)) = U_\lambda^{-1} \circ \pi_{\lambda^2 a}(n) \circ U_\lambda$

Proof: $[\pi_a(\exp(k^{-1}(v)+k^{-1}(z)))F](w) =$

$$= F(w+k^{-1}(v))e^{i<a,k^{-1}z> - \frac{1}{4}|a|(|v|^2 + 2(<w,k^{-1}v> - i<J_b w,k^{-1}v>))}$$

$$= F(w+k^{-1}v)e^{i<ka,z> - \frac{1}{4}|ka|(|v|^2 + 2(<kw,v> - i<J_{kb}kw,v>))}$$

$= T_k^{-1} \circ \pi_{k(a)}(\exp(v+z)) \circ T_k$, proving (i). The proof for (ii) is similar.

Given a closed, connected subgroup K of $A(\mathfrak{N})$, set

$$L^1_K(N) = \{f \in L^1(N) : f \circ k = f \quad \text{for all} \quad k \in K\} \quad .$$

This is a Banach subalgebra of $L^1(N)$ which can be identified with the algebra of K-biinvariant functions in $L^1(K \ltimes N)$, so that $(K \ltimes N, K)$ is a Gelfand pair precisely when $L^1_K(N)$ is commutative. For some natural choices of K we shall now give a necessary and sufficient condition for this to occur, which is purely algebraic and only involves the structure of $\mathfrak{v}$ as a $C(m)$-module.

Given an irreducible, unitary representation π of N, consider the family of operators $\{\pi(f) : f \in L^1_K(N)\}$ on the Hilbert space H_π. Then $L^1_K(N)$ is commutative if and only if this family is commutative for any such π. Assume now that

(2.2) $\qquad K$ is transitive on the unit sphere of $\mathfrak{z}$;

this is actually the case for $K = A_o(N)$ as well as for its subgroup Spin(m). Lemma (2.1) implies that $\pi_a(f) = T_k^{-1} \circ \pi_{k(a)}(f) \circ T_k$ for all $f \in L^1_K(N)$. Also, the map $f \longrightarrow f \circ D_\lambda^{-1}$ is an automorphism of this algebra and the same Lemma shows that $\pi_a(f \circ D_\lambda^{-1}) = U_\lambda^{-1} \circ \pi_{\lambda^2 a}(f) \circ U_\lambda$ for all $f \in L^1_K(N)$. We can then conclude that $L^1_K(N)$ is commutative if and only if the family of operators $\{\pi_a(f) : f \in L^1_K(N)\}$ is commutative for just *a single* unit vector $a \in \mathfrak{z}$.

Fix now such a unit vector $a \in \mathfrak{z}$ and let K_a be the stabilizer of a in K. It is clear that for any $f \in L^1_K(N)$ and $k \in K_a$, the operators $\pi_a(f)$ and T_k commute. Denote by $\mathcal{P}_{a,n} \subset \mathcal{F}_a$ the space of holomorphic polynomials on $(\mathfrak{v},J_a)$ of degree at most n; for any linear operator $S:\mathcal{P}_{a,n} \longrightarrow \mathcal{P}_{a,n}$, let $\tilde{S}:\mathcal{F}_a \longrightarrow \mathcal{F}_a$ denote the composition of S with the orthogonal projection of $\mathcal{F}_a$ onto $\mathcal{P}_{a,n}$.

(2.3) Lemma: Let S be a linear operator of $\mathcal{P}_{a,n}$ into itself such that $\tilde{S}$ commutes with $T_k:\mathcal{F}_a \longrightarrow \mathcal{F}_a$ for all $k \in K_a$. Then there exist $f \in L^1_K(N)$ such that $\pi_a(f) = \tilde{S}$.

Proof: Let $\hat{\pi}_\lambda$, $\lambda > 0$, denote the irreducible unitary representation of the Heisenberg group H_ℓ on the corresponding Fock space, such that $\hat{\pi}_\lambda(\exp(o + t)) = e^{it\lambda}$. Let $\{P_j\}$ be an orthonormal basis of monomials in that Fock space. Then it is known that

$$<\hat{\pi}_\lambda(\exp(v+t)))P_j,P_k> = \lambda^{k/2}\psi_{jk}(\sqrt{\lambda}v)e^{i\lambda t}$$

where $\{\psi_{jk}\}$ is an orthonormal system in $L^2(\mathbb{C}^\ell)$ consisting of functions in the Schwartz class $S(\mathbb{C}^\ell)$ (actually, built up with appropriate Laguerre functions). The notation here assumes the obvious realization of H_ℓ as a group of type H with $\mathfrak{v} = \mathbb{C}^\ell$ and $m = 1$.

Observe now that the kernel of the representation π_a of N is the orthogonal complement $a^\perp$ in $\mathfrak{z}$, so that π_a drops to an irreducible representation of the quotient group $N/a^\perp$, that is isomorphic (as a group of H-type) to H_ℓ, with $\ell = n/2$. The resulting representation is equivalent to $\hat{\pi}_1$

Fix a complex coordinate system in $(\mathfrak{v},J_a)$ and the corresponding orthogonal basis of monomials $\{P_j\}$ in $\mathcal{F}_a$. Let $\{a_{jk}\}$ be the matrix of $\tilde{S}$ relative to that basis and let $\eta \in C^\infty(\mathbb{R})$ be supported in a small neighborhood of 1 and such that $\eta(1) = 1$. Define f on N by

$$f(\exp(v+z)) = \sum_{j,\ell} a_{j\ell} \int_{\mathbb{R}} \int_K \eta(\lambda)\overline{\psi}_{j\ell}(\sqrt{\lambda}\, k(v)) e^{i\lambda<a,k(z)>} d\lambda dk$$

$$= \int_{\mathbb{R}} \int_K \operatorname{tr}(\tilde{S}\pi_{\lambda a}(k(\exp(v+z)))\eta(\lambda) d\lambda dk.$$

It is obvious that f is K-invariant. In order to see that $f \in L^1(N)$ we show that its Fourier transform along the central directions $\hat{f}(v,b) = \int_{\mathfrak{z}} f(\exp(v+z))e^{i<z,b>}dz$ is a Schwartz function on $\mathfrak{n} = \mathfrak{v} \oplus \mathfrak{z}$. With $b_1 = b/|b|$ we have $\hat{f}(v,b) = \sum_{j,\ell} a_{j\ell}\eta(|b|) \int_{\{k\in K:k(b_1)=a\}} \overline{\psi}_{j\ell}(\sqrt{|b|}\,k(v))dk.$

The function $\tilde{f}(v,\lambda,k) = \sum_{j,\ell} a_{j\ell}\eta(\lambda) \int_{K_a} \overline{\psi}_{j\ell}(\sqrt{\lambda}\, hk(v))dh,$ defined on $\mathfrak{v} \times \mathbb{R}_+ \times K$, is C^∞ with derivatives rapidly decreasing along $\mathfrak{v} \times \mathbb{R}_+$. Since $\hat{f}(v,\lambda\, k^{-1}a) = \tilde{f}(v,\lambda,k^{-1})$, we conclude that $f \in L^1(N)$ as claimed. Furthermore,

$$\hat{f}(v,a) = \sum_{j,\ell} a_{j\ell} \int_{K_a} \overline{\psi}_{j\ell}(kv)dk = \int_{K_a} \operatorname{tr}(\tilde{S}\pi_a(\exp kv)^{-1})dk$$

which, by (2.1), equals

$$\int_{K_a} \operatorname{tr}(T_k\tilde{S}\pi_a(\exp v)^{-1}T_k^{-1})dk = \operatorname{tr}(\tilde{S}\pi_a(\exp v)^{-1}) = \sum_{j,\ell} a_{j\ell}\overline{\psi}_{j\ell}(v).$$

Therefore,

$$<\pi_a(f)P_j,P_\ell> = \int_N f(\exp(v+z)\psi_{j\ell}(v)e^{i<a,z>}dvdz$$

$$= \int \hat{f}(v,a)\psi_{j\ell}(v)dv = a_{j\ell}\ ,$$

and the Lemma is proved.

Recall that $\mathcal{P}_a = \bigcup_n \mathcal{P}_{a,n}$, the space of holomorphic polynomials on $(\mathfrak{V}, J_a)$, is acted upon by the stabilizer $K_a \subset K$ via $(T_k P)(w) = P(k^{-1}w)$.

(2.4) Theorem. Assume that the subgroup $K \subset A(\mathfrak{n})$ acts transitively on the unit sphere in $\mathfrak{z}$. Then $L^1_K(N)$ is commutative if and only if for a unit $a \in \mathfrak{z}$, every irreducible representation of K_a occurs at most once in the representation T of K_a in $\mathcal{P}_a$.

Proof: Assume that $\mathcal{P}_a = \bigoplus_n E_n$, where each E_n is invariant and irreducible under K_a and the corresponding representations are all inequivalent. For any $f \in L^1_K(N)$, $\pi_a(f)$ intertwines T on K_a, so that $\pi_a(f)$ acts as a scalar on each E_n. This implies that $\{\pi_a(f),\ f \in L^1_K(N)\}$, and hence $L^1_K(N)$ itself, is commutative.

Conversely, assume that $\mathcal{P}_a$ contains two invariant, irreducible subspaces E_1, E_2, with equivalent actions; we may also assume that they are both contained in the same $\mathcal{P}_{a,n}$. But then the family $\{\tilde{S}: S: \mathcal{P}_{a,n} \to \mathcal{P}_{a,n}, \tilde{S}T = T\tilde{S}\}$ is not commutative. By (2.3) the same is true of $L^1_K(N)$, proving the Theorem.

§3. The cases $K = Spin(m)$ and $K = A_o(N)$. Throughout this section $N = N_m$ will denote the irreducible group of Heisenberg type with center of dimension m. We will discuss the commutativity of $L^1_K(N)$ for various values of m and the indicated choices of groups of rotations.

(3.1) <u>Proposition.</u> The algebra $L^1_{Spin(m)}(N_m)$ is commutative for $m = 5,6,7$. It is not commutative for $m \equiv 0,1,2,3,4 \pmod 8$.

<u>Proof:</u> Since $Spin(1) = 1$, we may assume $m > 1$. The stabilizer K_a of a fixed unit vector $a \in \mathfrak{z}$ in $K = Spin(m)$ is isomorphic to $Spin(m - 1)$. It is actually generated by the products $J_b J_c$ with a,b,c orthonormal (recall that for every unit $b \in \mathfrak{z}$, J_b can be regarded as an isometric automorphism of $\mathfrak{n}$, hence as a rotation of N). Similarly, the subalgebra of the Clifford algebra $C(m)$ generated by the J_b, a,b orthonormal, is isomorphic to $C(m - 1)$ and will be denoted accordingly.

Assume first that $m \equiv 0,1,2,4 \pmod 8$. Write $\mathfrak{n} = \mathfrak{v} \oplus \mathfrak{z}$ as in §1 and recall that $\mathfrak{v}$ is an irreducible module for $C(m)$. According to the dimensions table in §1, this cannot remain irreducible as a module over $C(m - 1)$; indeed, it must split as a sum of two irreducible $C(m - 1)$-modules. Let $\mathfrak{w} \subset \mathfrak{v}$ be one of them. Then $\mathfrak{w} \cap J_a\mathfrak{w}$ is invariant under $C(m)$ and therefore trivial. Since $J_a\mathfrak{w}$ is $C(m - 1)$ invariant,

$$\mathfrak{v} = \mathfrak{w} \oplus J_a\mathfrak{w} \tag{3.2}$$

is a complete reduction of $\mathfrak{v}$ as $C(m - 1)$-module. It also exhibits $\mathfrak{w}$ as a real form of $(\mathfrak{v}, J_a)$, invariant under $Spin(m - 1)$. Let $\{w_j\}$ be a complex basis for $(\mathfrak{v}, J_a)$ spanning $\mathfrak{w}$ over $\mathbb{R}$. Then

$\Sigma\langle w_j,\cdot\rangle^2$ is a holomorphic polynomial on $(\mathfrak{v},J_a)$, invariant under Spin(m - 1). Therefore the trivial representation appears with multiplicity >1 in the representation of Spin(m - 1) in P_a = holomorphic polynomials on $(\mathfrak{v},J_a)$. By (2.4), $L^1_K(N)$ cannot be commutative in this case.

Assume now $m \equiv 3 \pmod 8$. The natural embedding of Spin(m - 1) in C(m - 2) (cf. [H]) shows that $\mathfrak{v}$ again splits as a sum of two real irreducible subspaces under Spin(m - 1), although in this case each of them is J_a-invariant. In fact, let

(3.3) $$\gamma = J_{a_1} J_{a_2} \cdots J_{a_m}, \quad \{a_i\} = \text{orthonormal basis of } \mathfrak{z} .$$

This element of $\mathrm{Pin}(m) \subset C(m)$ is canonical up to sign. If $m \equiv 3 \pmod 8$, then γ commutes with all J_b, $b \in \mathfrak{z}$ and $\gamma^2 = 1$. Necessarily $\gamma = \pm 1$, since $\mathfrak{v}$ is irreducible under C(m). Letting $a = a_1$, $J_a = \pm J_{a_2} \cdots J_{a_m} \in \mathrm{Spin}(m-1)$. This shows that $(\mathfrak{v},J_a)$ itself splits as a sum of two complex, Spin(m - 1)-invariant subspaces. As modules over C(m - 2), hence over Spin(m - 1), these are known to be equivalent (cf. [R]). Therefore $L^1_K(N)$ cannot be commutative.

Finally we discuss the positive cases in (3.1). Since $\dim \mathfrak{v} = 8$ in all cases, we need to consider only $m = 5$, which is also of special interest (cf. §5). The element

$\gamma = J_{a_1} \cdots J_{a_5}$ is in this case a complex structure on $\mathfrak{v}$ which commutes with the action of $C(m)$, in particular with the complex structure J_a. Let $a = a_1$ and $\gamma_a = J_{a_2} \cdots J_{a_5}$. Since $\gamma_a^2 = 1$ but $\gamma_a J_{a_2} = - J_{a_2}\gamma_a$, γ_a defines a non-trivial splitting

$$\mathfrak{V} = \mathfrak{V}_+ \oplus \mathfrak{V}_- \tag{3.4}$$

with $\gamma_a = \pm 1$ on $\mathfrak{V}_\pm$. Each of these is γ-invariant, so they are complex subspaces of $(\mathfrak{V},\gamma)$ and, because $\gamma = J_a\gamma_a$, $\gamma = \pm J_a$ on $\mathfrak{V}_\pm$. Since J_{a_2}, for instance, interchanges them, $\dim_{\mathbb{C}} \mathfrak{V}_\pm = 2$. Recall now that $Spin(4) \cong SU(2) \times SU(2)$, being the stabilizer of a in $Spin(5)$, respects (3.4) and therefore its representation on $\mathfrak{V}_\pm$ is the projection on one of the two $SU(2)$-factors. It follows that the space of J_a-holomorphic polynomials on $\mathfrak{V}$ decomposes under $Spin(4)$ as the direct sum

$$P_a = \sum_{r,s} P^o_{a,r}(\mathfrak{V}_+) \otimes P^o_{a,s}(\mathfrak{V}_-)$$

where $P^o_{a,r}$ = J_a-holomorphic polynomials, homogeneous of degree r. Since the summands are non-equivalent, (2.4) implies now that $L^1_K(N)$ is commutative and the Proposition is proved.

Next we look at the case $K = A_o(N)$, the identity connected component of the full group of rotations of N. Recall that $A_o(N) = (U \times Spin(m))/(\text{finite subgroup})$, where U is the subgroup of elements acting trivially on the center of N.

(3.5) Proposition. The algebra $L^1_{A_o(N)}(N)$ is commutative for $m = 1,2,3,5,6,7$. It is not commutative for $m \equiv 0,1,4 \pmod 8$, $m > 7$ and for $m = 4$.

Proof: For $m \equiv 0,6,7 \pmod 8$ the group U is $\{\pm 1\}$, so that $A_o(N) = \mathrm{Spin}(m)$ and (3.1) gives the corresponding assertions.

Now let $m \equiv 1 \pmod 8$. Then $U = U(1)$ (indeed, $U = \{(\cos\theta)1 + (\sin\theta)\gamma\}$ where γ is as in (3.3)) and $A_o(N) = U(1) \times \mathrm{Spin}(m)$. For $m = 1$, N_1 is the 3-dimensional Heisenberg group and it is well known that $L^1_{U(1)}(N_1)$ is commutative in that case. Assume then $m > 1$ and fix a unit vector $a \in \mathfrak{z}$. Its stabilizer in $A_o(N)$ is $K_a \cong U(1) \times \mathrm{Spin}(m-1)$ and, as in the proof of (3.1), $\mathfrak{v}$ splits under $C(m-1)$ as the sum of two invariant (irreducible) subspaces of half the dimension. Set $\gamma_a = -J_a\gamma$; then $\gamma_a \in \mathrm{Spin}(m-1)$, $\gamma_a^2 = 1$, it commutes with J_a but it anticommutes with any J_b, $b \perp a$. Hence γ_a preserves any $C(m-1)$-invariant irreducible subspace $\mathcal{W} \subset \mathfrak{v}$ and induces in it a proper decomposition into (± 1)-eigenspaces $\mathcal{W} = \mathcal{W}_+ \oplus \mathcal{W}_-$. Now $\mathfrak{v} = (\mathcal{W}_+ \oplus J_a\mathcal{W}_+) \oplus (\mathcal{W}_- \oplus J_a\mathcal{W}_-)$. Since γ acts as $\pm J_a$ on $\mathcal{W}_\pm$, the group $K_a = \{(\cos\theta)1 + (\sin\theta)J_a\} \times \mathrm{Spin}(m-1)$ leaves $\mathcal{W}_\pm \oplus J_a\mathcal{W}_\pm$ invariant and the corresponding representations are contragradient of each other relative to the complex structure J_a. Therefore, if $\{u_j\}$, $\{v_j\}$ are orthonormal basis of $\mathcal{W}_+$ and $\mathcal{W}_-$ respectively, the polynomial $\sum_j \langle u_j,\cdot\rangle\langle v_j,\cdot\rangle$ is invariant under K_a. Hence $L^1_{A_o(N)}(N)$ is not commutative by (2.4).

Now let $m \equiv 4 \pmod 8$. In this case $\gamma^2 = 1$ and γ anticommutes with J_a for every $a \in \mathfrak{z}$. Hence $\mathfrak{v} = \mathfrak{v}_+ \oplus \mathfrak{v}_-$ with $\gamma = \pm 1$ on $\mathfrak{v}_\pm$ and $J_a \mathfrak{v}_\pm = \mathfrak{v}_\mp$. As in (3.2), $\mathfrak{v}_+$ is a real form of $(\mathfrak{v}, J_a)$ which is invariant under K_a, since both Spin(m - 1) and U(= SU(2)) commute with γ. The same argument as the one for K = Spin(m) shows now that $L^1_{A_o(N)}(N)$ cannot be commutative.

We now discuss the remaining positive cases of the proposition, m = 2,3,5. In the first two the real dimension of $\mathfrak{v}$ is four and $U \cong SU(2)$, so that $A_o(N) = U \times Spin(m)$ is transitive on the product of the unit spheres of $\mathfrak{v}$ and $\mathfrak{z}$. In particular, for every $n \in N$ there exist $k \in A_o(N)$ such that $k(n) = n^{-1}$. Hence $L^1_{A_o(N)}(N)$ is commutative in these cases.

<u>Remark</u>: m = 1,2,3 are the only cases where $L^1_U(N)$ is itself commutative.

When m = 5, again $U = U(1) = \{(\cos\theta)1 + (\sin\theta)\gamma\}$ and if $\mathfrak{v} = \mathfrak{v}_+ \oplus \mathfrak{v}_-$ as in (3.4), then U acts as $\{(\cos\theta)1 \pm (\sin\theta)\gamma\}$ on $\mathfrak{v}_\pm$. In particular a generic U-orbit in $\mathfrak{v}$ is contained in a generic Spin(4)-orbit. Therefore $L^1_{A_o(N_5)}(N_5) = L^1_{Spin(5)}(N_5)$ and commutativity follows from (3.1). This finishes the proof of (3.5).

§4. <u>The case of the full rotation group $A(N)$</u>. This group is not in general connected and, for the application of (2.4), additional information on the stabilizers of fixed $a \in \mathfrak{z}$ is needed.

Recall that $Pin(m)$ = group generated by the J_z's ($z \in \mathfrak{z} \cong \mathbb{R}^m$, $|z| = 1$ and J_z acting on $\mathfrak{z}$ as minus the reflection through $z^\perp$), U = group of rotations acting trivially on $\mathfrak{z}$ and that the natural homomorphism

$$(4.1) \qquad Pin(m) \times U \longrightarrow A(N)$$

has finite kernel and cokernel. If m is even, then $\det(J_z|_{\mathfrak{z}}) = -1$ and therefore $Pin(m)$ acts on $\mathfrak{z}$ as the full non-connected $O(m)$. If m is odd, $Pin(m)$ acts on $\mathfrak{z}$ as $SO(m)$; but there may still be an element

$$(4.2) \qquad \phi \in A(N) \quad \text{such that} \quad \phi|_{\mathfrak{z}} = -1$$

(in the standard realization of the Heisenberg group $N \cong \mathbb{C} \times \mathbb{R}$ such a ϕ is $\phi(z,t) = (z,-t)$). Since ϕ anticommutes with the J_z's, it commutes with the action of $Spin(m)$ on $\mathfrak{v}$.

When $m \equiv 3,7 \pmod 8$ there is no such ϕ. Indeed, the representation of $Spin(m)$ on $\mathfrak{v}$ is irreducible in this case - since is an irreducible $C(m-1)$-module - and the intertwining operators for it are the same as those for $C(m)$ (cf. table). Therefore ϕ would be in U contradicting the fact that $\phi|_{\mathfrak{z}} = -1$.

When $m \equiv 1,5 \pmod 8$ such a ϕ does exist. Assume first that $m \equiv 1 \pmod 8$. Then $\mathfrak{v}$ is no longer irreducible under Spin(m) but it decomposes as $\mathfrak{w} \oplus \gamma\mathfrak{w}$ (cf. §3). Since $\gamma J_z \in \mathrm{Spin}(m)$ for every unit $z \in \mathfrak{z}$, J_z interchanges $\mathfrak{w}$ and $\gamma\mathfrak{w}$; in other words, $\mathfrak{w}$ is a joint real form for all the complex structures J_z. Now let $\phi \in A(\mathfrak{n}) = A(N)$ be defined by $\phi = 1$ on $\mathfrak{w}$, $\phi = -1$ on $\gamma\mathfrak{w} \oplus \mathfrak{z}$; since ϕ anticommutes with the action of J_z on $\mathfrak{v}$, it gives the desired element. Consider now $m \equiv 5 \pmod 8$. Here $\mathfrak{v}$ is irreducible under Spin(m), but the ring of intertwining operators are the quaternions, while that for $C(m)$ is $\cong \mathbb{C}$, generated by 1 and γ. Therefore there must exist $\phi \in \mathrm{End}(\mathfrak{v})$ commuting with the action of Spin(m) but anticommuting with γ. This implies that ϕ anticommutes with the J_z's. Hence a ϕ as in (4.2) is obtained by extending ϕ linearly to $\mathfrak{n}$ with $\phi|_{\mathfrak{z}} = -1$.

Now fix a unit $a \in \mathfrak{z}$ and let K_a be its stabilizer in $A(N_m)$. Let $S_a \subset K_a$ be the image of the corresponding subgroup $\mathrm{Spin}(m-1) \times U$ under (4.1). Then K_a is generated by S_a, J_a and, if a ϕ as in (4.1) exists, $\phi J_{a'}$ with $a' \perp a$. Hence we have the cases:

m even: $\quad K_a/S_a = \{1, J_a\}$.

Notice that m odd $\Longrightarrow J_a S_a = S_a$.

$m \equiv 1,5 \pmod 8$: $K_a/S_a = \{1, \phi J_{a'}\}$

$m \equiv 3,7 \pmod 8$: $K_a/S_a = \{1\}$.

Consider now the algebra $L^1_{A(N)}(N)$, N irreducible group of H-type. Of course, (3.5) implies that it is commutative for $1 \leq m \leq 7$, $m \neq 4$. We can also see that it remains non-commutative for $m \equiv 1,3$ (mod 8), $m \neq 1,3$. The case $m \equiv 3$ also follows from (3.5) because $U \cong SU(2)$ is connected and therefore the discussion above shows that the stabilizer of an element $a \in \mathfrak{z}$ in $A(N)$ is the same as in $A_o(N)$.

For $m \equiv 1$ (mod 8), notice that $\phi J_{a'}$ is unitary on $(\mathfrak{v}, J_a)$ and interchanges $\mathcal{W}$ and $J_a\mathcal{W}$ (cf. §3). Therefore $\psi = J_a \phi J_{a'}$ preserves the real form $\mathcal{W}$. Also, ψ interchanges $\mathcal{W}_+$ and $\mathcal{W}_-$ because it anticommutes with γ_a. Now let $\{u_j\}$ be an orthonormal basis of $\mathcal{W}_+$. Since $\psi^2 = -1$, the polynomial

$$\left(\sum_j \langle u_j, \cdot \rangle \langle \psi u_j, \cdot \rangle \right)^2$$

is invariant under K_a, implying as before that $L^1_{A(N)}(N)$ cannot be commutative.

§5. Remarks on the algebra of invariants. The study of the spherical functions on $K \ltimes N/K$ naturally leads to the determination of the K-invariant polynomials on the Lie algebra $\mathfrak{n}$. For the Iwasawa groups, as well as in some other instances (e.g. $m = 6$), there are only two independent generators in the case $K = A(N)$, namely $v + z \longmapsto |v|^2$, $v + z \longmapsto |z|^2$ (the spherical functions are in these cases a combination of Bessel and Laguerre functions in these invariants [Ko1])

Consider now the case $m = 5$, $n = 8$, $K = \mathrm{Spin}(5)$. Fix a "generic" point $v + z \in \mathfrak{n} = \mathfrak{v} \oplus \mathfrak{z}$. The stabilizer of z in Spin(5) is $\mathrm{Spin}(4) \cong SU(2) \times SU(2)$. Then $\mathfrak{v}$ can be identified with $\mathbb{C}^2 \oplus \mathbb{C}^2$ with each factor of Spin(4) acting in the standard manner on one factor and trivially on the other. Therefore, the stabilizer of a generic $v \in \mathfrak{v}$ in Spin(4) and hence that of a generic element of $\mathfrak{n}$ in Spin(5), must be trivial. Since this implies dim(general orbit of Spin(5) in $\mathfrak{n}$) = dim Spin(5) = 10 but $\dim \mathfrak{n} = 13$, there must be three algebraically independent invariants. Two of these can always be taken to be $|v|^2$, $|z|^2$. As to the third, let γ be as in (3.3); then the invariant is

$$v + z \longrightarrow \langle v, \gamma J_z v\rangle .$$

Finally, we note that the discussion of the case $m \equiv 5$ in (3.5) implies that these polynomials are invariant under $A_o(N)$ as well.

References

[C] Cygan, J., "Subadditivity of homogeneous norms on certain nilpotent Lie groups", Proc. AMS 83 (1981), 69-70.

[H] Husemoller, D., "Fibre Bundles", Springer-Verlag (1966).

[Ka 1] Kaplan, A., "Fundamental solutions for a class of hypoelliptic PDE", Trans. AMS 258 (1980), 147-153.

[Ka 2] Kaplan, A., "On the geometry of groups of Heisenberg type", to appear in Bull. London Math. Soc.

[Ko 1] Koranyi, A., "Some applications of Gelfand pairs in classical analysis" Harmonic Analysis and Group Representations, C.I.M.E. (1980).

[Ko 2] Koranyi, A., "Geometric properties of Heisenberg type groups" to appear in Advances in Math.

[M] Metivier, G., "Hypoellipticité analytique dur des groupes nil-potents de rang 2", Duke Math. J., 47 (1980), 195-221.

[R] Riehm, C., "The automorphism group of a composition of quadratic forms", to appear in Trans. AMS.

[S] Seaman, W., "Hypersurfaces of constant mean curvature in euclidean spaces and groups of Heisenberg type", thesis, University of Massachusetts, Amherst (1981).

[TV] Tricerri, F. and Vanhecke, L., "Homogeneous structures", to appear in the Proceedings of the Special Year on Differential Geometry, College Park, Maryland (1982).

Department of Mathematics
University of Massachusetts
Amherst, MA 01003, USA

Dipartimento di Matematica
Politecnico di Torino
Torino, Italy

SURJECTIVITY OF THE CONDITIONALS EXPECTATIONS ON THE L^1 SPACES.

LUIGI ACCARDI (ROMA) CARLO CECCHINI (GENOVA).

Conditional expectations in von Neumann algebras are a useful tool in a variety of problems. In particular they have been succesfully applied to the armonic analysis on unimodular groups.

Recently the notion of conditional expectation associated to a state (or weight) on a von Neumann algebra has been extended and clarified so to make this tool applicable, among other things, to the harmonic analysis of non-unimodular groups. The definition of the conditional expectation associated to a given state, proposed in [1], was based on a "non-commutative characterization of the commutative conditional expectation".

In the present note we prove a more straightforward characterization of the conditional expectation based on the usual characterization of the classical conditional expectation. This approach will naturally lead to the study of the extension of the conditional expectation on the L^I-space associated to a given von Neumann algebra and a state and on such spaces, contrarily to what happens in the L^∞-case, the conditional expectation acts surjectively.

I) Let $\mathfrak{A}$ be a von Neumann algebra, $\mathfrak{B}$ a von Neumann sub-algebra of $\mathfrak{A}$. Denote $i : \mathfrak{B} \to \mathfrak{A}$ -the identity embedding and $i' : \mathfrak{A}_* \to \mathfrak{B}_*$ its dual which to a normal state on $\mathfrak{A}$ associates its restriction to $\mathfrak{B}$.

Any normal faithful state φ on $\mathfrak{A}$ defines an embedding $I_\varphi : \mathfrak{A} \to \mathfrak{A}_*$ characterized by the condition:

$$\langle I_\varphi (a), a_1 \rangle = (J \pi (a) J 1_\varphi , \pi (a_1) 1_\varphi) \qquad \text{(I.I}$$

for any $a, a_1 \in \mathcal{A}$ .Here $\langle \cdot , \cdot \rangle$ denotes the duality $\langle \mathcal{A}_*, \mathcal{A} \rangle$; $\{\mathcal{H}_\varphi, \pi, 1_\varphi\}$ is the GNS triple associated to $\{\mathcal{A}, \varphi\}$; $(.,.)$ is the scalar product in $\mathcal{H}_\varphi$;and J is the Tomita involution associated to $\{\mathcal{A}, \varphi\}$.

Denote φ_o the restriction of φ on $\mathcal{B}$ and $\{\mathcal{H}_{\varphi_o}, \pi_o, 1_{\varphi_o}\}, J_o, (.,.)_o$ the corresponding objects associated to the pair $\{\mathcal{B}, \varphi_o\}$.

We will identify $\mathcal{H}_{\varphi_o}$ with the sub-space $[\mathcal{B} . 1_\varphi]$ (=closure of $\mathcal{B} \cdot 1_\varphi$ in $\mathcal{H}_\varphi$)of $\mathcal{H}_\varphi$ and 1_φ with 1_{φ_o},and $P: \mathcal{H}_\varphi \to \mathcal{H}_{\varphi_o}$ will denote the orthogonal projection.

Theorem (I.I) There exists a unique map $E: \mathcal{A} \to \mathcal{B}$ characterized by the condition:

$$i' I_\varphi(a) = I_{\varphi_o}(E(a)) \; ; \forall a \in \mathcal{A} \tag{I.2}$$

Proof. Let $a \in \mathcal{A}$, $b \in \mathcal{B}$ .Denote $\chi(b) = \langle i' I_\varphi(a), b \rangle$.If $b \in \mathcal{B}_+$ then

$$|\chi(b)| \leq \|a\|_\infty \varphi(b)$$

Therefore,by the commutant valued Radon-Nikodim theorem and the Tomita isomorphism between $\pi(\mathcal{A})$ and $\pi(\mathcal{A})'$,there exists a unique element $E(a) \in \mathcal{A}$ such that

$$\chi(b) = (J_o \pi_o(E(a)) J_o 1_\varphi, \pi_o(b) . 1_\varphi) = \langle I_{\varphi_o}(E(a)), b \rangle \tag{I.3}$$

and this is equivalent to (I.2).

Because of (I.I) the identity (I.2) is equivalent to

$$(J \pi(a) J . 1_\varphi, \pi(b) 1_\varphi) = (J_o \pi_o(E(a)) J_o . 1_\varphi, \pi_o(b) 1_\varphi) \tag{I.4}$$

for any $b \in \mathcal{B}$ .Hence,if $\mathcal{A}$ is abelian (I.4) reduces to the identity:

$$\varphi(ab) = \varphi_o(E(a)b), a \in \mathcal{A}, b \in \mathcal{B} \tag{I.5}$$

which is the usual characterization of the classical φ-conditional expectation.In the general case,substituting $b_1^* b_2$ for b in (I.4), one easily proves the equivalence of this identity with:

$$\pi_o(E(a)) = J_o P J \pi(a) J P J_o \; , \quad a \in \mathcal{A} \tag{I.5}$$

Therefore the map $E: \mathcal{A} \to \mathcal{B}$ defined by Proposition (I.I) coincides with the φ-conditional expectation introduced in [1] .

A remarkable dofference between the classical and the non-commutative situation,is that in the latter case in general the φ-conditional expectation is not surjective,i.e.

$$\iota' I_{\varphi}(\mathfrak{A}) \subsetneq I_{\varphi_o}(\mathfrak{B}).$$

However,since the restriction map $\iota' : \mathfrak{A}_* \to \mathfrak{B}_*$ is surjective,then, for any b in $\mathfrak{B}$ there will be some state $\psi_b \in \mathfrak{A}_*$ such that:

$$\iota' \psi_b = I_{\varphi_o}(b)$$

Therefore any identification of the predual of a von Neumann algebra with a space of (possibly unbounded)operators yields a map which is natural to regard as an extension of the φ-conditional expectation;and,since this map will be surjective,this will provide a natural framework to study the properties of the range of the φ-conditional expectation.More precisely,independently on the definition or the realization of the L^{I} space associated to a von Neumann algebra, and a weight on it,let be given an identification

$$T_{\varphi} : \mathfrak{A}_* \to L^{I}(\mathfrak{A}, \varphi)$$

and the corresponding identification

$$T_{\varphi_o} : \mathfrak{B}_* \to L^{I}(\mathfrak{B}, \varphi_o)$$

Then one can define a surjective contraction

$\bar{E} : L^{I}(\mathfrak{A}, \varphi) \to L^{I}(\mathfrak{B}, \varphi_o)$through the commutative diagram

$$\begin{array}{ccc} \mathfrak{A}_* & \xrightarrow{\iota'} & \mathfrak{B}_* \\ T_{\varphi}\downarrow & & \downarrow T_{\varphi_o} \\ L^{I}(\mathfrak{A},\varphi) & \xrightarrow[\bar{E}]{} & L^{I}(\mathfrak{B},\varphi_o) \end{array} \qquad \text{(I.6}$$

Because of Theorem (I.I) the identity

$$\bar{E}\,(T_{\varphi}.\, I_{\varphi}\,(a)) = T_{\varphi_o}.I_{\varphi_o}\,(E\,(a));\ a \in \mathfrak{A} \qquad \text{(I.7}$$

takes place;therefore,if the L^{I}-spaces in (I.6)are realized as spaces of operators actingon $\mathcal{H}_{\varphi}$,it is natural to consider the map $\bar{E}$ as an extension of the φ-conditional expectation.

Now,the action of E on $\mathfrak{A} \cong L^{\infty}(\mathfrak{A}, \varphi)$is implemented on $\mathcal{H}_{\varphi}$ by a partial isometry (cf. the identity (I;5) and [1]).Can we say that the extension E of E is in some sense "implemented"?

The problem here arises from the fact that usually the embedding $a \in \mathfrak{A} \longrightarrow T_{\varphi} \cdot I_{\varphi}(a) \in L^{I}(\mathfrak{A}, \varphi)$ identifies the elements of $\mathfrak{A}$ with unbounded operators. There are several proposals of definition of the spaces $L^{I}(\mathfrak{A}, \varphi)$ due respectively to Haagerup [4] ,Connes [3] and Hilsum [5] ,Sherstnev [6] ,Araki [2] . In the folllowing we will discuss some preliminary results concerning the above mentioned implementation problem in the framework of the L^{I}-spaces defined by Connes [3] and Hilsum [5].

2) Let $\mathcal{M}$ be a von Neumann algebra; φ a normal, faithful, semifinite weight on $\mathcal{M}$;

$$\mathfrak{A} = \mathfrak{A}_{\varphi} = \{a \in \mathcal{M} : \varphi(a^{*} a), \varphi(a\, a^{*}) < +\infty\} \qquad (2.I$$

$\mathcal{M}(\varphi) = \{$linear combinations of the elements of the form $x^{*}x$ with $\varphi(x^{*}x) < +\infty\}$ (2.2

Let $\dot{\varphi}$ be the linear extension of φ on $\mathcal{M}(\varphi)$ and $\mathcal{H}_{\varphi}$ the completion of $\mathfrak{A}$ for the scalar product:

$$\langle a_{I}, a_{2}\rangle = \varphi(a_{I}^{*} a_{2}) ; a_{I}, a_{2} \in \mathfrak{A} \qquad (2.3$$

$\mathfrak{A}$ is a w^{*}-dense left ideal of $\mathcal{M}$ and the map $\pi : \mathcal{M} \to \mathcal{B}(\mathcal{H}_{\varphi})$

$$\pi(m).a = m.a \quad ; m \in \mathcal{M} ; a \in \mathfrak{A} \qquad (2.4$$

is a normal $*$-isomorphism. The operator $S = S_{\mathfrak{A}} = S^{\varphi}_{\mathfrak{A}}$ denotes the closure of the involution $a \in \mathfrak{A} \longrightarrow a^{*} \in \mathfrak{A}$; and its polar decomposition will be denoted $S = J\Delta^{1/2}$.

Denote $\mathcal{H}^{*}_{\varphi}$ the completion of $\mathfrak{A}$ for the scalar product:

$$\langle a_{1}, a_{2}\rangle_{*} = \langle a^{*}_{1}, a^{*}_{2}\rangle ; a_{1}, a_{2} \in \mathfrak{A} \qquad (2.5$$

Arguing as in Lemma (6) of Connes' paper [3] one shows that for any positive linear functional ψ on $\mathcal{M}$ the quadratic form q_{ψ} on $\mathcal{H}^{*}_{\varphi}$ defined by: (1) dom $q_{\psi} = \mathfrak{A}$ (2) $q_{\psi}(a) = \psi(a^{*}a)$; is lower semicontinuous. Hence there exists a positive operator $T_{\varphi}(\psi)$ on $\mathcal{H}^{*}_{\varphi}$ characterized by the following properties:

1) dom $T_{\varphi}(\psi)^{1/2} \supseteq \mathfrak{A}$

2) $T_{\varphi}(\psi)$ is the largest positive self-adjoint operator on $\mathcal{H}^{*}_{\varphi}$ such that:

$$\|T_{\varphi}(\psi)^{1/2} a\|^{2}_{*} = \psi(a^{*}a) ; a \in \mathfrak{A} \qquad (2.6$$

The map $a \in \mathfrak{A} \subseteq \mathcal{H}_{\varphi} \to a^{*} \in \mathcal{H}^{*}_{\varphi}$, extends to a anti-unitary isomorphism $R: \mathcal{H}_{\varphi} \longrightarrow \mathcal{H}^{*}_{\varphi}$,and the operator

$\tilde{T}_\varphi(\psi) = R^+ . T_\varphi(\psi) . R : \mathcal{H}_\psi \longrightarrow \mathcal{H}_\psi$

has a dense domain($\supseteq \mathfrak{A}$) and satisfies

$$\|\tilde{T}_\varphi(\psi)^{1/2} a\|^2 = \psi(aa^*) \;; a \in \mathfrak{A} \qquad (2.7$$

The operator $d\psi/d\varphi'$ ($\varphi' = \varphi . j$ =the weight induced by φ on the commutant of $\pi(\mathfrak{A})$ through Tomita's isomorphism),defined by Connes in [3] ,is the largest positive self-adjoint operator T on $\mathcal{H}_\varphi$ satisfying

$$\|T^{1/2}\xi\|^2 = \psi(|R_\varphi(\xi)^+|^2) \qquad (2.8$$

for any φ'-bounded vector ξ (here we are using the notations and terminology of [3]).

Therefore,since the elements of $\mathfrak{A}$ are φ'-bounded:

$$\tilde{T}_\varphi(\psi) \geq d\psi/d\varphi' \qquad (2.9$$

The converse inequality follows from the characterization of $d\psi/d\varphi'$ and the following Lemma.

<u>Lemmma 2.1)</u> Let T be a positive self-adjoint operator on $\mathcal{H}_\varphi$ such that dom $T^{1/2} \supseteq \mathfrak{A}$ and

$$\|T^{1/2}a\|^2 = \psi(a\,a^*) \qquad (2.10$$

then dom $T^{1/2}$ includes the φ'-bounded vectors.

<u>Proof</u>. Let ξ be a φ'-bounded vector in $\mathcal{H}_\varphi$.For ξ to be in the closure of the graph of $T^{1/2}$ it is sufficient to show ,in view of (2.10),that there is a sequence (a_n) in $\mathfrak{A}$ such that $a_n \longrightarrow \xi$ in $\mathcal{H}_\varphi$ and (a_m) is a Cauchy sequence for the norm $a \longrightarrow \psi(|a^*|^2)$.

Let (e_α) be a net in the unit ball of $\mathfrak{A}$ converging to 1 in the strong-topology of $\mathcal{M}$;such a net exists by Kaplansky's density Lemma.Denote,for each α, $a_\alpha = \pi(e_\alpha).\xi \in \mathcal{H}_\varphi$ .Then clearly $a_\alpha \longrightarrow \xi$ in $\mathcal{H}_\varphi$.Let us show that $a_\alpha \in \mathfrak{A}$.Since a_α is φ'-bounded we need onlyto show that a_α is in the domain of $S_\mathfrak{A}$ .Now,if $a' \in \mathfrak{A}'$,one has:

$$\langle F_\mathfrak{A} a', a_\alpha \rangle = \langle \pi(e_\alpha^*) F_\mathfrak{A} a', \xi \rangle = \langle \pi'(F_\mathfrak{A} a')\, e_\alpha^*, \xi \rangle = \langle e_\alpha^*, \pi'(a')\xi \rangle$$

and $a_\alpha \in$ dom $S_\mathfrak{A}$ follows since ξ is φ'-bounded.Finally,since $\pi(a_\alpha) = \pi(e_\alpha) R_{\varphi'}(\xi)$, one has

$$|\pi(a_\alpha^* - a_\beta)|^2 = \pi(e_\alpha - e_\beta) R_{\varphi'}(\xi) R_{\varphi'}(\xi)^+ \pi(e_\alpha^* - e_\beta^*) \leq \|R_{\varphi'}(\xi)^+\|^2 \pi(|e_\alpha^*$$

Hence (a_α) is a Cauchy net for the norm $a \longrightarrow \psi(|a^*|^2)$ .From this the the-

-sis follows. Thus one has the equality

$$R^{*}.T_{\varphi}(\psi) . R = d\psi/d_{\backslash}\varphi' \qquad (2.11$$

Let us now look at the representation of the conditional expectation in these two (isomorphic) realizations of the space $L^1(\mathcal{M},\varphi)$.

More explicitely, let $\mathcal{N}$ be a von Neumann sub-algebra of $\mathcal{M}$ such that $\varphi_0 = \varphi|\mathcal{N}$ is semi-finite; let $\mathcal{B} \subseteq \mathcal{N}$ be defined in analogy with $\mathcal{O}\mathcal{l}$ ($cf.$(2.1)); and let $\mathcal{H}_{\varphi_0}$ be the completion of $\mathcal{B} \subseteq \mathcal{O}\mathcal{l}$ for the scalar product (2.3); we identify $\mathcal{H}_{\varphi_0}$ to a sub-space of $\mathcal{H}_\varphi$ (and denote P: $\mathcal{H}_\varphi \rightarrow \mathcal{H}_{\varphi_0}$ the orthogonal projection. Similarly $\mathcal{H}^*_{\varphi_0}$ is defined as the completion of $\mathcal{B}$ for the scalar product (2.5) and identified to a sub-space of $\mathcal{H}^*_\varphi$ and Q: $\mathcal{H}^*_\varphi \rightarrow \mathcal{H}_{\varphi_0}$ denotes the orthogonal projection Then denoting for any $\psi \in \mathcal{M}_*$, $\psi_0 = \psi|\mathcal{N}$, by definition $T_{\varphi_0}(\psi_0)$ is the largest positive self-adjoint operator on $\mathcal{H}_{\varphi_0}$ satisfying

$$\| T_{\varphi_0}(\psi_0)^{1/2}\, b \|^2_* = \psi_0(b^* b); b \in \mathcal{B} \qquad (2.12$$

On the other hand one has also

$$\| | T_\varphi(\psi)^{1/2} Q | \|^2_* = \psi_0\, (b^* b)$$

therefore:

$$T_{\varphi_0}(\psi_0) = \bar{E}\, (T_\varphi(\psi)\,) \geq Q T_\varphi(\psi)\, Q \qquad (2.13$$

and, with a similar argument:

$$d\psi_0/d\varphi_0' = \bar{E}\, (d\psi/d\varphi') \geq P\, d\psi/d\varphi'\, P \qquad (2.14$$

An unrestricted validity of the equality in (2.14), or equivalently (2.13), would mean that at the L^1-level the exact analogue of the classical abelian formula for the φ-expectation takes place, a thing which at the L^∞-level holds only under very restrictive conditions. We conjecture that in (2.14) the equality takes place, at least in the case in which the condition:

$$\mathcal{B}' = \mathcal{H}_{\varphi_0} \cap \mathcal{O}\mathcal{l}'$$

is satisfied.

BIBLIOGRAPHY.

I. L.Accardi,C. Cecchini.Conditional expectations in von Neumann algebras and a theorem of Takesaki.J.Funct. Anal.45 (I982) p. 245-273.

2. H.Araki,T. Masuda.Positive cones and L_p spaces for von Neumann algebras. (I98I) preprint.

3. Connes A. On the spatial theory of von Neumann algebras.J. Funct. Anal. 35 (I980) p. I53-I64.

4. Haagerup U. L^p spaces associated with an arbitrary von Neumann algebra. Coll int. C.N.R.S. 1979 (274) p. 175-184.

5. Hilsum M. Les espaces L^p d'une algebre de von Neumann definies par la derivée spatiale.J.Funct.Anal. 40 (I98I)p. I5I-I69.

6. Sherstnev. A general measure and integration theory in von Neumann algebras (in Russian).Matematika 8 (I982)p. 20-35.

Luigi Accardi
Dipartimento di Matematica
Università di Roma II
Torvergata, Roma
Italy.

Carlo Cecchini
Istituto di Matematica
Università di Genova
via L.B.Alberti 4,Genova
Italy.

GENERALISATIONS OF HEISENBERG'S INEQUALITY

by

Michael Cowling
Università di Genova
16132 Genova
Italia

and

John F. Price
University of New South Wales
Kensington 2033
Australia

In our paper [3] we prove inequalities of the type

(1) $$\|f\|_2 \leq \text{const.}(\|vf\|_p + \|w\hat{f}\|_q)$$

or $$\|f\|_2 \leq \text{const.}\|vf\|_p^{\theta}\, \|w\hat{f}\|_q^{1-\theta}$$

(where $\hat{f}$ is the Fourier transform of f defined by $\hat{f}(y) = \int_{\mathbb{R}} f(x)e^{-2\pi ixy}dx$) under various hypotheses on the weight functions v and w. Roughly speaking, we may summarise these as follows. Given p in $[1,\infty]$, define $p^{\#} = 2p/(p-2)$ if $p > 2$ and $p^{\#} = \infty$ otherwise. We always require that

(2) $$\|\chi_{\mathbb{R}\setminus E}\, v^{-1}\|_{p^{\#}} < \infty \text{ , and}$$

($\hat{2}$) $$\|\chi_{\mathbb{R}\setminus F}\, w^{-1}\|_{q^{\#}} < \infty$$

for subsets E and F of $\mathbb{R}$ of finite measure. In this case we can prove (1) for all f in L^2. If we also suppose that E and F are bounded, or that $\chi_E v^{-1}$ and $\chi_F w^{-1}$ belong to some Lebesgue space L^r, with r in $(0,\infty)$, then (1) holds for all f in S' (with f and $\hat{f}$ locally integrable functions on the appropriate sets).

In [3] we give examples to show ways in which conditions (2) and ($\hat{2}$) are essential. Here we begin with a further example which shows that something more than (2) and ($\hat{2}$) is needed if (1) is to hold for all f in S'. (We are grateful to J.J. Benedetto and R.S. Strichartz for bringing this example to our attention.) Suppose that $f = \Sigma_{n\in\mathbb{Z}}\,\delta_n$; then $\hat{f} = \Sigma_{n\in\mathbb{Z}}\,\delta_n$. Let

$$E = F = \cup_{n\in\mathbb{Z}}\,(n-(n^2+1)^{-1},\ n + (n^2+1)^{-1})$$

and $$v(x) = w(x) = x^2\, \chi_{\mathbb{R}\setminus E}(x) \quad \text{for } x \text{ in } \mathbb{R} .$$

Then (2) and $(\hat{2})$ are satisfied and $\|vf\|_p = \|w\hat{f}\|_q = 0$ for all p and q, while $\|f\|_2 = \infty$.

This gives rise to the following problem: under what conditions on E and F does the question below have an affirmative response?

<u>QUESTION</u> f in S' with $\text{supp}(f) \subseteq E$ and $\text{supp}(\hat{f}) \subseteq F$ implies $f = 0$?

We also showed that if v and w grow very rapidly then the finiteness of $\|vf\|_p$ and $\|w\hat{f}\|_q$ implies that $f = 0$. Hence (1) is valid, but only in a trivial sense. Here we improve the result we gave, and obtain a best possible theorem. In order to enunciate our theorem, we introduce a little notation. By e_a we denote the following function:

$$e_a(x) = \exp(ax^2) \qquad \text{for} \quad x \in \mathbb{R}.$$

<u>THEOREM</u> Suppose that $p,q \in [1,\infty]$ with at least one of them finite. Suppose also that $a,b \in \mathbb{R}^+$ and that v and w are functions satisfying

$$v(x) \geqslant \lambda e_a(x); \quad w(x) \geqslant \mu e_a(x)$$

for $|x|$ sufficiently large and constants $\lambda,\mu > 0$.

If $ab \geqslant \pi^2$, then the only f in S' satisfying $\|vf\|_p + \|w\hat{f}\|_q < \infty$ is $f = 0$. In contrast, if $ab < \pi^2$ there are infinitely many nonzero functions f in S satisfying $\|e_a f\|_p + \|e_b \hat{f}\|_q < \infty$.

<u>REMARKS</u> The case in which $p = q = \infty$ is covered by Hardy's theorem ([5]; see also [4], pp. 155-158). It asserts that if f satisfies $\|e_a f\|_\infty + \|e_b \hat{f}\|_\infty < \infty$, then $f = 0$, or f is a constant multiple of e_{-a}, or there are infinitely many such f according as $ab > \pi^2$, $ab = \pi^2$ or $ab < \pi^2$.

The proof of our theorem is a variant of the proof of Hardy's theorem together with a simple observation from [2]. Hardy's theorem itself relies on Phragmén-Lindelöf arguments and Liouville's theorem. We begin our proof with a lemma which is an L^p-version of the Phragmén-Lindelöf methods.

The statement and proof of the lemma are facilitated by the following notation. By Q_θ we denote the sector in the complex plane:

$$Q_\theta = \{\rho e^{i\psi}: \rho \in \mathbb{R}^+, \psi \in (0,\theta)\},$$

and by Q we denote the quadrant $Q_{\pi/2}$. The usual closure of Q_θ will be denoted by $\bar{Q}_\theta$.

<u>LEMMA</u> Suppose that g is analytic in Q and continuous on $\bar{Q}$. Suppose also that for $p \in [1,\infty)$ and a constant A,

$$|g(x+iy)| \leqslant A \exp(\pi x^2) \qquad (x+iy \in \bar{Q})$$

and

$$\left(\int_0^\infty |g(x)|^p \, dx \right)^{1/p} \leqslant A .$$

Then
$$\int_\sigma^{\sigma+1} |g(\rho e^{i\psi})| d\rho \leqslant A \max\{e^\pi, (\sigma+1)^{1/p}\}$$

for $\psi \in [0,\pi/2]$ and $\sigma \in \mathbb{R}^+$.

<u>Proof</u> Fix temporarily θ in $(0,\pi/2)$ and ε in $(0,\pi/2-\theta)$, and define $h = h_{\theta,\varepsilon}$ on $\bar{Q}_\theta$ by the formula

$$h(z) = g(z) \exp\{i\varepsilon e^{i\varepsilon} z^{(\pi-2\varepsilon)/\theta} + i\pi\cot(\theta) z^2/2\} .$$

Let s be a simple function on $[\sigma,\sigma+1]$ of L^∞-norm 1. It will suffice to show that

(4) $$\left| \int_\sigma^{\sigma+1} s(\rho) h(\rho e^{i\psi}) d\rho \right| \leqslant A \max\{e^\pi, (\sigma+1)^{1/p}\}$$

for all ψ in $[0,\theta]$. (The desired result follows from (4) by first letting ε tend to 0, then letting θ tend to $\pi/2$, and finally taking the supremum over all such s.)

In order to prove (4) we first observe that, if $\psi \in [0,\theta]$, then

(5) $$|h(\rho e^{i\psi})|$$

$$= |g(\rho e^{i\psi})| . \exp\{\mathrm{Re}(i\varepsilon e^{i\varepsilon} \rho^{(\pi-2\varepsilon)/\theta} e^{i(\pi-2\varepsilon)\psi/\theta} + i\pi \cot(\theta) \rho^2 e^{2i\psi}/2)\}$$

$$\leqslant A \exp\{\pi\cos^2(\psi)\rho^2 - \varepsilon\sin(\varepsilon+(\pi-2\varepsilon)\psi/\theta)\rho^{(\pi-2\varepsilon)/\theta} - \pi\cot(\theta)\sin(2\psi)\rho^2/2\}$$

$$\leqslant A \exp\{\pi\rho^2 - \varepsilon \sin(\varepsilon+(\pi-2\varepsilon)\psi/\theta)\rho^{(\pi-2\varepsilon)/\theta}\}.$$

Now $(\pi-2\varepsilon)/\theta > 2$ and $0 < \varepsilon + (\pi-2\varepsilon)\psi/\theta < \pi$ so that $h(\rho e^{i\psi})$ is bounded on $\bar{Q}_\theta$ and tends to 0 uniformly in ψ in $[0,\theta]$ as $\rho \to \infty$. Define k on $\bar{Q}_\theta$ by the formula

$$k(\rho e^{i\psi}) = \int_\sigma^{\sigma+1} s(\tau)h(\rho\tau e^{i\psi})d\tau \ ;$$

clearly k is also analytic in Q_θ, continuous and bounded in $\bar{Q}_\theta$ and tends to 0 uniformly in ψ in $[0,\theta]$ as $\rho \to \infty$.

From (5) we deduce that for $\rho > 0$,

$$|h(\rho e^{i\theta})| \leqslant A \exp\{\pi \cos^2(\theta)\rho^2 - \varepsilon \sin(\varepsilon)\, \rho^{(\pi-2\varepsilon)\psi/\theta} - \pi \cot(\theta) \sin(2\theta)\rho^2/2\}$$

$$\leqslant A \ ,$$

so that $$|k(\rho e^{i\theta})| \leqslant \int_\sigma^{\sigma+1} 1.A \, d\tau = A \ .$$

Similarly, if $\rho \in [0,(\sigma+1)^{-1}]$,

$$|k(\rho)| \leqslant \int_\sigma^{\sigma+1} |h(\rho\tau)|d\tau \leqslant \sup\{|h(\rho)|: \rho \leqslant 1\} \leqslant A\, e^{\pi}$$

by (5), while if $\rho > (\sigma+1)^{-1}$, then

$$\begin{aligned} |k(\rho)| &\leqslant \int_\sigma^{\sigma+1} |h(\rho\tau)|d\tau \\ &\leqslant \rho^{-1/p}\Big(\int_\sigma^{\sigma+1} |h(\rho\tau)|^p \rho d\tau\Big)^{1/p} \\ &\leqslant (\sigma+1)^{1/p}.A. \end{aligned}$$

Now the maximum principle implies that

$$|k(e^{i\psi})| \leqslant A.\max\{e^{\pi},(\sigma+1)^{1/p}\}$$

for ψ in $[0,\theta]$ from which the lemma follows. □

We remark that if $p = \infty$, then the above estimates may be refined so that $\max\{e^{\pi},(\sigma+1)^{1/p}\}$ is replaced by 1.

<u>Proof</u> (<u>of the theorem</u>) Under the hypotheses of the theorem, f and $\hat{f}$ are in fact continuous functions. For f may be written as the sum of a compactly

supported distribution and an integrable function, so that $\hat{f}$ is continuous. Analogously, f is continuous. It follows that both $\|fe_a\|_p$ and $\|\hat{f}e_b\|_q$ are finite.

Assume that $ab \geqslant \pi^2$. By dilating if necessary, we may assume that

$$\|fe_\pi\|_p < \infty$$

and

$$\|\hat{f}e_\pi\|_q < \infty$$

Further, by interchanging f and $\hat{f}$ if need be, we may and shall assume that $p < \infty$.

We recall that f extends to an entire function:

$$f(z) = \int_{\mathbb{R}} \hat{f}(\xi)\exp(2\pi i z\xi)d\xi \qquad (z \in \mathbb{C}),$$

and observe that

$$\begin{aligned} |f(x+iy)| &\leqslant \int_{\mathbb{R}} |\hat{f}(\xi)| \exp(-2\pi y\xi)d\xi \\ &= \exp(\pi y^2) \int_{\mathbb{R}} |\hat{f}(\xi)|\exp(\pi\xi^2) \exp(-\pi(\xi+y)^2)d\xi \\ &\leqslant \exp(\pi y^2) \|\hat{f}e_\pi\|_q \|e_\pi^{-1}\|_{q'} . \end{aligned}$$

We denote by g the entire function given by the rule

$$g(z) = \exp(\pi z^2)f(z) \qquad (z \in \mathbb{C}) .$$

It is now obvious that there is a constant A such that

$$(6) \qquad |g(x+iy)| \leqslant A \exp(\pi x^2) \qquad (x+iy \in \mathbb{C}) ,$$

$$(7) \qquad \left(\int_{\mathbb{R}} |g(x)|^p dx\right)^{1/p} \leqslant A .$$

By applying the lemma to the functions $g(z)$, $g(-z)$, $\bar{g}(\bar{z})$ and $\bar{g}(-\bar{z})$, we deduce that for any ψ in $[0,2\pi]$ and any sufficiently large σ in $\mathbb{R}^+$,

$$\int_\sigma^{\sigma+1} |g(\rho e^{i\psi})|d\rho \leqslant B(\sigma+1)^{1/p} ,$$

for some constant B.

Now by Cauchy's integral formula,

$$g^{(n)}(0) = n!(2\pi)^{-1} \int_0^{2\pi} g(\rho e^{i\psi})\ (\rho e^{i\psi})^{-n}\ d\psi$$

so that
$$|g^{(n)}(0)| \leqslant n!(2\pi)^{-1} \int_0^{2\pi} |g(\rho e^{i\psi})|\ \rho^{-n}\ d\psi\ .$$

Consequently, for sufficiently large σ in $\mathbb{R}^+$,

$$|g^{(n)}(0)| \leqslant n!(2\pi)^{-1} \int_\sigma^{\sigma+1} \int_0^{2\pi} |g(\rho e^{i\psi})|\rho^{-n}\ d\psi\ d\rho$$

$$\leqslant n!(2\pi)^{-1}\ \sigma^{-n} \int_0^{2\pi} \int_\sigma^{\sigma+1} |g(\rho e^{i\psi})|\ d\rho\ d\psi$$

$$\leqslant B\ r!\ \sigma^{-n}(\sigma+1)^{1/p}\ .$$

This implies that $g^{(n)}(0) = 0$ if $n \geqslant 2$, so $g(z) = Cz + D$, for some constants C and D. From (6), $C = 0$; from (7), $D = 0$ also. Therefore $f = 0$, as required.

On the other hand, if $ab < \pi^2$, choose α in $(a,\pi^2/b)$ and let $f(x) = \phi(x)\exp(-\alpha x^2)$ for x in $\mathbb{R}$ where ϕ is a polynomial. Then $\hat{f}(y) = \psi(y)\exp(-\pi^2 x^2/\alpha)$ for y in $\mathbb{R}$ where ψ is also a polynomial Evidently $\|e_a f\|_p$ and $\|e_b \hat{f}\|_q$ are both finite, which completes the proof. □

In conclusion, we remark that further developments of (1) with applications to quantum mechanics may be found in [6]. A different approach to local uncertainty principles is contained in [1].

REFERENCES

[1] J.J. Benedetto, A local uncertainty principle. Submitted for publication.

[2] M. Benedicks, Positive harmonic functions vanishing on the boundary of certain domains in $\mathbb{R}^{n+1}$. In Harmonic Analysis in Euclidean Spaces, 345-348, Amer. Math. Soc., Providence, 1979.

[3] M. Cowling and J.F. Price, Bandwidth versus time concentration: The Heisenberg-Pauli-Weyl inequality. Submitted for publication.

[4] H. Dym and H.P. McKean, Fourier Series and Integrals, Academic Press, New York and London, 1972.

[5] G.H. Hardy, A theorem concerning Fourier transforms, J. London Math. Soc. 8(1933), 227-231.

[6] J.F. Price, Inequalities and local uncertainty principles, J. Math. Phys. (to appear).